Ethics of Science and Technology Assessment

Founding Editor

Carl Friedrich Gethmann

Volume 55

The series *Ethics of Science and Technology Assessment* focuses on the impact that scientific and technological advances have on individuals, their social lives, and on the natural environment. Its goal is to cover the field of Science and Technologies Studies (STS), without being limited to it. The series welcomes scientific and philosophical reviews on questions, consequences and challenges entailed by the nature and practices of science and technology, as well as original essays on the impact and role of scientific advances, technological research and research ethics. Volumes published in the series include monographs and edited books based on the results of interdisciplinary research projects. Books that are devoted to supporting education at the graduate and post-graduate levels are especially welcome.

Davy van Doren · Bert Droste-Franke ·
Gabriele Fohr
Editors

Towards Energy System Resilience

Theory, Governance and Practice

Springer

Editors
Davy van Doren
IQIB—Institut für qualifizierende
Innovationsforschung und -beratung GmbH
Bad Neuenahr-Ahrweiler, Germany

Bert Droste-Franke
IQIB—Institut für qualifizierende
Innovationsforschung und -beratung GmbH
Bad Neuenahr-Ahrweiler, Germany

Gabriele Fohr
IQIB—Institut für qualifizierende
Innovationsforschung und -beratung GmbH
Bad Neuenahr-Ahrweiler, Germany

ISSN 1860-4803 ISSN 1860-4811 (electronic)
Ethics of Science and Technology Assessment
ISBN 978-3-032-03186-0 ISBN 978-3-032-03187-7 (eBook)
https://doi.org/10.1007/978-3-032-03187-7

This work was supported by BMWK—Bundesministerium für Wirtschaft und Klimaschutz (FKZ: 03EI1020A).

This Springer imprint is published by the registered company Springer Nature Switzerland AG
The registered company address is: Gewerbestrasse 11, 6330 Cham, Switzerland

Preface

Existing trends of climate change indicate an increased probability and intensity of extreme weather conditions. However, recent events have made clear that existing instruments related to risk assessment and crisis management often appear inadequate. In addition, post-catastrophic restoration activities have indicated significant tensions between 'build-back-soon' versus 'build-back-better' approaches. Consequently, there has been political direction to go beyond the drafting of traditional risk management plans by designing systems from the start onwards—as good as possible—that can prevent, resist, absorb, and rapidly recover from disruptive incidents.

In times of increasing crises and disasters, regions are faced with the challenge of strengthening their resilience and shaping the transformation to a sustainable and liveable future. To support such regions, we as IQIB—a non-profit organisation and subsidiary of the German Aerospace Center (DLR) and formerly known as the European Academy of Technology and Innovation Assessment—are oriented towards guiding future technical developments within Germany and abroad. As a research and consulting company based in Bad Neuenahr-Ahrweiler, we experienced the flood disaster in 2021 and are actively contributing our expertise to solutions for practice-oriented, climate-adapted reconstruction. With our trans- and interdisciplinary research projects and practical initiatives, we contribute to strengthening resilience and climate adaptation regionally, nationally, and internationally. For us, strengthening resilience means increasing the resistance, recovery and transformation capabilities of socio-technical systems.

The IQIB-coordinated project on Resilience Monitoring for the Digitalisation of the Energy Transition—or ReMoDigital—can be placed within this increasing awareness of developing and managing energy systems resilience. Its main purpose was to develop tools applicable for supporting the stepwise development towards a future low-carbon energy system that is climate- and cyber-resilient. Such management tools particularly aim to reduce energy system vulnerability through improving services in relation to fluctuating renewable energy supplies, the realisation of sector coupling, the introduction of new actors and roles, as well as the digitalization of decentralized energy production, storage, and consumption patterns.

To develop transparent, rationally justified and socially acceptable recommendations for dealing with new scientific and technical developments within the project ReMoDigital, a so called 'project group' was formed. Project groups are composed and organised in such manner that different arguments and perspectives can be discussed and reflected upon based on an interdisciplinary and multi-expertise basis. This semi-structured and deliberative-discursive activity can then be used in the analysis of mutual recognised and prioritised problems, which finally can result in condensed consensual conclusions, pragmatic if-then-recommendations, or even context-specific solutions. The project group approach was fundamentally developed in the 1990s at the European Academy as a standard method for rational technology assessment and has since then evolved (see e.g. Decker and Grunwald 2001; Grunwald 2013; Gethmann et al., 2015).[1] In the past three decades, around 35 project groups of this kind have been carried out, of which the results were primarily published in the Springer series 'Ethics of Science and Technology Assessment', within which this book represents the latest entry.

The ReMoDigital-project group particularly provided a platform to reflect fundamental challenges in relation to resilient energy systems, as well as to explore options for system analysis. In addition, the output of the project group provided a valuable basis for reflection in relation to the potential landscape of 'resilience analysis' that can be integrated within stress-test- and monitoring-tools, and as such supported methodological elaboration and allowed for the definition of future avenues for the developed instrument within the context of ReMoDigital.

The included chapters in this book have been initialised, reflected upon, and finalised by a broad range of experts involved in this project group. From the ReMoDigital consortium—apart from those involved from IQIB—these were from the DLR Institute of Networked Energy Systems (DLR-VE) Hans Christian Gils, Karsten von Maydell, Karen Derendorf, and Henning Wigger; from the DLR Institute of Transport Research (DLR-VF) Tudor Mocanu and Christian Winkler; from OFFIS Sebastian Lehnhoff and Michael Brand; and from the Center for Interdisciplinary Risk and Innovation Studies (ZIRIUS) Sigrid Prehofer and Wolfgang Weimer-Jehle. In addition, various experienced and relevant scientists were involved as external members, of whom some already had previous experience regarding the project group format. These were Prof. Ortwin Renn (Research Institute for Sustainability RIFS), Prof. Arnim von Gleich (University of Bremen), Prof. Matthias Kaiser (University of Bergen), Prof. Martin Wickel and Prof. Jörg Knieling (HafenCity University Hamburg), and Prof. Thomas Ziesemer (United Nations University—Maastricht Economic and Social Research Institute on Innovation and Technology).

[1] Decker, M. and Grunwald, A. (2001) 'Rational Technology Assessment as Interdisciplinary Research', in *Interdisciplinarity in Technology Assessment*. Springer, Berlin, Heidelberg, pp. 33–60. https://doi.org/10.1007/978-3-662-04371-4_4; Gethmann, C. F. et al. (2015) *Interdisciplinary Research and Trans-disciplinary Validity Claims*. Cham: Springer International Publishing (Ethics of Science and Technology Assessment). https://doi.org/10.1007/978-3-319-11400-2; Grunwald, A. (2013) *Rationale Technikfolgenbeurteilung: Konzeption und methodische Grundlagen*. Edited by A. Grunwald. Springer Berlin Heidelberg.

Starting in 2020, the project group of ReMoDigital—supervised by Bert Droste-Franke—was coordinated and organised by Gabriele Fohr, which included preliminary talks, regular meetings and review rounds, as well as the composition and layout of the book. In December 2023, coordination was taken over by Davy van Doren, which included the inter-chapter harmonisation of content, the abstraction of aggregate recommendations, and the overall management of book publication.

Additional to the main experts of the ReMoDigital project group—a core that also represent the main authors of this book—there has been involvement of additional external experts. Here, we like to acknowledge all those that have provided valuable feedback and contributions in relation to the developed concepts and solutions within the project group, including Davood Babazadeh (Hamburg University of Technology), Chris Cummings (US Army Corps of Engineers), Marco Krüger (University of Tübingen), Daniel Lichte (DLR Institute for the Protection of Terrestrial Infrastructures), Miranda Schreurs (Technical University of Munich), Frank Sill Torres (DLR Institute for the Protection of Maritime Infrastructures), and Cathrin Zengerling (University of Freiburg).

We want to sincerely thank Michael Quante—the active series editor of 'Ethics of Science and Technology Assessment'—for his positive feedback on the preliminary ideas and content of this book, which has finally resulted in the version here at hand. In addition, we also like to acknowledge our IQIB colleagues—in particular Markus Voge who carried out the main work of stress-test tool realisation in the systems analysis part of the project—for their various engaging organisational, technical, and scientific support. Finally, we want to thank the German Federal Ministry for Economic Affairs and Climate Action, which funded the project ReMoDigital within the context of the 7th Energy Research Programme.[2]

Bad Neuenahr-Ahrweiler, Germany

Davy van Doren
Bert Droste-Franke
Gabriele Fohr

[2] Funding IDs (FKZ) 03EI1020A, 03EI1020B, 03EI1020C, 03EI1020D and 03EI1020E. See also BMWK—7th Energy Research Programme of the Federal Government.

Acknowledgements

The project "Resilience Monitoring for the Digitalisation of the Energy Transition (ReMoDigital)" was supported by the German Federal Ministry for Economic Affairs and Climate Action (Bundesministerium für Wirtschaft und Klimaschutz)—with the funding ID (FKZ) 03EI1020A, 03EI1020B, 03EI1020C, 03EI1020D and 03EI1020E—within the context of the 7th Energy Research Programme.

The content of the book is the sole responsibility of the involved authors.

Contents

About the Editors

Dr. Davy van Doren is a scientific researcher at IQIB GmbH. In 2006, he obtained a M.Sc. in biology (Leiden University), for which he performed an internship at the Naturalis Biodiversity Center (Leiden, the Netherlands) focussed on the analysis of marine biodiversity in the Southern Caribbean. In 2008, he joined the *Building with Nature* initiative as an environmental engineer for the dredging company Boskalis (Papendrecht, the Netherlands), for which he conducted applied research in relation to the monitoring and preservation of marine ecosystems in the United Arab Emirates. In reflection on this experience, he has written how sustainable development could be realised through the harmonisation of people, planet, and profit in large-scale development projects (https://doi.org/10.1108/978-1-83549-250-520241009). In 2010, he obtained a M.Sc. in sustainable development (Utrecht University), for which he analysed—as part of an internship at the Port of Rotterdam (the Netherlands) and the Rotterdam Climate Initiative (RCI)—standardisation and certification strategies to facilitate sustainable trade and supply of biofuels. In 2011, he joined the Fraunhofer Institute for Systems and Innovation Research ISI (Karlsruhe, Germany), where he focussed on the development and management of the emerging techno-scientific domain synthetic biology. In this regard, he has published in relation to the conceptualisation of multidisciplinary activity (https://doi.org/10.4337/9781782547884.00019), the activity and dynamic of R&D (https://doi.org/10.1007/s11693-013-9121-7), evolving practices of technology assessment (https://doi.org/10.1093/scipol/scu034), and organisational management approaches in relation to external technology commercialisation (https://doi.org/10.1080/09537325.2021.1895103), leading to a doctorate obtained at the Open Universiteit (Heerlen, the Netherlands). Since joining the European Academy in 2016, he has worked on a diversity of research projects related to the simulation of innovation system-based technology diffusion, the assessment of energy system resilience and associated policy measures, as well as the modelling of intra- and inter-organisational innovation and transfer processes (https://doi.org/10.1007/978-3-030-34127-5_44).

Dr. Bert Droste-Franke is head of the subject area "System Evaluation and Societal Future Viability" at IQIB GmbH. His focus is on system, innovation and technology impact analyses for decision support based on theoretical work, inter-/transdisciplinary expert groups, quantitative analyses and modelling. Bert Droste-Franke studied physics with a minor in economics at the Universities of Göttingen and Heidelberg. In 1996 Diploma obtained with a thesis on "Aircraft-borne acetone measurements in the upper troposphere" at the Max Planck Institute for Nuclear Physics in Heidelberg. In 2004 Doctorate in engineering sciences with a thesis on the quantification of environmental damage as a contribution to environmental economic accounts at the University of Stuttgart. From 1996 to 2006 he was a scholarship holder and research assistant at the Institute for Energy Economics and Rational Use of Energy (IER) at the University of Stuttgart (specialising in Integrated Assessment Modelling, impact path analyses, technology assessment, external costs). Since 2006, he has been a project coordinator and scientist/senior scientist at the EA (now IQIB GmbH) in interdisciplinary projects to analyse framework conditions and options for sustainable innovation systems, with a focus on "energy systems". In 2015, he became Head of the "Energy" research area, in 2019 he became Head of the Research business area, and since 2023 he is Head of the subject area "System Evaluation and Societal Future Viability". Examples of written works include "Balancing Renewable Electricity" (https://doi.org/10.1007/978-3-642-25157-3) "Improving Energy Decisions" (https://doi.org/10.1007/978-3-319-11346-3), contributions to methodologies for technology assessment ("Methoden für die Technikfolgenabschätzung…" (https://link.springer.com/book/10.1007/978-3-658-48491-0)), the German flooding 2021—from rebuilding to climate resilience ("Die Flutkatastrophe 2021—Vom Wiederaufbau zur Klimaresilienz" (upcoming)) (all Springer), and a handbook on technology assessment (https://doi.org/10.5771/9783748901990).

Gabriele Fohr is a city and regional planner whose current areas of research include renewable energy, climate adaptation, and other sustainability issues—with a special focus on the design of suitable spatial planning processes, such as participative modelling. She received her engineering degree at Technical University of Berlin (TU Berlin) with a diploma thesis on Innovative Milieus at the German Institute for Economic Research (DIW Berlin). Before joining EA European Academy, she was a research associate at Potsdam University of Applied Sciences, where she was also holding a Ph.D. scholarship on the applicability of resilience or respectively adaptive management, to sustainable urban development. In addition, she helped advance student art and sustainability projects, which were acclaimed as "exzellente Lehre" (excellent teaching). As a senior researcher at EA and IQIB GmbH, she initiated, proposed, led, and contributed to numerous multidisciplinary research projects in the field of energy and system evaluation. Her works particularly cover analyses of regional and technological innovation systems (https://doi.org/10.1007/978-3-319-43940-2_4, https://doi.org/10.14512/tatup.28.3.34) as well as the development of concepts (https://doi.org/10.1186/s12889-024-19025-4), methods (https://doi.org/10.3390/ijerph18136874) and models (https://doi.org/10.1007/978-3-030-61503-1_46). She was scientific coordinator of the project group of ReMoDigital from 2020 to 2023, and is an author-editor of this book.

Part I
Introduction

Chapter 1
Shaping and Monitoring Resilient Energy Systems—An Introduction

Davy van Doren

Abstract Although recent years have tragically demonstrated the effect of climate change on the probability and intensity of extreme weather-related incidents, it has been argued that associated systemic collapse can often be partly attributed to inadequate risk assessment and management practices. As such, there is need for improved governance approaches that are better suited to deal with the uncertain scope and timing of high-stress situations, as well as with the general unpredictable dynamic being characteristic for complex systems. In this introduction, we focus on the form and function of monitoring and digitalisation solutions within the context of developing resilient energy systems. For this, we identify associated main themes and characterise these along the dimensions of understanding, governing, and operationalising resilience in energy systems, as well as introduce various contributions contained in this work that elaborate in-depth on specified topics.

Keywords Energy systems · Systemic stress · Risk assessment · Resilience · Monitoring tools · Digitalisation · Governance

1.1 Introduction

Despite blinking warning lights on July 13th, in the summer of 2021, even scientists were stunned by the chaos that would hit Western Germany four days later (Cornwall 2021).

In the aftermath of the flooding in Rhineland-Palatinate and North Rhine-Westphalia, over 180 casualties were counted and an estimated building and infrastructural damage of approximately 40 billion Euro (Birkmann et al. 2023). Of the ten-thousands of citizens that were affected by the natural disaster, many were abruptly pushed into a state of survival. Critical infrastructure functioning came to a sudden

D. van Doren (✉)
IQIB - Institut für qualifizierende Innovationsforschung und –beratung GmbH, Bad Neuenahr-Ahrweiler, Germany
e-mail: davy.vandoren@iqib.de

D. van Doren et al. (eds.), *Towards Energy System Resilience*, Ethics of Science and Technology Assessment 55, https://doi.org/10.1007/978-3-032-03187-7_1

stop and victims were instantly deprived from fundamental services due to lost electricity, interrupted gas pipes, broken-down fresh water supply, and died down mobile networks. Mobility in and around the valley came under tremendous stress as many hurried—often over heavily affected roads—with helping hands towards the Ahr-Region, while others rushed to escape this new dangerous reality of a shockingly transformed valley. Hundreds of patients were evacuated from destructed clinics and non-functioning hospitals in the first days after the flooding (Roggenkamp and Herget 2024).

Now, years later, much effort has been put into understanding the conditions leading to these devastating floods. After rain became excessive—falling on saturated soils from previous precipitation around the deep structured valley—the recorded maximum water level of 3.71 m from 2016 of the Ahr was exceeded within just 48 h, with a major share of rain falling in approximately 10 h (Schäfer et al. 2021). After the measurement station broke down at a water level of 5.05 m, model calculations from the Rhineland-Palatinate State Environment Agency have estimated the water level to have reached around seven meters (Trenczek et al. 2022), while SGD Nord estimated local water levels of even above 10 meters during the flooding.[1]

Extreme weather conditions, astonishingly, were not identified as the only driving force behind the systemic collapse of the Ahr-valley. Research has identified that the malfunctioning of existing warning systems caused many people—particularly those living along smaller streams—to be caught by surprise. One study showed that a majority of inhabitants of the affected areas did not receive any warning, as well as that the majority of those that were warned did not expect severe flooding (Thieken et al. 2023). In addition, a large share of affected people lacked situational knowledge on protective behaviour, potentially due to flawed media reports and official warnings failing to provide adequate recommendations to people at risk (Thieken et al. 2023).

In this chapter, some of the key challenges in relation to risk assessment and management approaches—issues that also seem to have strongly influenced pre- and post-catastrophic processes within the Ahr-valley context—are introduced. In relation to this, a general argument is provided regarding the need for resilience integrated in governance approaches, with a particular focus on energy systems. In addition, we introduce the overall investigation presented in this book, as well as summarise the included contributions that focus on specific themes. We conclude with some remarks on the potential value of resilience monitoring and digitalisation tools for better coping with extreme weather conditions.

[1] https://sgdnord.rlp.de/fileadmin/sgdnord/Abteilung_3/Dokumente/Ahr/Bezugswasserstaende_Ahr_A3_Teil1.pdf

1.2 Systemic Collapse in Relation to Risk Assessment and Management Approaches

Based on the status quo, existing trends of climate change indicate an increased probability and intensity of extreme weather conditions (Liu et al. 2023). This fact is highlighted by various tragic weather-related catastrophes—both within and outside Europe—that have occurred over the past decade (Dottori et al. 2023).

Often, however, associated systemic collapse is not caused by weather conditions exclusively. As recent events made clear, applied risk assessment and management approaches appear often inadequate. Since the initiation of the European Floods Directive in 2007, all EU countries are required to create and update hazard and risk maps, primarily using input from geography-based assessments and the modelling of flood scenarios and associated potential adverse consequences to citizens and assets (European Parliament 2007). These maps are then intended to be used in drafting management plans that include concrete and coordinated measures to reduce the risks associated to flooding.

However, recent floodings have highlighted that the operational value of such plans might be limited. For example, one study showed that significant historical data was not considered in the drafting of risk management plan for the Ahr-Valley, resulting in biased assessment (Schäfer et al. 2021). Here, the events of 2021 showed similarity—from a hydrological perspective—with a second flooding in 1910 with assessed flow rates around 500 m^3/s. Another flooding in 1804, over two centuries ago, was previously estimated by the Bonn University to have a flow rate of 1100 m^3/s. In 2021, however, both 1804 and 1910 floodings were not included in risk assessment. Although these historic events were primarily excluded due to a lack of ´scientific sound measurements´ prior 1947, the then active plans might have been significantly different when such significant events had been considered.

Additional to these limitations of forecasting and warning procedures, post-catastrophic restoration activities have also indicated significant drawbacks. A main aspect regards the balance between ´build-back-soon´ versus ´build-back-better´, or to what extent the conceptualisation and implementation of alternative solutions potentially reducing the risk of reoccurrence should be considered. In 2022, one year after the disaster, the mayor of Bad Neuenahr-Ahrweiler Guido Orthen stated that bureaucracy has been unable to deal responsively with innovative protective measures that could keep residents safe from future floods (Agence France Presse and Associated Press 2022). In addition, he stated that growing mountains of paperwork had exhausted the limited capacity of local officials in charge of handling reconstruction cases, leading to large delays in the handing out of available government aid. This ultimately resulted in an overall feeling of disenchantment and powerlessness among residents, causing many of them to leave town. Such experience strongly opposes the general assumed functionality of bureaucracy to compensate for the misfits of free market behaviour (Carnis 2009).

1.3 Need for Improved Governance

1.3.1 Integration of Resilience

Unfortunately, this tragic flooding cannot be considered as an isolated incidence, but one part of a widespread trend across Europe (Copernicus Climate Change Service and World Meteorological Organization 2024).

To better deal with the extreme events that can cause the breakdown of complex systems, there has been political direction to go beyond the drafting of traditional risk management plans. In this light, the Critical Entities Resilience Directive by the European Commission—which entered into force on 16 January 2023—has created a framework to support Member States in preventing, resisting, absorbing and recovering from disruptive incidents related to natural hazards, terrorism, insider threats, sabotage, or public health (The European parliament and the council of the European Union 2022).

As the title implies, the concept of resilience stands central in this directive. Holling—widely acknowledged as resilience´s founding father—observed already decades ago that ecosystems do not evolve to a single stable climax state, but instead undergo periodic cycles of change. This led him to the inspiration that, contrary to more classical system-in-equilibrium notions, ecosystems contain a certain resistance to cope with stressors, as well as entail a strong element of adaptation in the light of changing conditions (Holling 1973; Martin-Breen and Anderies 2011).

Since then, the concept of resilience has extended its reach into various socio-economical, socio-technical and socio-ecological contexts (Folke et al. 2010), including agriculture (Munawar et al. 2020), healthcare (Iflaifel et al. 2020; Nemeth et al. 2011), information and communications technology (Bertschek et al. 2019), supply chains (Sáenz and Revilla 2014; Rezapour et al. 2017; Mensah and Merkuryev 2014), urban planning (Chelleri 2012), education (Naidu 2021), employment and economy (Kahl and Hundt 2015), organisations (Mcmanus et al. 2007; Jain 2024), politics (Grove 2013; McDonnell 2020; Vale 2014), and society in general (Davidson-Hunt and Berkes 2003). Also, resilience has been studied in relation to the generation, distribution and consumption of energy (Jasiūnas et al. 2021).

1.3.2 ReMoDigital—Resilience Monitoring and the Digitalisation of Energy Transition Processes

The project ReMoDigital—a project focussed on resilience monitoring for the digitalisation of the energy transition—should be placed in this prioritisation of developing and managing resilience in energy systems.[2] Such management tools particularly

[2] See https://www.iqib.de/de/remodigital-resilienz-monitoring-fur-die-digitalisierung-der-energiewende.html; accessed 2 October 2025.

aim to facilitate more robust services of energy systems through monitoring the effects of far-reaching structural changes—mainly caused by the generally stronger fluctuation of renewable energy supplies, the realisation of sector coupling, and consequently the introduction of new actors and non-traditional roles (Chang et al. 2021). The aim of ReMoDigital was to develop a method that allows—in the context of large-scale turbulence and shocks—the control and monitoring of system functionality and adaptability. In addition, it placed a particular focus upon digitalization, which potentially can make technical energy system more decentralized and orchestrate complex energy conversion, storage, and consumption patterns, but also adds systemic vulnerabilities (Clifton et al. 2023).

The aim of this book is to provide a fundament that addresses theoretical and operational issues and objectives specified within the context of ReMoDigital. To achieve this, resilience is here defined and conceptualised as follows:

- ***Services***: *Resilience is the ability of a socio-technical system to sustain, or quickly recover, its services, even under the condition of major stress.*
- ***Prevention and recovery***: *Resilience includes a preventive aspect by means of sustaining system services and damage limitation, as well as a crisis-management aspect by means of recovering quickly.*
- ***Characterisation and Improvement***: *Resilience can be used to descriptively characterize a system, but also to normatively improve a system towards being more resilient.*
- ***Design and Performance***: *Resilience includes technical design criteria, as well as criteria related to organizational and institutional performance.*
- ***Learning***: *Resilience implies a dynamic approach based on social and institutional learning, which can facilitate the transformation of systems.*

1.4 Realising Resilience of Energy Systems—An Outline of Topics

To build a fundament for the conceptualisation and development of a resilience monitoring tool for energy systems, this book has been structured in three main parts. The book concludes with a synthesis, in which a condensed summary is given from the main lessons-learned, as well as in which key recommendations distilled from all provided contributions are provided.

1.4.1 Understanding of Resilience

In this part, a theoretical basis is provided regarding resilience in relation to complex systems.

In Chap. 2 *by Renn et al.*, a systematic exploration is made regarding conceptual and operational aspects of resilience engineering and resilience monitoring. For this, insights are drawn from complexity research in relation to central capabilities and design elements for resilience development, including potential implications of such insights in relation to risk management and monitoring processes.

In Chap. 3 *by von Gleich and Giese*, the concept of biomimetic resilience is explored and reviewed in relation to other established resilience and risk management concepts. Based on assessed benefits and identified restrictions associated to biomimetic resilience, potential solutions are discussed how this concept could anchor precautionary principles and be transferred from natural role models to socio-techno-economic implementations.

In Chap. 4 *by Weimer-Jehle and Prehofer*, problems fundamentally associated to analysis of system-context are discussed, as well as how context scenarios could be developed and applied to overcome associated problems. A discussion is provided how the system context problem manifests itself in the application of the resilience concept for energy systems, and also includes a reflection on how context scenarios have been applied in practice.

1.4.2 Governance of Resilience

In this part, the predominant focus rests on governance issues associated to resilience.

In Chap. 5 *by Renn*, an in-depth review is made of the precautionary principle in the context of risk assessment and risk evaluation. It also explains risk management strategies developed by the International Risk Governance Council and reflects on how such strategies connect to ongoing debates on precaution and resilience.

In Chap. 6 *by Kaiser*, a conceptual discussion is provided on how existing value-landscapes inform and influence behaviour within the context of crisis situations, which can arise resulting from abrupt events or tipping-point based slow-burners. It also reflects on the role of science-for-policy for crisis management and complex societal disruptions, as well as investigates the importance of neighbourhood networks and learning processes for creating social resilience.

In Chap. 7 *by Knieling*, scientific discourses on infrastructure planning related to complexity, insecurity, and social acceptance are considered. Based on three comparative case studies on different participation policies introduced in Germany, the Netherlands, and the United Kingdom, an empirical based discussion is provided regarding potential adequate solutions to integrating public participation into the implementation process of renewable energy projects.

In Chap. 8 *by Fohr*, the drawbacks of existing energy system models in measuring resilience are elaborated, particularly due to limited observed emphasis on social, behavioural, and political drivers. To overcome this, an argument is made that decision-making processes could be improved and made more legitimate through complementing analytic findings with deliberation processes, including suggestions how deliberative elements could be integrated into energy resilience monitoring processes.

In Chap. 9 *by Wickel*, the conceptual embedment of resilience in infrastructure law is investigated, using exemplary specific parts of the electricity supply infrastructure. With a specific focus on German federal law, it examines which structural features of infrastructure law could be concentrated upon for regulating resilience.

In Chap. 10 *by Ziesemer*, the costs of obtaining resilience are introduced as a problem of resilience maximization, leading to the development of a resilience index covering vulnerability and prevention. In comparison to established indices from engineering, geography, and economics, a simple, growth-rate-stable, empirical vector-error correction and German data-based model is provided to enable resilience shock analysis, including some reflections on relevant incentive systems and policy approaches for driving resilience.

1.4.3 Operationalisation of Resilience

In this part, there is an emphasis on how resilience has been—or can be—operationalised within various contexts of energy systems.

In Chap. 11 *by Wigger* et al., insights are provided into the general characteristics of indicators, as well as into fundamental requirements with respect to the measuring and monitoring of resilience for energy systems. Also, some exemplary indicators are reviewed in-depth regarding their value for the assessment and management of resilience in energy systems.

In Chap. 12 *by Brand* et al., it is argued that the increased diffusion of distributed and often weather dependent energy resources requires digitalisation solutions to flexibly balance electricity generation and consumption. Regarding associated vulnerabilities of such solutions, four different measures—i.e. enhanced situational awareness, virtualization, flexibilization, and distributed black-start—are analysed in relation to their impact on systemic resilience.

In Chap. 13 *by Gils and Yeligeti*, special attention is given to the resilience of future national energy systems. In doing so, the impact of the energy system transformation in relation to resilience is discussed. Regarding the evaluation of future energy system resilience, both possibilities and limitations are evaluated for systematic stress-test analyses that are based on the application of existing energy system optimisation models.

In Chap. 14 *by Derendorf* et al., the driving force of renewable energy growth on the diffusion of information and communications technology in electricity supply—including related advantages and disadvantages—is discussed. With a focus on the effects of small and interconnected grid participants, various events are reviewed that can potentially disturb the stability of energy systems, as well as which measures could be implemented to improve the operation and associated resilience of distribution grids.

In Chap. 15 *by Mocanu* et al., a review of resilience in transportation is provided in the light of increased electrification, required charging infrastructures, vehicle-to-grid (V2G) systems, and emerging opportunities and challenges for sector coupling. Based on a developed roadmap towards a transport system resilience assessment framework, relevant research topics are identified and mapped in the context of the energy transition and digitalization.

In Chap. 16 *by Droste-Franke*, a discussion is given on analytical challenges and opportunities for providing support in designing resilient and digitalized energy systems, with a particular focus on monitoring catastrophic impacts on system-internal infrastructures, environment, and society, including potential effect propagation to other systems. Suggestions are made for how a stress-test- and monitoring-tool—based on contextualized framework conditions, prepared scenarios, and selected digital resilience measures—can be developed and applied in decision making processes.

In Table 1.1, an overview of the main chapters and associated keywords is provided. In the final chapter by van Doren et al., a structured synthesis is provided of all main chapters, including recommendations how resilience in energy systems can be realised.

1.5 Conclusion

The observed increase of extreme weather-related incidents, in combination with the often inadequate response from public authorities responsible for risk assessment and management, has stressed the need for improved governance tools. With a focus on energy systems, a generalised research approach was here introduced regarding existing aims to realise resilient energy systems. In addition, an overview of topics relevant for achieving such aims was provided, including a particular focus on the potential value of resilience monitoring and digitalisation tools. The provided contributions show that there are still considerate challenges that need to be overcome to better cope with systemic stress, but also highlight potential steps that can be taken towards developing and managing resilient energy systems.

Table 1.1 Overview of book chapters and associated keywords

Chapter	Title	Keywords
Resilience in Complex Systems		
2	Coping with complexity—A systems approach to resilience	Resilience, vulnerability, recovery, complexity, transformative, adaptive, biomimetic
3	Biomimetic Resilience—Performance Claim, Concept and Design Principles	Biomimetic resilience, risk management, design principles, unknown unknowns, precaution
4	System and Context	Resilience, energy systems, context sensitivity, context scenarios, cross-impact balances
Governance of Resilience		
5	Risk, Resilience and precaution	Complexity, uncertainty, ambiguity, precautionary principle, risk assessment, risk management
6	Resilience and values—a conceptual approach	Societal resilience, social identities, value landscapes, norm-driven decision making, public engagement
7	Governance of Energy Infrastructures: Comparing Planning Cultures in Germany, the Netherlands, and the United Kingdom	Infrastructure planning, complexity and insecurity, social acceptance, public participation policy
8	Resilience Monitoring as a Participative Process	Energy resilience, resilience monitoring, analytic-deliberative approach, multi-stakeholder process, participative scenario-building, social resilience, self-organization
9	Resilience in the System of Infrastructure Law	Infrastructure law, resilience, emission control law, energy law, planning approval
10	Resilience, shock prevention, and interacting systems: Economic aspects of REMOdigital	Resilience costs, interacting systems, vector-error-correction model, shock analysis
Operationalisation of Resilience		
11	Indicators for the resilience monitoring of energy systems	Indicators, assessment, energy system, resilience, reliability, indicator selection, monitoring
12	Resilience of Digitalized Power Systems Challenges and Solutions	Digitalised power systems, resilience, anticipate, react, recover
13	Using optimising infrastructure planning models for the resilience monitoring of future integrated energy systems	Energy systems modelling, optimisation, sector coupling, stress testing, resilience
14	Implementing resilience strategies in future digitalized distribution grids	Resilience, stability, distribution grid, fallback, disturbance in controllability, disaster prevention, disaster management

(continued)

Table 1.1 (continued)

Chapter	Title	Keywords
15	Assessing the resilience of the current and future transportation system	Transportation, transportation system, transportation resilience, digitalization, sector coupling, climate change, travellers' behaviour, assessment framework
16	Systems Consideration for Resilience Monitoring	Energy systems, transport systems, ICT systems, systems analysis, resilience modelling, systems theory, stress test tool, Ahr-flood disaster

References

Agence France Presse and Associated Press: After the floods: Germany's Ahr valley then and now – in pictures. The Guardian (2022). https://www.theguardian.com/world/2022/jul/13/floods-then-and-now-photographs-germany-ahr-valley-flooding-disaster-july-2021. Accessed 2 Oct 2025

Bertschek, I., Polder, M., Schulte, P.: ICT and resilience in times of crisis: evidence from cross-country micro moments data. Econ. Innov. New Technol. **28**(8), 759–774 (2019). https://doi.org/10.1080/10438599.2018.1557417

Birkmann, J., et al.: Strengthening resilience in reconstruction after extreme events – Insights from flood affected communities in Germany. International Journal of Disaster Risk Reduction **96**, 18 (2023). https://doi.org/10.1016/j.ijdrr.2023.103965

Carnis, L.: The economic theory of bureaucracy: insights from the Niskanian model and the Misesian approach. Q. J. Aust. Econ. **12**(3), 57–78 (2009). https://www.researchgate.net/publication/228421658. Accessed 2 Oct 2025

Chang, M., et al.: Trends in tools and approaches for modelling the energy transition. Appl. Energy **290**, 116731 (2021). https://doi.org/10.1016/j.apenergy.2021.116731

Chelleri, L.: From the «Resilient city» to urban resilience. a review essay on understanding and integrating the resilience perspective for urban systems. Doc. D'analisi Geogr. **58**(2), 287–306 (2012). https://doi.org/10.5565/rev/dag.175

Clifton, A., et al.: Grand challenges in the digitalisation of wind energy. Wind Energy Sci. **8**(6), 947–974 (2023). https://doi.org/10.5194/WES-8-947-2023

Copernicus Climate Change Service and World Meteorological Organization: European State of the Climate 2023 (2024). https://climate.copernicus.eu/europe-experiences-widespread-flooding-and-severe-heatwaves-2023

Cornwall, W.: Europe's deadly floods leave scientists stunned. Science **373**(6553), 372–373 (2021). https://doi.org/10.1126/science.373.6553.372

Davidson-Hunt, I. J. and Berkes, F.: Nature and society through the lens of resilience: toward a human-in-ecosystem perspective. In: Berkes, F., Colding, J., Folke, C. (eds.) Navigating Social-Ecological Systems: Building Resilience for Complexity, pp. 53–83. Cambridge University Press (2003)

Dottori, F., et al.: Cost-effective adaptation strategies to rising river flood risk in Europe. Nat. Clim. Chang. **13**(2), 196–202 (2023). https://doi.org/10.1038/s41558-022-01540-0

European Parliament (2007) DIRECTIVE 2007/60/EC OF THE EUROPEAN PARLIAMENT AND OF THE COUNCIL of 23 October 2007 on the assessment and management of flood risks. Off. J. European Union, p. 8. https://eur-lex.europa.eu/eli/dir/2007/60/oj. Accessed 2 Oct 2025

Folke, C., et al.: Resilience thinking: integrating resilience, adaptability and transformability. Ecol. Soc. **15**(4), 9pp. https://www.jstor.org/stable/26268226. Accessed 2 Oct 2025

Grove, K.: On resilience politics: from transformation to subversion. Resilience **1**(2), 146–153 (2013). https://doi.org/10.1080/21693293.2013.804661

Holling, C.S.: Resilience and stability of ecological systems. Annu. Rev. Ecol. Syst. **4**, 1–23 (1973)

Iflaifel, M., et al.: Resilient health care: a systematic review of conceptualisations, study methods and factors that develop resilience. BMC Health Serv. Res. **20**(1), 1–21 (2020). https://doi.org/10.1186/S12913-020-05208-3/TABLES/2

Jain, A.: Organizational resilience to knowledge loss in biotechnology research. IEEE Trans. Eng. Manag. **71**, 5627–5640 (2024). https://doi.org/10.1109/TEM.2024.3365802

Jasiūnas, J., Lund, P.D., Mikkola, J.: Energy system resilience – a review. Renew. Sustain. Energy Rev. **150**, 111476 (2021). https://doi.org/10.1016/J.RSER.2021.111476

Kahl, J., Hundt, C.: Employment performance in times of crisis: a multilevel analysis of economic resilience in the German biotechnology industry. Compet. Rev. **25**(4), 371–391 (2015). https://doi.org/10.1108/CR-12-2014-0038/FULL/XML

Liu, W., et al.: A probabilistic assessment of urban flood risk and impacts of future climate change. J. Hydrol. **618**, 129267 (2023). https://doi.org/10.1016/J.JHYDROL.2023.129267

Martin-Breen, P., Anderies, J.M.: Resilience: a literature review (2011)

McDonnell, S.: Other dark sides of resilience: politics and power in community-based efforts to strengthen resilience. Anthropol. Forum **30**(1–2), 55–72 (2020). https://doi.org/10.1080/00664677.2019.1647828

Mcmanus, S., et al.: Resilience management: a framework for assessing and improving the resilience of organisations. Resilient organisations research report. http://hdl.handle.net/10092/2810. Accessed 2 Oct 2025

Mensah, P., Merkuryev, Y.: Developing a resilient supply chain. Procedia Soc. Behav. Sci. **110**, 309–319 (2014). https://doi.org/10.1016/J.SBSPRO.2013.12.875

Munawar, S., et al.: Role of Biotechnology in Climate Resilient Agriculture', Environment, Climate, Plant and Vegetation Growth, pp. 339–365 (2020). https://doi.org/10.1007/978-3-030-49732-3_14

Naidu, S.: Building resilience in education systems post-COVID-19. Distance Educ. **42**(1), 1–4 (2021). https://doi.org/10.1080/01587919.2021.1885092

Nemeth, C., et al.: Resilience is not control: Healthcare, crisis management, and ICT. Cogn. Technol. Work **13**(3), 189–202 (2011). https://doi.org/10.1007/S10111-011-0174-7/METRICS

Rezapour, S., Farahani, R.Z., Pourakbar, M.: Resilient supply chain network design under competition: a case study. Eur. J. Oper. Res. **259**(3), 1017–1035 (2017). https://doi.org/10.1016/J.EJOR.2016.11.041

Roggenkamp, T., Herget, J.: Flood reconstruction – the unexpected rather frequent event at River Ahr in July 2021. Global Planet. Change **240**, 104541 (2024). https://doi.org/10.1016/J.GLOPLACHA.2024.104541

Sáenz, M.J., Revilla, E.: Creating more resilient supply chains. MIT Sloan Manag. Rev. (2014). https://sloanreview.mit.edu/article/creating-more-resilient-supply-chains/. Accessed 2 Oct 2025

Schäfer, A.A.M., et al.: Hochwasser Mitteleuropa, Juli 2021 – Bericht Nr. 1 „Nordrhein-Westfalen & Rheinland-Pfalz" (2021). https://doi.org/10.5445/IR/1000135730

The European parliament and the council of the European Union: Directive (EU) 2022/2557 of the European Parliament and of the Council of 14 December 2022 on the resilience of critical entities and repealing Council Directive 2008/114/EC. Off. J. Eur. Union **L333**(164), 35pp. (2022)

Thieken, A.H., et al.: Performance of the flood warning system in Germany in July 2021 - insights from affected residents. Nat. Hazard. **23**(2), 973–990 (2023). https://doi.org/10.5194/NHESS-23-973-2023

Trenczek, J., et al.: Kosten durch Klimawandelfolgen - Schäden der Sturzfluten und Überschwemmungen im Juli 2021 in Deutschland: Eine ex-post-Analyse. Düsseldorf (2022)

Vale, L.J.: The politics of resilient cities: whose resilience and whose city? Build. Res. Inf. **42**(2), 191–201 (2014). https://doi.org/10.1080/09613218.2014.850602

Part II
Understanding of Resilience

Chapter 2
Coping with Complexity—A Systems Approach to Resilience

Ortwin Renn, Matthias Kaiser, and Arnim von Gleich

Abstract Resilience in this chapter is defined as the ability of a socio-technical system to sustain or quickly recover its services to society even under the condition of major stress. Resilience includes a preventive aspect (sustain service functionality) as well as a crisis-management aspect (recover quickly). It may be used to characterize a system (descriptive), but it may also serve as a normative concept to improve a system towards a more resilient stage, such as building back better. It includes technical design criteria as well as organizational and institutional performance. Finally, resilience implies a dynamic approach based on social and institutional learning The paper distinguishes four different modes of resilience: (i) building back better; (ii) adaptive; (iii) transformative and (iv) biomimetic. Based on complexity theory, the paper advocates a dynamic, comprehensive and transformative perspective when analysing or designing energy systems.

Keywords Resilience · Vulnerability · Recovery · Complexity · Transformative · Adaptive · Biomimetic

2.1 Introduction

The concept of resilience has been used in many disciplines for being able to respond adequately when the system is under stress (for an overview see Comfort et al. 2010). It has been widely applied in ecological research and denotes the resistance of natural ecosystems to cope with stressors (Holling 1973). In psychology, resilience denotes

O. Renn (✉)
Research Institute for Sustainability / at GFZ (RIFS), Potsdam, Germany
e-mail: ortwin.renn@rifs-potsdam.de

M. Kaiser
University of Bergen—Centre for the Study of the Sciences and the Humanities, Bergen, Norway
e-mail: Matthias.kaiser@uib.no

A. von Gleich
University of Bremen, Bremen, Germany
e-mail: gleich@uni-bremen.de

D. van Doren et al. (eds.), *Towards Energy System Resilience*, Ethics of Science and Technology Assessment 55, https://doi.org/10.1007/978-3-032-03187-7_2

the ability to cope successfully when being exposed to stress conditions (Fletcher and Sarkar 2013) and in the field of engineering resilience it is used to describe the capability of a socio-technical system to resist, restabilise and reconfigure in times of crisis and stress (Coaffee 2008; Comes et al. 2020). In more general terms: Resilience is focused on the ability and capacity of systems to resist shocks and to have the capability to deal with and recover from threatening events (Rose 2007, Jackson and Ferris 2017). This idea of resistance, recovery and adaptation can also be applied to social systems (Review in Norris et al. 2008; Adger 2000). The main emphasis there is on organizational learning and institutional preparedness to cope with stress and disaster. The US Department of Homeland Security (DHS) uses this definition: "Resilience is the ability of systems, infrastructures, government, business, and citizenry to resist, absorb, and recover from or adapt to an adverse occurrence that may cause harm, destruction, or loss [that is] of national significance" (cited after Longstaff et al. 2010: 19). Hutter (2013) added to this analysis the ability of systems to respond flexibly and effectively when a system is under high stress from unexpected crisis. Pulling from an interdisciplinary body of theoretical and policy-oriented literature, Longstaff et al. (2010) regard resilience as a function of resource robustness and adaptive capacity.

The Stockholm Resilience Centre (2015) advocates an understanding of resilience that integrates the human societies as part of ecological systems, believing that humans and nature are strongly interlinked. Based on this understanding, ecological and social dynamics determine the resilience of a socio-ecological system. The main objective of resilience is to cope with crises while staying within planetary boundaries. Folke defines resilience as the "capacity of an SES (socio-ecological system) to continually change and adapt yet remain within critical thresholds" (Folke et al. 2010: 1). Continuous adaptability and transformability are thus major components of resilience as a dynamic rather than a static objective. The ability to engage in transformations that improve the level of sustainability over time goes beyond adaptation, it includes the ability to anticipate future stress situations and improve the preparedness of systems to cope with future challenges without compromising targets for sustainable development (Brown 2014, Bahadur and Tanner 2014, Keck and Sakdapolrak 2013).

The governance framework suggested by the International Risk Governance Council (IRGC 2005, 2017) depicts resilience as a normative goal for risk management systems to deal with highly uncertain events or processes (surprises). It is seen as a property of risk-absorbing systems to withstand stress (objective resilience) but also the confidence of risk management actors to be able to master crisis situations (subjective resilience). The attempt to build resilient systems as part of risk management is called resilience engineering.

For this chapter we define *resilience as the ability of a socio-technical system to sustain or quickly recover its* services to society *even under the condition of major stress* (see similar in Nationalakademien 2017). Resilience includes a preventive aspect (sustain service functionality) as well as a crisis-management aspect (recover quickly). It may be used to characterize a system (descriptive), but it may also serve as a normative concept to improve a system towards a more resilient stage, such as building back better (Hynes et al. 2022, SAPEA 2022). Finally, our understanding of resilience incorporates technical design criteria as well as organizational and institutional performance. Finally, our understanding of resilience implies a dynamic approach based on social and institutional learning. Therefore, resilience includes the ability to adapt, reconstruct, recover and evolve.

In this paper we will explore conceptual as well as operational aspects of resilience engineering and monitoring from a systems perspective. The first section introduces the systemic perspective drawing on the insights from complexity research. We then apply these insights to the various concepts of resilience with focus on capabilities and design elements (part 3), review the implications of these concepts to risk management and monitoring (part 4) and end the paper with some general conclusions (part 5).

2.2 A Systemic Perspective on Complexity

2.2.1 Complex Systems

Societies are complex systems and processes of transformations such as the energy transition can therefore only be understood if they are regarded as dynamic structural elements in complex systems. The common cause-effect models are less and less effective in a systemically networked world, which is so interconnected in its subareas by the technical possibilities that the effects of interventions are impossible or extremely hard to predict (Peletier et al. 2018: 72ff.). This applies to the world as a whole, but also to its numerous subsystems, especially to energy or agriculture. Complex systems consist of many interacting parts, are typically characterized by non-linear functions,[1] and are capable of spontaneous formation of new macroscopic structures that cannot be estimated by simple linear extrapolation from the properties of their elementary parts. Therefore, it is increasingly appropriate to focus on systemic interactions between and among elements instead of linear aggregation of individual data sets. This applies in principle to all systems and processes, be they representation of phenomena in nature, technology, medicine, business or society (Lucas et al. 2018).

All systems are defined by a set of elements and a set of relations (or functions) between the elements. They are scoped by defining its boundaries, i.e. by defining which elements are inside and which are outside: <F, Ä>.[2] Most real systems are open systems, i.e. they exchange information with outside systems; closed systems do not do this. When we model the world around us, we build them on the basis of well-defined systems, where we characterize the state of the system and the interactions

[1] A simple example of a nonlinear function is: $\frac{dx}{dt} = f(x) = x(x^2 - \alpha)$.. The parameter a depicts a slowly varying quantity. The system has three stable states, exactly those where f(x) = 0, namely: x = 0, x = Öa, and x = -Öa. This feature makes the function non-linear.

[2] This is a generic description of a simple system where F depicts the ontology, i.e. the objects which are in our system, given by a description, and A depicts the relation between these objects.

by mathematical functions (sometimes merely by simple arrows indicating positive or negative enforcement).

What are the lessons of complexity research for improving our understanding of resilience? How do societies deal with complex situations and how can we learn from the findings of complexity science developed in the natural sciences about the fundamental characteristics of complex systems? How can we avoid being overwhelmed by complexity and frozen in awe? How can we use the results of complexity research to develop effective orientations for coping with systemic risks and their complex interactions? One of the possible ways to design resilient systems is through resilience engineering.

2.2.2 Interdisciplinary Perspective

Dynamic structure formation processes are subjected to an ordering analysis requiring a variety of interdisciplinary methods of complexity science. They make it possible to understand the dynamics of transformations such as the energy transition or the mobility turn. The basis for such an analysis is provided by the stock of theory and methods that has developed over the past 50 years from the study of dynamic processes, first in physics and chemistry, and later, detached from scientific peculiarities and limitations, transferred to many other systems (Lucas et al. 2018, Lucas and Renn 2020). The general question here is how the interaction of many elements, be they molecules in materials, cells in organisms, machines and apparatuses in technical plants or finally people in markets and organizations, can create dynamic structures, but also destroy them. The temporal development of the structure formation via self-organization, and thus the core task of the analysis, is recorded by differential or difference equations, by time series analyses or also by direct simulation of the interactions between the elements. Computer models and their supporting algorithms play an important role. In complex systems with intentionally acting people as agents, behavioural probabilities are constitutive for dynamic structure formation; there are no fixed deterministic laws that one can apply in this interdisciplinary stock of theory and methods, but only stochastic relationship patterns, i.e. limited to probabilities and possibilities (Peletier et al. 2018: 15f).

2.2.3 Construction Plans of Complex Systems

Complex systems are usually open systems. They interact with their environment beyond their system boundaries. This exchange includes matter, energy and information. Flow patterns in liquids are initiated by the supply of energy across system boundaries, chemical patterns by the supply of suitable chemicals and energy. In societies, the supply of information from the environment leads to the formation of new structures within socio-technical systems. They emerge as a result of technical developments with organizational and individual behaviour in the context of collective culture and political decision-making.

Complex systems and the dynamic structures generated in them are agent-based, i.e., there are elements in the system that take over control functions for the system as a whole, but at the same time interact with each other (Renn and Lucas 2021). Structures are created by the properties of and interactions between the elementary parameters of a system, which are called agents in the language of complexity theory. In the area of dead matter, for example, such effect variables can be molecules such as in flow patterns or oscillating chemical reactions, or sand grains in dune patterns.

In socio-technical systems, too, structures are created by the characteristics and interactions of agents. The development of social etiquette, an established fashion for clothes, the emergence of new lifestyles or even a public opinion on a certain topic are further examples of this dynamic pattern. Complex socio-technical systems and dynamic structures are communication-based, i.e. they arise through local communication, i.e. interactions between agents (Cairney 2012). Macroscopic structures can differ from structures based on interactions between their agents by several orders of magnitude. What is decisive is that it is not a central control system that generates the dynamic structures, but the interactions between the agents. Thus, higher structural orders are achieved without recognizable external controlling elements. This is the principle of self-organization as a fundamental principle of complexity science (McKelvey 2004). In many, if not all, systems it is impossible to control the rules of interactions by appropriate measures and interventions from outside the system.

The interactions between the agents are non-linear. This has the consequence that small causes can show unexpectedly large effects which are inherently unpredictable. As a result, the dynamics of the entire system cannot be derived from the dynamics of its components. The temporal development of a system is therefore difficult to predict, in some cases counterintuitive, and harbours hardly assessable risks or opportunities for its future. The emergence of macroscopic structures from a chaotic multitude of elementary interactions is due to selection through competition. Of the macroscopic patterns generated by the actions of the agents, the so-called modes, one or the other affect the actions of the agents in such a way that they are selectively amplified and thus stabilize the associated macroscopic structure (Levin et al. 2013). This so-called circular causality, together with nonlinearity, is constitutive for dynamic structure formation and can be observed in all domains in which elements interact in complex relationships with each other (Henrickson and McKelvey 2002).

Complex systems and dynamic structural formations finally have emergent properties, i.e. the ability to generate structures on the way of self-organization, which cannot be explained by the isolated behaviour of the system components. Thus, individual molecules move chaotically and do not form ordered vortices, whereas a whole molecular system of countless molecules does so if the external and internal conditions are favourable. Also, the collective dynamics of traffic flow, e.g., a sudden traffic jam without recognizable external cause, is an emergent phenomenon, which results from an interaction between many vehicles. Gentrification in large cities can be seen as an emergent feature based on seemingly chaotic and undirected choices of inhabitants. And while individuals often have moderate political views, they can turn into radical and sometimes irrational hysteria at mass gatherings under appropriate internal and external influences.

What does complexity theory tell us about resilience and risk? The linear model of a risk agent being released by a natural event, technology or human activity and causing harm to a risk absorbing system may be too simplistic to describe how threats to the various targets that humans value emerge and promulgate. Natural, technical and behavioural agents interact with each other, and the results of their interactions are often not predictable; they may pose unexpected or unseen consequences on the risk absorbing system. The risk absorbing system may become an agent in itself by channelling the incoming risk agents towards another absorbing system thus entering into a cascade of vagabond risk agent movements that meander within and between systems until they find the most vulnerable entry point into the targeted system (Lucas et al. 2018). This oscillating pattern is well documented in cybersecurity where viruses or malware circulate around potential targets until the firewall of one target is breached and the target itself becomes a weapon against other targets.

Similar patterns apply to energy systems as a manifestation of a socio-technical system, or more precisely the transformation of the same, the energy transition. Thus, socio-economic and socio-cultural aspects, societal forces and institutions act as risk agents in conjunction with technologies and natural forces such as floods, earthquakes or extreme weather events. Resilience includes all the capacities of any risk absorbing system to be adaptive to dynamic changes of potential risk agents in a world of mutual interactions. Risk analysis is not confined to technical failures and their management but include the influence of many factors within large socio-technical systems, including state governance, organizations, market developments, civil society actors, participation, guiding principles, acceptance and acceptability, social cohesion, trust (credibility), usability, understandability and the scandalization potential of topics (e.g. food vs. fuel debate debate) (IRGC 2017).

2.2.4 Obstacles to Resilience in Complex Socio-Technical Systems

2.2.4.1 Obstacles in the Political System

One can hold that modern society is much better prepared to meet some significant existential threats than in earlier human history. A lot (but not all) of this increased resilience is due to scientific advances in many spheres of life. Many communities of the past were self-sustaining with little communication with the outside world. Seasonal instabilities or pests affecting harvests of food were a recurring problem in the past, and had huge societal consequences on a local, regional and sometimes even national level. With techniques of modern agriculture and global supply chains of food these shocks have been significantly alleviated (even if climate change now amplifies the risks again through global effects). Floods, fires, wars, and various diseases regularly hit these communities with large harms and without the ability to respond effectively to the shocks. The occurrence of Black Death / the Plague had devastating effects in much of Western Europe for several centuries. With modern scientific and technological advances, we have now much better control over these risks, and we have built up a degree of resilience towards these risks. We even succeeded in eliminating some of such global risks like smallpox.

But the example of COVID-19 (and some others: e.g. food and energy supply, or the war in Ukraine) has taught us that we are still far from being prepared to so-called Black-Swan events, i.e. very rare events with potentially huge negative impact. Experts in epidemiology had long warned that pandemics were to be expected regularly, possibly several of them during a century. But even expert knowledge did not prepare us for the COVID-19 pandemic when it occurred; in fact, the “experts” may have given us a false sense of preparedness and security. An example was the Global Health Security Index (GHSI), published roughly half a year before the outbreak of the pandemic (cf. Kaiser et al. 2021), which gave good marks for preparedness precisely to those countries which arguably fared worst during the actual pandemic. The reason behind this was a lack of understanding how different societal systems interact with each other, and how political, national and cultural diversity add complexity to such a degree that reliable predictions on preparedness are virtually impossible. The lesson from this should be that reliance on an expert community from science and technology is not a fool-proof shield against black-swan events. Assuming that some experts are “dealing with the issue” is not bringing us closer to societal resilience (Hynes et al. 2020).

But even if our scientific and technological experts would be better at what they are doing in regard black-swan events, one quickly realizes that this is still not enough to build up societal resilience. We would need a political system that responds in the adequate manner to this expert knowledge. We are then looking at the science-politics interface which is commonly recognized as often strained and beset with misunderstandings from each party regarding the abilities and constraints of the other party. Good brokerage is an ideal but still rarely realized (Gluckman et al. 2021).

The typical problem is that expert warnings are drowned by the day-to-day politics, and typically end up on the backburner of the political agenda. For a very long time this was the situation in regard to climate change, just now slowly changing due to the magnifying focus of extreme weather events. Several authors have identified the structural obstacles in the political system to effectively deal with preparedness and resilience, regardless of what political system we live under (cf. Gluckman and Bardsley 2021; Letwin 2020). Here is a list of some structural obstacles in the political system:

(1) A government thinks in terms of its time in office, the timespan to the next general election. This is typically 3–5 years, in Germany 4 years (there are plans for extending it to 5 years). Any government needs to point to measures that have effectively solved a problem occurring during this time and handled

during this time. In other words, long-term thinking does not pay-off when the next election occurs.

(2) Related to this is the point that there are always more "pressing issues" on the political agenda than those that require long-term thinking and that are highly complex. Ministers, parliamentarians and senior officials are always overloaded with problems from their constituency which demand immediate attention. This implies eliminating those issues from the prioritized agenda which assumedly can wait since they are assumed not to happen today or tomorrow.

(3) Certainty always trumps uncertainty in the sense that any issue or measure which politicians / decision makers are sure will happen will sideline other issues which only might happen. Black-swan events are rare and inherently unpredictable. Events expected as immediate consequences of a current system (or disturbance thereof) and deduced by extrapolation from interfering factors will take priority over the more complex issues which lack the detectable causal chain leading up to them.

(4) Increasing societal resilience may provide invisible benefits which are discounted in the world of political merits (the so-called prevention paradox). If a government "succeeds in preventing a disaster from occurring, the wider population will probably be unaware that it might have occurred" (Letwin 2020: 22). In the world of politics, doing something for which you never will be thanked is a waste of your efforts. Invisible benefits are discounted.

If you still engage in building up resilience and prevention of major risks, you might never know if your measures actually worked, if the worst risks were avoided due to your actions or due to other external factors. The only thing you know for sure is that if your measures do not work in the case of the shock occurring, you will be blamed for doing a bad job and people will make you accountable for experienced harm. This is not much of a political incentive to design long-term strategies for risk reduction and increased resilience.

In order to break through these obstacles, the science-politics interface needs to develop strategies that can bring other (than expert) voices of society to the front which are highly esteemed by larger societal segments in order to give weight to the issue. Nudging and participatory governance can be strategies.

2.2.4.2 Obstacles Based on Mental and Evolutionary Biases

Black-swan events were discussed by philosophers (Hume, Mill, Popper) in relation to the problem of induction in logic, but Nassim Taleb brought the metaphor to a wider audience, specifying it in the sense that he claimed that these unexpected events of large magnitude and high impact play an important role in the course of history, more so than regular occurrences. Statistically they may be outliers and unpredictable but in terms of transformational power in societal systems they are a major driver. But why are they unexpected?

Meyer and Kunreuther addressed this question in their book *The Ostrich Paradox* (2017). Their answer is that our minds are hard-wired to fall prey for a number of biases, and they work to suppress highly uncertain impacts with large negative consequences. Thus, they claim that psychology comes into the picture when we try to understand why we do not perform better in preparing for disasters. The issue of bias is well-known in risk studies, early advanced by Kahnemann and Tversky, and with special relevance when dealing with uncertainty in a complex world (Renn 2017). Gluckman and Bardsley (2021) discuss them as well. Meyer and Kunreuther identify six such biases as especially important in relation to the question why we underprepare for uncertain, rare but inevitable events with high societal impact. These mental biases include myopia, amnesia, optimism, inertia, simplification, and herding:

(1) Myopia is here the tendency that our mind treats rewards that are immediately present different from those that are in the future, discounting the future even when rationality would demand a higher weight on future events.
(2) Amnesia describes the tendency to remember good outcomes and repeating actions that led to these outcomes in the past. But with respect to rare events this strategy may lead us to downplay strategies to "see value in costly actions that carry few observable rewards" (Letwin 2020: 73), since the "benefit" of success lies in avoiding losses rather than accumulating positive benefits.
(3) The optimism bias describes the tendency to believe that bad events occur to others rather than us. There is strong optimism that one can find a way out of it, even if things turn out bad for most others. As a consequence, people also underestimate the effects of cascading risks where risks do not stay localized but extend to various other spheres.
(4) Inertia is the tendency to be easily overwhelmed by the magnitude of the necessary tasks and to prefer not to make decisions or act at all. Instead, one hopes for default solutions that will deal with the issue at hand by itself.
(5) Simplification denotes the tendency of our brain to concentrate on cues which are large and should therefore have our immediate attention. Small probabilities will then naturally take a back seat despite their consequences potentially being very large.
(6) Herding is a well-known phenomenon that people follow the crowds rather than acting on their own. People seek guidance in the behaviour of others which often intensifies the biases mentioned above. Herding leads to heuristics which are quick in response but slow in learning.

These are just some of the mental biases which together might explain why we as societies are unprepared for black swans and lack strong and effective efforts to build up resilience against these rare events. Dealing with resilience in regard complex and uncertain risks / shocks should recognize these mental biases and the structural obstacles of political decisionmakers if we aim at a more satisfactory governance of such risks.

2.2.5 A Model for Dealing with Complex System Interactions

In innovation and transformation research, there are numerous research approaches and models on these complex interactions between drivers and impacts on a variety of diverse trajectories. Innovation research has long been dominated by the question of innovative capacity. For transformation research, the question of the direction of innovation and the corresponding possibilities to influence it is gaining in importance. One example to study these opportunities is the RESYSTRA project focused on how innovation processes can be influenced towards a desired direction (Brand, von Gleich 2015). It worked with the 'turtle model' (Fig. 2.1), with its representation of six selected drivers for directed system transformations (three push factors: regulatory push, scandalization push, technology push, three pull factors: incentive pull, vision pull, market pull). Based on this, the following representation of drivers for a transformation within complex energy systems was developed as a starting point for agent-based modelling of possible pathways to a successful market implementation of synthetic fuels (Schnülle et al. 2020). All agent-based modelling needs to start with pragmatic decisions on how to carve out segments in the pool of basic system interactions.

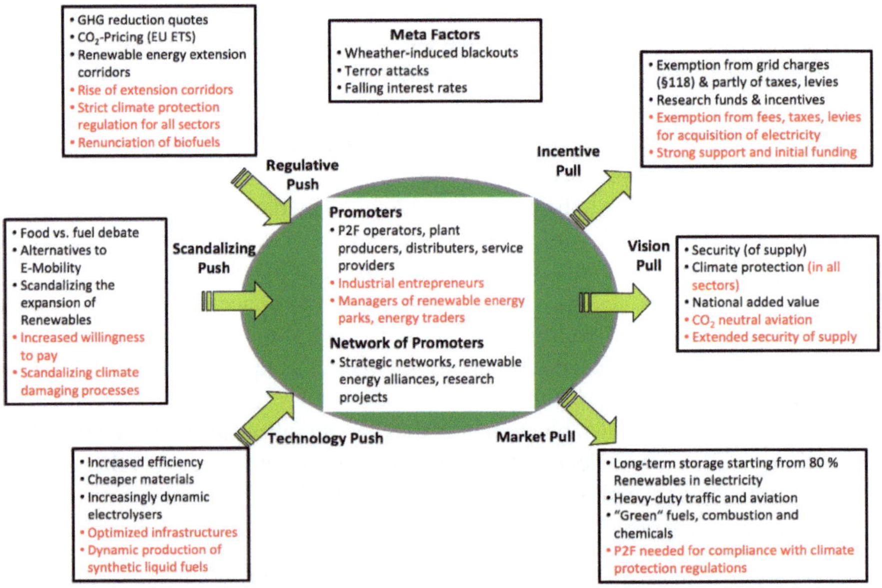

Fig. 2.1 Direction-giving drivers for the establishment of power-to-fuel plants. Directional impulses already present black, red: necessary in the future (Brand et al. 2017)

2.3 A Systemic Perspective on Resilience

2.3.1 Resilient Systems as Adapting and Evolving Systems

As explained in the introduction, resilience is focused on the ability and capacity of systems to resist shocks and to have the capability to deal with and recover from threatening events and even to adapt and reconstruct. For the purpose of resilience engineering, for designing a systemic concept of resilience that can be applied to energy systems, it seems appropriate to emphasize the dynamic aspect of resilience, i.e. the systemic capacity to change as a result of circumstances that push the system beyond the boundaries of its competence envelope. The system may have to amend some, or even all of its goals, procedures, resources, roles, or responsibilities. As a result of those changes, the system under study then expresses a revised competence envelope. In effect, it becomes a different system (Hoffman and Hancock 2017).

This dynamic and systemic understanding relates strongly to the extent of internal changes in affected socio-technical systems (adaptation, improvisation, learning, innovation, evolution) in response to disturbances and failures.

2.3.2 Resilience and Sustainability

Exploring the relationship between resilience and sustainability, it may be useful to distinguish three major concepts relating specifically to the role of resilience in the quest for sustainability (cf. reviews in Reid and Botterill 2013, Xu 2018, IRGC 2018). The first school claims that resilience is independent of sustainability; resilience refers to a continuation of a desired service, such as electricity, heat or fresh water, in times of stress (Jackson and Ferris 2012; Haimes 2009). Whether such a resilient strategy of designing more robust systems for vital services actually meets the goals of sustainability, such as environmental quality, circular economy or peaceful resolution of conflicts, is not included in this understanding. In the extreme, this concept might imply that one can strive for a resilient weapon industry, which needs to continue to provide service to armies even in times of peace.

The second concept basically claims that resilience is a part of sustainability. A major advocate of this view is the Stockholm Resilience Centre (Folke 2006; Folke et al. 2016; Biggs et al. 2015; Rockström et al. 2023). The main claim here is that resilience is a way to sustain the functionality of services that are crucial to meeting human needs. So, both sustainability and resilience have "something to maintain" according to this understanding of the terms. The main goal is to sustain a development that evolves within the ecological and societal planetary boundaries.

The third concept takes a more radical view: resilience is in conflict with sustainability. Sustaining the status quo and making it more resilient defies the concept of sustainability (van der Leeuw 2020: 81ff.). Resilience in this understanding is directed towards maintaining the given structures (not services) rather than engaging in a dynamic system change where sustainability is a dynamic goal to guide change not to preserve stability. There are also conflicts between resilience and sustainability in resilience enhancing measures e.g., when more diversity and redundancy come into conflict with economy and efficiency in resource use.

Similar thoughts can be extended to the concepts of sustainability in relation to resilience (review in Savaget et al. 2019). There is the well-established distinction between the three pillars: ecological, social, and economic (sometimes augmented with cultural or ethical). There are proponents for keeping sustainability focused on ecologic quality while others advocate more normative visions of sustainability striving for social justice among and between the generations (Vasconcellos-Oliverra 2018). In contrast to the sectoral and normative distinction, many scholars stress functional concepts of sustainability such as maintaining ecological services to humankind, providing economic and social services to all people regardless of whether they live today or the future, or finding an equitable distribution rule for environmental resources, public and private goods and social opportunities (Renn et al. 2009). Lastly, sustainability may be associated with processes of decision-making and communication, stressing equal opportunities for learning and adaptation (van Mierlo and Beers 2020). Some sustainability definitions explicitly refer to resilience, but most do not; however, they may implicitly include this concept when they focus on sustaining "something".

This common feature of "maintaining or sustaining something" may be the key to a more coherent understanding of the relationship between the two terms (Faber et al. 2020). Almost all definitions of resilience and sustainability include the notion of sustaining something over time. With respect to resilience, it is the functionality of critical services that should be preserved even when the system is under stress; with respect to sustainability, it is the continuation of humane living conditions for present and future generations (within a co-evolution of social and natural systems). The construction of an ideal resilient system architecture may in principle avoid normative statements. Normativity especially comes into play with the definition and configuration of the services to be maintained. While an abstract understanding of resilience, like the one represented in Fig. 2.2, does not specify the type of services und functions that need to be maintained or quickly recovered, sustainability adds purpose to the services: they should be directed towards humane living conditions. These include respecting the boundaries of natural ecosystems and resources, meeting basic needs of all human beings, and ensuring peaceful means of conflict resolution (Robertson 2017: 3–4).

Given this overlap, the combination of resilience and sustainability could be framed as a requirement that the conditions for providing ecological services, for ensuring economic well-being and enabling quality of life with its essential social services are to be maintained over time in full awareness of multiple constraints and conflicting goals such as present and future resource capacities/limitations (planetary boundaries) and the occurrence of unexpected or unlikely stressors. Meeting this goal does not imply sustaining current structures and practices but rather fostering technical and social innovations that make it more likely that these human conditions

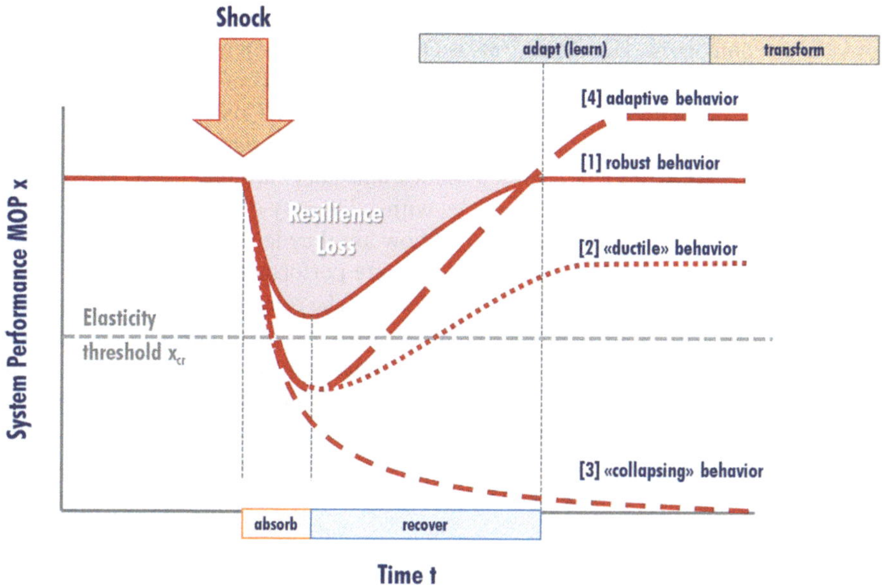

Fig. 2.2 System performance after shock (Taken from: Kröger 2017)

can be sustained and extended from the present to future generations. In this respect resilience and sustainability share the same common vision (cf. Fiksel 2003).

But the question of who defines and determines the services to be maintained remains essential. It is about the criticality of resources with regard to their relevance for society, about the quality, quantity and characteristics of the services to be maintained. What are the benchmarks for assessing their quality? It is also about minimum standards when restrictions become unavoidable. What must be maintained in what quality and quantity at any cost? However, sustainability is not only about promoting resilience related to maintaining humane living conditions. There are two additional aspects that go beyond resilience. First, sustainability is based on the normative goal of intra- and intergenerational justice (Doorn 2019). Societies might experience a very resilient system; but such a system may operate on the expense of the poor or any other disadvantaged group. One might include distributional justice and fairness under the rubric of humane living conditions, but it may be better to keep it as a separate component. Orienting oneself on the average level of welfare or mean quality of life obscures the fact that essential goods and services are often unfairly distributed among the affected groups and individuals. The Pareto optimality criterion that has inspired most of our economic thinking is blind for distributional effects. Therefore, for a development to be called sustainable it is not sufficient that crucial services are sustained over time, but that these services are also equally or fairly distributed among the respective populations. That postulate does not only apply for humans but also, to different degrees, for other creatures that share our planet (Washington et al. 2017).

The third and last implication of sustainability is the need to focus on processes not only on outcomes (Hendriks 2009). Decisions often imply painful trade-offs between conflicting goals or moral principles. The dilemma of policymakers to find the best route between protection against the spread of Covid-19 infections and recovery of economic and social activities illustrated this point (Collins et al. 2020; Hynes et al. 2020). Conflicts between values and goals are inevitable in complex decision situations in democratic societies with a diversity of value-landscapes. The question is: who assigns the trade-offs and how are they legitimized? A sustainable policy framework implies that such trade-offs are performed in a most democratic way based on ethical grounding (Voss et al. 2009). The relevant ethical concepts and approaches may to a certain extent differ among the various cultures and traditions in the world, while some principles like Human Rights may be expected to apply globally. However, a sustainable way to resolve such conflicts is to employ due processes based on both democratic decision-making institutions and rational ethical argumentation. This requirement includes a public debate about the moral principles that should be applied and a process of stakeholder and public engagement in assigning and justifying trade-offs.

2.3.3 Four Modes of Resilience Engineering

Based on the conceptual clarifications of resilience as a dynamic and systemic feature of socio-technical systems within a transformation towards more sustainable practices, one can distinguish specific objectives, capabilities and design principles that should be implemented through resilience engineering:

A broadly accepted view of the performance of resilient systems aftershocks is shown in Fig. 2.2.

Three types of resilience can be identified in this graph:

- **Bounce back resilience** (ductile and robust behaviour), which focuses on system preservation, resistance, robustness, and the fastest possible return to the original state of the system (roughly Holling's engineering resilience, Holling 1996).
- **Adaptive resilience** which includes the capability to flexibly respond to shocks and thereby even to increase the systems performance through a learning process (e.g. Holling's ecological resilience), and
- **Transformative resilience** in which a far-reaching reconstruction of the system is possible in order to maintain or improve the services (Asadazeh et al. 2022).

In the graph, nothing is said about the character of shocks and about the service to be maintained. The characterization of confounding factors is an important issue when operating within the stimulus–response paradigm. The characteristics of disruptive factors have a great influence and can imply a variety of impacts. Internal failures, external stress, anomalies, accidents, negligence, oversight, abuse, changes in framework conditions are all candidates for disruptive factors. There may be acute, slow-burners or insidious disruptive factors, specific or non-specific disturbances or stress at weak points, for example within the system's governance structure (Hollnagel 2012). It can be cumulative shock events and multiple failures, social amplification (ex. BSE) and wrong decisions in the system itself (ex. Chernobyl). While conventionally, critical infrastructure resilience focuses on single infrastructures—such as food, energy, or health—shocks ripple beyond traditional boundaries of each system. These cascading effects can have a positive or negative impact on the resilience of adjacent systems. Importantly, the indirect damages to other systems may be larger than the impacts on the system, in which the crisis originally occurred (Merz et al. 2013). Catastrophes (ex. wars) have often led to quite rapid social transformations.

But resilience focuses on the capabilities of the affected system to sustain its services. Our thesis is that system responses to stress are based on their capabilities, which are supported and generated by system design (the system's architecture) (Hollnagel et al. 2011; Brown et al. 2017; Shen et al. 2020). In this respect, what we are aiming at is resilience engineering, the construction of more resilient systems (Hollnagel 2016). We identify basic capabilities and try to assign design elements that support these capabilities.

Three basic capacities of resilient systems have been widely consensual in the resilience debate to date (Folke et al. 2010; Rockström et al. 2023): (a) resistance (absorptive coping capacity), (b) adaptivity (adoptive and anticipatory capacity) and (c) transformative capacity (innovativeness). But the discussion is far from being concluded. In the following we show capabilities and supporting design elements for the above mentioned three types of resilience, supplemented by a fourth type with focus on evolvability (Figs. 2.3, 2.4, 2.5 and 2.6).

Bounce back resilience is probably the most common understanding of resilience. And widespread ideas about the design of resilient systems often only include redundancy and diversity.[3]

The three modes of resilience engineering (bounce back, adaptive and transformative) should be complemented by a fourth mode, called biomimetic resilience engineering.[4] It goes beyond the stimulus–response paradigm and tries to implement a more far-reaching understanding of precaution. This means it is detached from the search for conceivable disruptive factors and designs that prepare for them (mode of what if?), and instead tries to orient itself on extensive precaution (mode of whatever may come!). Preparing against surprises can be implemented by learning from the process of evolution which has lasted for millions of years, by recurse to evolutionarily successful strategies and system designs. Such an orientation may be misinterpreted as just an extreme extension of randomly generated perturbation events within the "what if?" paradigm. But the advantage of the 'evolution strategy' towards designing resilient systems is not the generation of designs as a 'reaction' to innumerable perturbation events. That would be a Lamarckian misunderstanding of evolution. In Darwinian evolution theory the random creation of innumerable designs comes first and then these designs are tested in natural selection by countless

[3] E.g. Roege et al (2014), Molyneaux et al (2016).

[4] Cf. von Gleich, Giese (2019) and next paper in this volume.

disruptive elements. In this way, beyond the stimulus–response paradigm, a cornucopia of structures and processes is created that are often outside our imagination. Adopting these 'innovations', that are successfully tested by selection, the surviving systems evolve gradually towards the ability to deal with almost anything that may come along. The specific capabilities and designs for biomimetic resilience are often described by concepts of modern systems theories (e.g. self-organization, nested systems, homeostasis, autopoiesis).

The four Modes of resilience with their capabilities and design principles to support them are summarized in Table 2.1.

Now that we have discussed objectives, disruptive factors, capabilities and recommendations for the design of resilient systems, the question of services that must be sustained remains. Regarding the design of resilient systems, the determination of relevance and urgency of services or functions to be preserved becomes particularly important. Who defines the strongly normative statements about services or functions to be preserved? It may be necessary to identify basic services (crucial or key services). One part of this task is defined by parliamentarian laws and procedures, everything else must be determined deliberatively in a participatory way. Combined with the desirable capabilities of a resilient energy system to be strived for, an outline must be deliberatively developed, describing which services are to be preserved in which quality, in which form and to what extent and costs. In combination with the characterization of disruptive events this will have consequences for the specific design of the resilient energy system.

Resources are limited and even a resilient system will not always be able to maintain all services in all qualities. This also raises the socio-economic question of effort and return (Pareto optimum). In addition, distribution effects need to be considered even if a Pareto optimal solution was reached (Nicola 2013). Thus, the question is about fair distribution. Who benefits from resilience measures and who pays for them? Who bears the risks and who can seize the opportunities?

Table 2.1 Capabilities and design principles of resilient systems

Resilience mode	Capabilities	Design principles
Bounce back	Resistance, absorption, improvisation, mitigation, preparedness	Structural stability, redundancy, diversity, buffers, organizational skills
Adaptive	Learning, flexibility, anticipation regeneration	Smartness, plasticity, slack, substitutes, organizational learning, self-healing
Transformative	Precaution (what if?), reconstruction	Dynamic stability, loose coupling, variability, subsidiarity
Biomimetic	Precaution (whatever may come!), safe-to-fail, homeostasis, autopoiesis	Self-organization, high variability, abundance, hierarchical design, nested systems, productive instability

2.3.4 *Cyber Security as a Role Model?*

Resilience engineering, the construction of more resilient systems, can on the one hand be undertaken in a general way based on the outline of an ideal resilient system which will serve as a guiding concept. On the other hand, it is about designing real, socio-technical, socio-economic or societal, systems according to this guiding concept based on their actual situation. In doing so, the circumstances in each case must be taken as a prerequisite (expected disruptive impulses, existing system elements and relations, services to be maintained).

One of the most elaborate concepts of resilience engineering as well as resilience metrics and scoring has emerged in the field of information and communication technologies, especially regarding cyber security. This is not surprising considering the high technological standards and the smartness already implemented in the field of information and communication technology, nor is the engagement of military research in this area. Military systems have to function in the unreduced complexity of the battleground and always reckon with surprises. Given the increasing merging of the energy system with the information and communication system, the experiences and strategies from within the information and communication system are invaluable.

Highly elaborated is the Cyber Resiliency Engineering Framework published by MITRE (Bodeau et al. 2018) and adopted by the US National Institute of Standards and Technology (NIST). This concept is strongly aimed at resilience engineering and hereby not only follows the objective of adaptability, but beyond this also of evolvability.[5]

Figure 2.7 shows the overall framework with extended objectives of 'what' (including transformation[6] and redesign[7]), 'why' (risk management strategy) and 'how' (techniques, approaches and design principles).

The focus on implementation (how) with the help of strategic and structural design principles is remarkable.

Explicit design principles are shown in Fig. 2.8 with reference to security (brown), survivability (dark green), evolvability (light brown) and the consideration of advanced cyber threats (light green).

Resilience engineering has often focused on socio-technical aspects. In the MITRE approach the majority of design principles has its focus on social and professional actors, their knowledge and consciousness (trust, situational awareness), tasks (handling of resources) and behaviour (assume, expect, control, determine, limit, contain/exclude, manage, maintain, change) compared to technical designs (layer and partition defences, architecture for adaptability, reduction, change or disruption of attack surfaces, redundancy, diversity, dispersion). More detailed information on these principles and still more interesting design principles can be found in NIST 2018

---

[5] With reference to Ricci et al (2014).

[6] Transform: "Modify mission or business functions and supporting processes to handle adversity and address environmental changes more effectively" (NIST 2018: 15).

[7] Re-architecture: "Modify architectures to handle adversity and address environmental changes more effectively" (NIST 2018: 15).

(based on Jackson and Ferris 2012), such as loose coupling,[8] inter-node interaction,[9] complexity avoidance, drift correction,[10] reorganization and repairability.

2.3.5 *Increased Digitization of Society as Posing Increased Challenges*

Digitalisation is probably a historical watershed in the technology's past, present and future. In the past a certain technology could pose a substantial risk when subjected to a major shock but typically the technological system-failure would affect only a limited number of other systems and fallback solutions. Even nowadays in some lesser industrialized countries a breakdown of, for instance, the electricity grid does not lead to a complete standstill in all sectors of public life.

But with increased digitization of our societies these realities will be quite different. The digitized society implies inter-connectedness of virtually all spheres of public life. The electricity grid, computing, science, the internet, the financial markets, the industrial production- and supply-lines, satellite positioning systems, mobility on sea, land and air, mobile communications, health care systems, banking, smart homes and international relations, all these systems depend in various ways on each other. The end-result is basically one gigantic network of other networks, all digitized and exchanging information with each other. The benefits are obvious to many people (as also the scientific response to COVID-19 pandemic has shown), but the downside is the magnitude of potential vulnerability: if one of the systems fails, many or all others will also fail. And it will be hard to detect a cause which is easily and quickly repaired.

Currently we see the trend towards digitization gaining much momentum. Large investments are made to create this inter-connectedness in the digitized world. At the same time, we see the downgrading and dismantling of the old non-digitized systems. Fallback solutions as response to system failure are less and less available. 20 or 30 years ago a failure in mobile communication could easily have resulted in larger use of still existing landlines. In Western European societies this is less and less the case as virtually all communication is already by satellite and smartphones. In sum: digitized societies pay the prize for smarter technological solutions to basic functions of society through radically increased vulnerability in scale and scope and the possibility of shocks which quickly spread through all segments of society. While

[8] "The loose coupling principle states that the system should have the capability to limit the ability of failures to propagate from one component to the next in a system of many components" (Jackson and Ferris 2012: 7).

[9] "The inter-node interaction principle states that every node, or element, of a system should be capable of communicating, cooperating, and collaborating with every other node" (Jackson and Ferris 2012: 7).

[10] "The drift correction principle states that if the system is drifting towards the boundary of resilience and there is evidence that a failure is approaching, then either measures can be taken to avoid the threat or it can be diminished through corrective action" (Jackson and Ferris 2012: 7).

some sub-systems like the military sector aim to maintain fall-back structures that are independent of digitized systems, this is less and less the case in civil society. And the energy sector is prime in its rapid development towards all-encompassing digitization. Assuming that the current development of artificial intelligence (AI) will get an even stronger foothold in how we manage diverse sectors of public life and base them on AI services, the vulnerability of societies to shocks and attacks on the core elements of the digitized networks (like satellites, the storage and connection centres at Network Service Providers, etc.) is further increased.

If we aim at increased resilience of our digitized energy systems, we need more wisdom in choosing our development paths. We need to convert one-dimensional strategies into multi-dimensional strategies which can offer us fallback solutions even when our digitized world is under direct attack or subject to black-swans like e. g. space weather events (Letwin 2020).

2.4 Resilience Monitoring and Assessment

There is a difference between monitoring and measurement. Monitoring is more about observing changes over time in a particular system or area. A measurement process rather aims at absolute values on a scale. The quest for an absolute measure of resilience is understandable. Not least if one follows the widespread opinion in management literature that one can only manage what one can measure.[11] And there are remarkable approaches in trying to measure resilience.[12] A rather elaborate example is integrated in the Cyber Resiliency Engineering Framework presented above.[13] A less demanding and prerequisite step towards measuring resilience may be a pairwise comparison of systems. Such access to resilience monitoring should be sufficient to fulfil the task of identifying progress or regress on the way to building more resilient systems. The focus of a monitoring approach pursued in ReMoDigital should not be on measuring resilience as such, but on the evaluation of the impacts of design interventions in the energy system that are intended to increase its resilience (Béné et al. 2015, GIZ 2014, Hills et al. 2015). A comparative assessment should be able to show that one system appears to be more resilient than the other against the guiding principle of an ideal system[14] and/or against the background of a series of criteria.

[11] The often quoted saying probably goes back to the book by Kaplan and Norton on the Balanced Scorecard (Kaplan and Norton 1996).

[12] Molyneaux et al (2016), Béné (2013).

[13] The full title of the MITRE publication is: "Cyber Resiliency Metrics, Measures of Effectiveness, and Scoring. Enabling Systems Engineers and Program Managers to Select the Most Useful Assessment Methods". For a more conventional approach of resilience measurement regarding the electricity sector see EPRI (2017).

[14] See Fig. 2.1 in Chap. 3.

2.4.1 Vulnerability and Resilience

Even if one sticks to the goal of measurability, measuring vulnerability seems to be more promising than measuring resilience (von Gleich et al. 2010). Based on studies of developing countries, it has a much longer tradition (Naudé et al. 2009). But even in vulnerability analysis a comparative strategy may be more appropriate. The distinction between resilience and vulnerability needs to be clarified in the context of resilience monitoring and resilience-related system analysis (Béné et al. 2015). There is a case for seeing resilience more as a guiding principle and focusing the monitoring process on the evaluation of steps towards the mission statement.

As long as one remains at the level of a resilience concept that focuses on disruptive events and their possible consequences, there may not be much difference between a resilience and a vulnerability analysis. But if one proceeds in precaution from a 'what if?' towards a 'whatever may come' approach without identifiable disturbing factors, a clear difference opens in terms of vulnerability analysis. Then we should distinguish between an event-based vulnerability analysis (EVA) and a structural vulnerability analysis (SVA). The different elements and steps are shown in Fig. 2.9.

2.4.2 Incident-Related (Event Based) and Incident-Independent (Structural) Vulnerability Analysis

Regarding event-based VA it is important to note that exposure and sensitivity (and—based on an impact model—the associated potential impacts) are contrasted with adaptive capacity. In the area of adaptive capacity, available resources, task forces, disaster training and the like play an important role. Thus, the event-based VA has great similarity with methods that are currently being discussed as 'stress tests' (Anderson 2016).

Structural VA, on the other hand, abstracts from possible external disruptive events and focuses on possible weak points within the (affected) system itself. Structural VA is more about socio-technical reliability, which also includes misbehaviour within the system. Structural VA asks the question at which relations or nodes the system will give way if it comes under severe stress. This approach can also be understood as a kind of stress test. However, the stress tests carried out so far, which are mainly located in the financial system (Quagliariello 2009; Borio et al. 2012) and in software security, seem to focus primarily on external disruptive factors. Structural VA is closer to the FMEA (failure mode and effect analysis) and FTA (fault tree analysis), methods known from the engineering sciences aiming at more reliable systems (DIN 2006; Eberhardt 2012).

Remarkable in both approaches are the right strands in Fig. 2.9 leading to adaptive capacity, which is part of but not identical with resilience (Gallopin 2006). The EVA draws its reasons and traditions from classical risk management. SVA is more in the mechanical engineering tradition of designing more reliable systems. EVA tries to eliminate identifiable vulnerabilities with regard to external confounding factors. SVA focuses on vulnerabilities within the systems, without necessarily referring to any specific external disruptive factor. Based on the results of both approaches one can focus on the elimination or mitigation of identified vulnerabilities by a more resilient design of the system based on design principles displayed in Figs. 2.3 and 2.4. But the quest for more resilient systems should not be reduced to the avoidance of vulnerabilities. If we aim at a strong understanding of precaution, following the 'what-ever-may-come' paradigm, resilience must be more than the elimination of identifiable vulnerabilities. Resilience then is more than avoiding or minimizing identifiable risks. And it is not only about defence but also about the capability to seize new opportunities. In this way, the construction of more resilient systems should include design principles of transformative and biomimetic resilience displayed in Fig. 2.5 and 2.6. As in the biological evolutionary process, based on random mutations, these principles can creatively contribute to the fitness of the system, i.e. open up new opportunities, in many cases without any reference to identifiable vulnerabilities. But it must be noted that there are trade-offs between several design principles, for example between the number of redundancies, the amount of storage and buffers on the one hand and energy efficiency on the other.

2.4.3 Comparative Resilience Assessment and Model-Based Stress Tests

An event based and structural VA should be carried out as the first steps in resilience monitoring, as an initial analysis before possible interventions to improve resilience come into play.

When working with the 'what-ever-may-come' precautionary approach, two further steps are necessary for a resilience analysis or resilience monitoring based on the capabilities and design principles displayed in Figs. 2.3, 2.4, 2.5 and 2.6 and the outline of an ideal resilient system.[15] A comparative resilience assessment oriented at such a guiding concept of an ideal resilient system should be applied in form of a 'presence-absence judgement'. Which one of the compared systems has implemented more and/or more important design principles and is therefore closer to the ideal? But even this approach does not adequately reflect the complexity of resilience. Helmrich et al. can be agreed with when they write: "resilience is not a designed characteristic of a system but an emergent property of a collection of design choices" (Helmrich et al. 2020: 4 with reference to Park et al. 2013). In the next step a stress test (as a certain type of incident-related VA) can be applied in which a model of the (energy) system considered to be more resilient is confronted with known or possibly randomly generated disturbance factors (or selected energy transition scenarios) in order to find out whether the system under investigation is

[15] See Fig. 2.1 of Chap. 3.

Bounce Back Resilience

Capabilities	Design Principles
Resistance[a], system preservation	Structural stability
Absorption of disruption	Redundancy (structural, functional)[d]
Ability to cope with performance deficits	Availability and diversity of resources (personal, financial, technical)[e]
Robustness[b], fail safety	Dispersion, segmentation
Mitigation, minimization of impact	Buffer / storage capacity[f]
Repair, recovering, restoring[c]	Minimum attack surface
Preparedness (training)	Organizational skills (emergency institutions, task forces)

Fig. 2.3 Capabilities and Design Principles of Bounce Back Resilience. **a** "Continue essential mission or business functions despite adversity" (NIST 2018: 15). **b** Describes the ability to withstand a shock including protective measures such as 'winterization' or flood proofing of assets and infrastructures (McPhail et al. 2018). Especially for frequently re-occurring and predictable risks (such as seasonal flooding or wildfires) increasing robustness is a promising management strategy. However, in the case of extreme and outlier events, systems that are robust against smaller events may have a reduced flexibility and suffer from catastrophic losses. **c** "Restore mission or business functions during and after adversity" (NIST 2018: 15). **d** Allows for systems to function even though a specific element or line has been disrupted—for instance in terms of energy lines or road segments. **e** Refers to spreading the risk by having a variety of independently operating options. Examples are multiple sources of electricity, using different suppliers in different geographical regions, or a diversified portfolio of sources for electricity. **f** Buffer capacities include stockpiles of vital supplies ranging from food, fresh water and medical supplies to oil and gas.

able to withstand at least the already imaginable disturbance factors (Linkov et al. 2022).

But all steps must be based on explicit ideas of what services the resilient energy system should provide (in terms of form, quality and quantity).

2.5 Risk Governance of Systemic Risks Integrating Resilience Design Principles

A prerequisite for the conventional way of making systems more resilient is a prudent and effective policy for reducing potential sources of risk as well as reducing the vulnerability of the risk target. One can regulate the drivers of risk, such as pollution, waste disposal, use of explosive material, the handling of meat markets, etc. or

Adaptive Resilience

Capabilities	Design Principles
Adaptive learning[a]	Diversity (structural, functional)[e]
Flexibility[b]	Structural flexibility, smartness
Improvisation with capabilities and resources[c]	Slack, availability of free resources
Coping with secondary impacts institutional, political, psychological	Substitutes for damaged structures and processes
Anticipation, monitoring[d], awareness perception of early warnings	Specialized institutions for early warning and monitoring
	Organizational learning (cooperatives)
	Inclusiveness and equity[f]

Fig. 2.4 Additional capabilities and design elements of adaptive resilience. **a** Capacity to learn from the disaster and to improve the future resilience of the remodeled system. **b** The ability to act upon better information versus creation of path-dependencies and lock-ins (Baharmand et al. 2019). The concept relates to the adaptive capacity of a system to learn from experience and to improve performance by adaptation. **c** Improvisation is the flexible use of available resources and skills (Gößling-Reisemann et al. 2013). **d** The sensitization and vigilance of a socio-technical system with regard to possible disruptive events improves its ability to react appropriately (Gleißner 2017; Hollnagel 2016). **e** Stirling distinguishes between different forms of diversity, variety, disparity and balance (Stirling 2010). **f** Resilience is not only—possibly not primarily—a technical issue. Inclusiveness and equity can largely determine the adaptive capacity of socio-technical and socio-economic systems. Adger 2000, Saja et al. 2021, Rockström et al. 2023. Even before the resilience debate, this topic was primarily dealt with in vulnerability analysis (disaster readiness)

one can make the targets less susceptible to these risks, such as constructing buildings fire-proof, modifying ecosystems to make them more resistant against climate change, or making people less receptive to viruses. Interventions to eliminate, reduce or redistribute risks are covered by the term "Risk Management". The management of systemic risks poses challenges with respect to complexity, uncertainty and ambiguity. An important first step in the sense of a vulnerability analysis is mapping and analysing the critical dependencies between different sectors and socio-economic-environmental systems (Comes et al. 2011, SAPEA 2022, Liu and Renn 2025). For critical infrastructures, most common approaches focus on identifying interdependencies at a structural level (Rinaldi et al. 2001) or simulating and modelling how the impacts affect different infrastructure systems (Ouyang 2014). For supply chains and economic analyses, conventionally, macro-economic input–output (Kunz et al. 2013) or general equilibria models are used (Ma and Chen 2022). Yet, to expand from there, we also need to predict and integrate the continuous dynamics and adaptation at the micro-level in coupled socio-economic and technical systems that consider

Transformative (evolutionary) Resilience

Capabilities	Design Principles
Precaution based on experience (‚what if?‘)	Dynamic stability based on feedback loops
Openness for system innovation	Connectivity, loose coupling, modularity
Reconstruction to better built facilities	Variability
Evolvability[a]	Multi scalarity[b]
Taking advantage of opportunities	Interconnectedness
	Self-organization (technical, social, institutional)[c]
	Subsidiarity, polycentric governance[d]

Fig. 2.5 Additional capabilities and design elements of transformative (evolutionary) resilience. **a** Which implies the gradual restructuring of the system based on variability and selection. **b** Is especially used in spatial analysis (e.g. Palazzo 2019) and refers to the interdependence of different spatial scales—local, regional, national, and international levels. **c** Self-organization is the basic competenz to form new and sometimes emergent stuructures. It refers to the ability of a socio-technical system to adapt and respond with limited external intervention (Comes 2016). Examples range from volunteering in flood-ridden areas, citizen initiatives for new energy cooperatives, or rescue operations after an earthquake. Importantly, self-organisation in crises will occur, yet is often ad-hoc, and disconnected from the professional and institutional response system (Nespeca et al. 2020). Therefore, appropriate mechanisms are needed to coordinate, connect, support, and manage the interplay of both the volunteer and the institutional system. **d** Subsidiarity means to solve problems at the lowest possible level. Thus it has an overlapp with decentralization which may reduce vulnerability and increases the ability to respond appropriately at the respective system level (Biggs et al. 2012)

Biomimetic Resilience

Capabilities	Design Principles
Precaution against unknown unknowns 'whatever may come'	Error proof designs (safe to fail)
Safe to fail[a]	High variability of designs[c]
Self-healing based on self-organization	Hierarchical structure[d]
Autopoiesis[b]	Cellularity, nested systems[e]
	Productive instability[f]

Fig. 2.6 Additional capabilities and design principles of biomimetic resilience include. **a** In case of failure the system falls into a stable state (Ahern 2011). **b** Ability to create and preserve emergent solutions and structures (Maturana and Varela 1980). **c** Which opens new path to fulfill services. **d** Which enables various responses on different scales resp. layers. **e** Which can cushion the disruptive impulse, offer flexible responses and open up new path (Carpenter et al. 2012). **f** Which improves the ability to adaptively rebuild systems (Schmidt 2008)

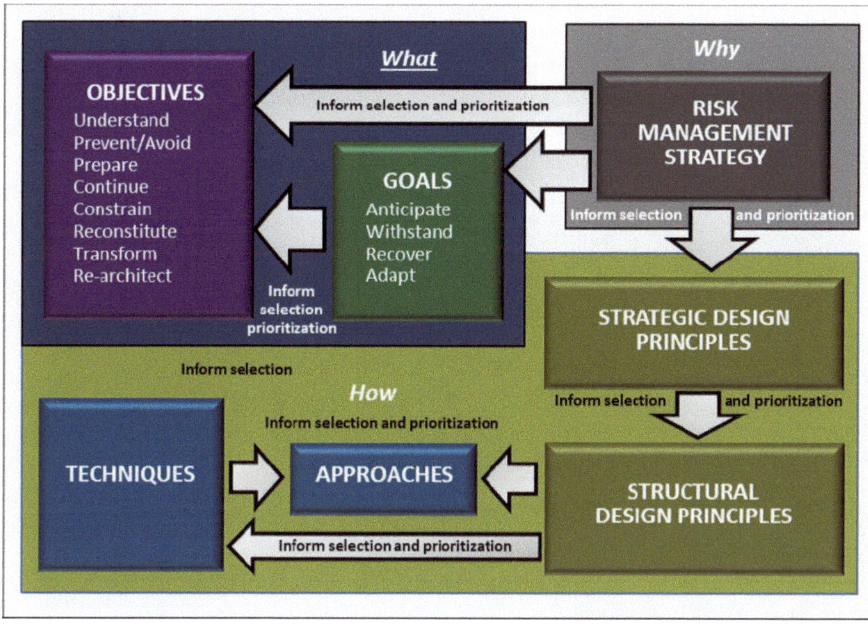

Fig. 2.7 Cyber Resiliency Engineering Framework: Mapping the Cyber Resiliency Domain (taken from Bodeau et al. 2018: V) ©2018 The MITRE Corporation. All Rights Reserved. Reprinted with permission of The MITRE Corporation

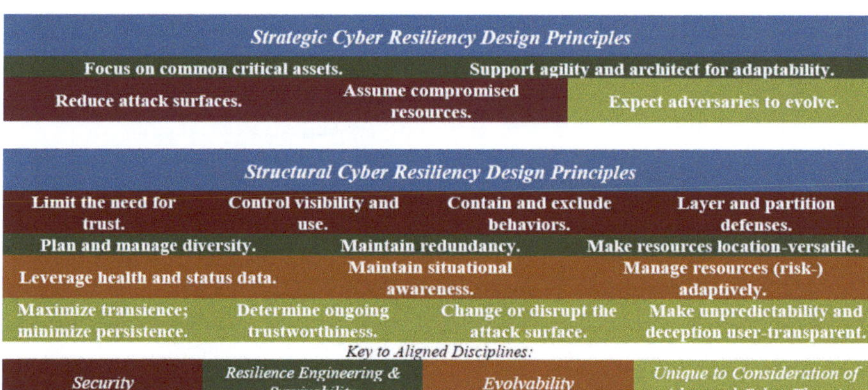

Fig. 2.8 Cyber Resilience Design Principles (taken from: Bodeau et al. 2018: 24) ©2018 The MITRE Corporation. All Rights Reserved. Reprinted with permission of The MITRE Corporation

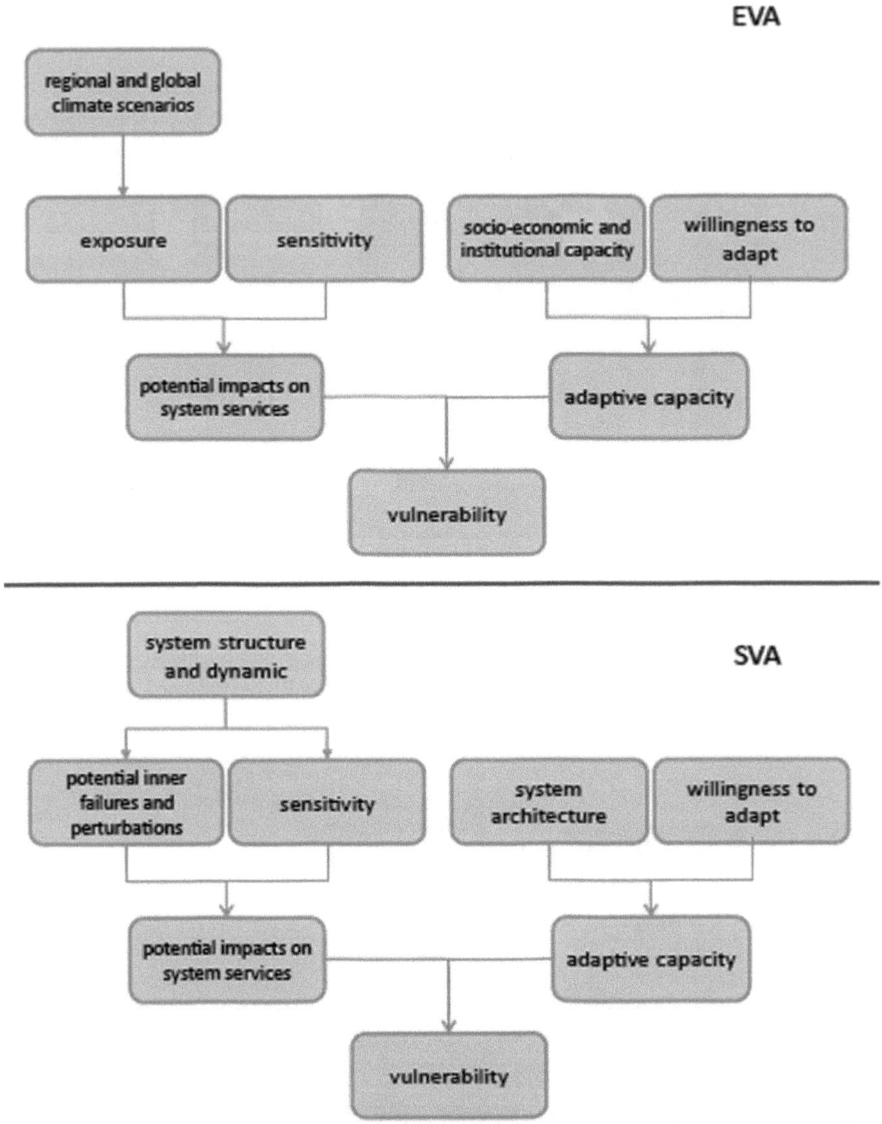

Fig. 2.9 Theoretical framework for the vulnerability assessment. EVA = event-based vulnerability assessment; SVA = structural vulnerability assessment (taken from Gößling-Reisemann et al. 2013)

shifts in human behaviour, the decision-making process, governance arrangements, economic adaptation, and environmental consequences.

Risk governance denotes both the institutional structure and the policy process that guide and restrain collective activities of a group, societies or international communities to regulate, reduce or control risk problems (IRGC 2018). During the last decades, risk governance has been shifted from traditional state-centric approaches with hierarchically organized governmental agencies as the dominant space of power to multi-level systems, in which the political authority for handling risk problems is distributed to separately constituted public bodies with overlapping jurisdictions (cf. Skelcher 2005, Klinke and Renn 2021). Institutional diversity can offer considerable advantages when systemic risk problems need to be addressed because, first, risk problems that impact several domains can be managed at different levels, second, an inherent degree of overlap and redundancy makes non-hierarchical adaptive and integrative risk governance systems more resilient, and third, the larger number of actors facilitates experimentation and learning (Renn 2008: 177ff.) Disadvantages refer to the possible compartmentalization of risk, the fragmentation of the risk governance process, costly collective risk decision-making, and the potential loss of democratic accountability.

Given the challenges mentioned above, risk management is trapped between two contrasting dimensions:

- integration versus specification
- leadership versus popular support

The first dimension addresses the trans-sectoral and transboundary aspect of systemic risks and its complex intricate structure. On the one hand, advanced interdisciplinary expertise is required for understanding, modelling and governing each systemic risk separately and, on the other hand, each risk cannot be fully grasped without understanding its connection with its environment. To be literate and effective in both tasks may overburden any assessment team or risk management agency. The second dimension relates to the democratic legitimization that is required to promulgate and enforce risk regulations. If systemic risks are underrated in the public and trial and error methods prove to be unfit for taking regulatory actions, there is not much chance for getting widespread approval for any risk reduction program. However, if political leaders step forward without public support and take on a leadership role they may be ousted from their office in the next election.

But these dilemmas are not insurmountable. First, the very feature of systemic risk regarding interconnectivity is not only a problem but also a great opportunity. If risk experts can locate intervention points that have impact on many related risk domains at the same time, they can trigger a positive risk reduction cascade that reinforces itself (Sharpe and Lenton 2021). Such powerful trigger points are, for example, located at the nexus between energy, water and food. Using solar energy and digital control measures for pumping and reusing water resources for irrigation of land that produces climate adjusted food with little emission potential could benefit climate, world nutrition, planetary health and biodiversity at the same time. It would produce income for farmers and could provide support for countries of the Global South.

Turning to the societal challenge of justifying ambitious measures for risk reduction, we have witnessed that representatives of the younger generation press for more action and are willing to reduce their comfort if risks are seriously addressed. There is no guarantee that the drive will not fade over time, but the movement shows that mobilization of the willing and convincing communication campaigns are powerful instruments to overcome the wicked combination of common pool resource dilemma and lack of immediate disaster experience. However, in the long run more concrete benefits of alternatives to foregoing current comfort probably are needed to maintain this drive.

Given these opportunities, effective risk management requires a systems approach that makes a major investment into a better understanding of the fabrics of complexity and stochasticity in order to identify powerful intervention points with positive risk reduction cascades. New tools developed by experts of complexity sciences and systemic risk analysts include black swan scenarios for testing the robustness of risk absorbing systems (i.e. the targets that are affected by risks), morphological-permutation studies that systematically explore possible interactions between the causal factors and test the sensitivity of each scenario, and reverse stress tests that start with the analysis of multiple vulnerabilities as a conceptual tracer for potential interventions.

These considerations have direct implications for resilience. As this concept reflects the confidence of all actors to deal with even uncertain outcomes, it provides a mental guideline for the negotiations between beneficiaries and potential victims of risks (IRGC 2017). Furthermore, it includes a discourse about coping capacity and compensation schemes if the worst were to happen. The degree of coping capacity that is regarded as sufficient or justified for approving a new risk agent or a disaster management plan to become enacted depends therefore on a discourse between the directly affected groups of the population. Such a reflective involvement of policy makers, scientists, stakeholders and public interest groups can be accomplished through a spectrum of different procedures such as negotiated rulemaking, mediation, roundtable or open forums, advisory committees, and others (see Beierle and Cayford 2002; Benn et al. 2009; Renn 2015).

2.5.1 Coping with High Ambiguity

If risk problems are associated with high ambiguity, it is not enough to demonstrate that risk regulation addresses the public concerns of those directly affected by the impacts of the risk source. In these cases, the process of evaluation and management needs to be open to public input and new forms of deliberation. This corresponds with the participative aspect of resilience (Lorenz 2013). Such discursive activities should start with revisiting the question of proper framing. Is the issue really a risk problem or is it an issue of lifestyle or future vision? Often the benefits are contested as well as the risks. The debate about 'the use of crops for generating energy' may illustrate the point that observers may be concerned not only about the risks of technical feasibility and economic viability but also about the acceptability of using food for energy production. Thus, the controversy is often much broader than dealing with the direct risks only. The aim here is to find an overlapping consensus on the dimensions of ambiguity that need to be addressed in comparing risks and benefits, and balancing pros and cons. High ambiguity would require the most inclusive strategy for involvement because not only directly affected groups but also those indirectly affected should have an opportunity to contribute to this debate.

These situations are typically addressed in transdisciplinary project work, based on partnerships including non-academic actors and open to a diversity of knowledge systems and value-landscapes. It poses new challenges to research as scientific and disciplinary knowledge no longer can assume primacy but needs to engage in dialogues in order to produce knowledge that is fit-for-purpose. This situation is often also described as post-normal science (Funtowicz and Ravetz 1993). One feature is that the partnerships are not defined through a given pre-defined and well described problem but engage in a common effort to agree how to delimit the problem they will work on. Reflexivity of all partners is an essential feature of this kind of transdisciplinarity (Kaiser Gluckman 2023). Resolving ambiguities in risk debates necessitates the participatory involvement of the public to openly discuss competing arguments, beliefs and values. Participatory involvement offers opportunities to resolve conflicting expectations through a process of identifying overarching common values, and to define options that will allow a desirable lifestyle without compromising the vision of others. Critical to success here is the establishment of equitable and just distribution rules when it comes to common resources and a common understanding of the scope, size and range of the problem, as well as the options for dealing with the problem (Renn and Schweizer 2009, 2020). Unless there is some agreement on the boundaries of what is included, there is hardly any chance for a common solution. Such a common agreement will touch upon the coping capacity of systems to deal with different frames of risks and not only with the physical impacts of risks. There are various social constructions of resilience that the participants associate with the management options. The set of possible procedures for involving the public includes citizen panels or juries, citizen forums, consensus conferences, public advisory committees and similar approaches (see Rowe and Frewer 2000, Beierle and Cayford 2002, Renn 2008: 284ff., Wong 2018, Radtke et al. 2018).

2.6 Conclusion

Resilience is based on the ability of a system to cope with surprises. This means that a far-reaching concept of precaution is pursued. The most far-reaching conception of resilience relies not only on the capacities for resistance and adaptation, but also on the capacities for transformation and evolution of the affected systems. Resilience includes a descriptive and normative function, and it provides a framework for generating design criteria. These criteria and principles need to be in line with the complex nature of the socio-technical system itself but also with the environment in which it is embedded. The dynamic formation of structures in diverse socio-technical contexts is subject to the self-organization of relations between agents and institutional forms of organization. So far, the function and effectiveness of self-organization has been underestimated in the literature on resilient structures and design. The potential for self-organization is based on interactions between the elementary relationships between technology, human action and organizational governance and generates a macroscopic dynamic. The result depends on the rules and processes of these interactions. In socio-technical systems, in contrast to most other domains, this normative guidance can be influenced by social decision-making processes. In particular, there is a need for guidance in resilience engineering that steers the potential of self-organization in the formation of structures in the direction of more resilient systems. If normative guidance is missing or not enforced, resilience will not be accomplished leading to high vulnerabilities, inefficiencies, social inequality, etc. Above all, the still tolerable range of variation in the relationships between the elements of systems can be exceeded, thus triggering massive instabilities in times of stress.

A procedure in which feasible design interventions in socio-technical systems (such as energy systems) are developed based on fundamental principles that were outlined in this chapter and then tested in real life experiments is both scientifically compatible as well as practicable for innovation and transformation processes. In addition, these principles should include dynamic criteria and guiding principles for measuring progress towards resilience and sustainability as major goals for societal transformation. As the climate debate makes clear, a comparative assessment of the costs of such interventions and the costs of disasters, i.e. of interventions that are not carried out, also plays an important role in the legitimization of resilience-enhancing interventions.

Such a comparative assessment can be combined with a multi-criteria evaluation and optimization (e.g., LCA, SWOT, cost–benefit, trade-off, multi-attribute decision making, and evaluation by stakeholders) that builds on the normative assumptions of the requirements for the services or functionalities to be maintained. These requirements go far beyond the security aspects addressed in actual debates about resilience. Resilience includes an active learning process in which all actors monitor the system-environment-interactions and readjust the corresponding socio-technical system in accordance with the feedback received. Such an approach requires an inclusive governance framework that is open for input from civil society as well as affected citizens. Finally, it is to be hoped that the 'new normality', which was so often discussed in connection with corona, will prevail as a guiding principle for designing and governing innovative technologies or critical infrastructures.

References

Adger, W.N.: Social and ecological resilience: are they related? Prog. Hum. Geogr. **24**(3), 347–364 (2000)

Ahern, J.: From fail-safe to safe-to-fail: sustainability and resilience in the new urban world. Landsc. Urban Plan. **100**(4), 341–343 (2011)

Anderson, R.W. (ed.): Stress Testing and Macroprudential Regulation. A Transatlantic Assessment. CERP Press, London (2016)

Asadazeh, A., Khavarian-Garmsir, A.R., Sarifi, A., Salehi, P., Kötter, T.: Transformative resilience: an overview of its structure, evolution, and trends. Sustainability **14**(22), 15267 (2022). https://doi.org/10.3390/su142215267

Bahadur, A., Tanner, T.: Transformational resilience thinking: putting people, power and politics at the heart of urban climate resilience. Environ. Urban. **26**(1), 200–214 (2014). https://doi.org/10.1177/0956247814522154

Baharmand, H., Comes, T., Lauras, M.: Defining and measuring the network flexibility of humanitarian supply chains: insights from the 2015 Nepal earthquake. Ann. Oper. Res. **283**(1–2), 961–1000 (2019)

Benn, S., Dunphy, D., Martin, A.: Governance of environmental risk: new approaches to managing stakeholder involvement. J. Environ. Manag. **9**(4), 1567–1579 (2009)

Beierle, T.C., Cayford, J.: Democracy in Practice. Public Participation in Environmental Decisions. Washington D.C. Resources for the Future (2002)

Béné, C.: Towards a quantifiable measure of resilience. IDS Working Paper 434, Brighton IDS (2013)

Béné, C., Frankenberger, T., Nelson, S.: Design, monitoring and evaluation of resilience interventions: conceptual and empirical considerations. Institute of Development Studies IDS Working Paper 459 Brighton (2015)

Biggs, R., Schlüter, M., Schoon, M.L.: Principles for building resilience. In: Sustaining Ecosystem Services in Social-Ecological Systems. Cambridge University Press, Cambridge, UK (2015)

Biggs, R., Schlüter, M., Biggs, D., Bohensky, E.L., BurnSilver, S., Cundill, G., Dakos, V., Daw, T.M., Evans, L.S., Kotschy, K., Leitch, A.M., Meek, C., Quinlan, A., Raudsepp-Hearne, C., Robards, M.D., Schoon, M.L., Schultz, L., West, P.C.: Toward principles for enhancing the resilience of ecosystem services. Annu. Rev. Environ. Resour. **37**(1), 421–448 (2012). https://doi.org/10.1146/annurev-environ-051211-123836

Bodeau, D.J., Graubart, R.D., McQuaid, R.M., Woodill, J.: Cyber resiliency metrics. Measures of effectiveness, and scoring. enabling systems engineers and program managers to select the most useful assessment methods. MITRE TECHNICAL REPORT MTR180314 Bedford MA (2018)

Borio, C., Drehmann, M., Tsatsaronis, K.: Stress-testing macro stress testing: does it live up to expectations? J. Financ. Stab. **12**, 3–15 (2012)

Brand, U., von Gleich, A.: Transformation toward a secure and precaution-oriented energy system with the guiding concept of resilience - implementation of low-exergy solutions in northwestern Germany. Energies **8**(7), 6995–7019 (2015). https://doi.org/10.3390/en8076995, https://www.researchgate.net/publication/281116516_Transformation_toward_a_Secure_and_Precaution-Oriented_Energy_System_with_the_Guiding_Concept_of_Resilience-Implementation_of_Low-Exergy_Solutions_in_Northwestern_Germany

Brand, U., Giese, B., von Gleich, A., Heinbach, K., Petschow, U., Schnülle, C., Stührmann, S., Stührmann, T., Thier, P., Wachsmuth, J., Wigger, H.: Auf dem Weg zu Resilienten Energiesystemen! Resiliente Gestaltung der Energiesysteme am Beispiel der Transformationsoptionen „EE-Methan-System“ und „Regionale Selbstversorgung“ Schlussbericht des vom BMBF geförderten Projektes RESYSTRA (FKZ:01UN1219A-B), Bremen, Berlin 2017 (2017). https://media.suub.uni-bremen.de/handle/elib/5842

Brown, K.: Global environmental change I: a social turn for resilience? Prog. Hum. Geogr. **38**(1), 107–117 (2014). https://doi.org/10.1177/0309132513498837

Brown, C., Seville, E., Vargo, J.: Measuring the organizational resilience of critical infrastructure providers: a New Zealand case study. Int. J. Crit. Infrastruct. Prot. **18**, 37–49 (2017)

Cairney, P.: Complexity theory in political science and public policy. Polit. Stud. Rev. **10**(3), 346–358 (2012)

Carpenter, S., Arrow, K.J., Barret, S., Biggs, R., Brock, W.A., Crépin, A., Engström, G., Folke, C., et al.: General resilience to cope with extreme events. Sustainability **4**(12), 3248–3259 (2012)

Coaffee, J.: Risk, resilience, and environmentally sustainable cities. Energy Policy **36**(12), 4633–4638 (2008). https://doi.org/10.1016/j.enpol.2008.09.048

Collins, A., Florin, M.F., Renn, O.: COVID-19 risk governance: drivers, responses and lessons to be learned. J. Risk Res. **2**, 1760332 (2020). https://doi.org/10.1080/13669877.2020.1760332
Comes, T.: Designing for networked community resilience. Procedia Eng. **159**, 6–11 (2016). https://doi.org/10.1016/j.proeng.2016.08.057
Comes, T., Hiete, M., Wijngaards, N., Schultmann, F.: Decision maps: a framework for multi-criteria decision support under severe uncertainty. Decis. Support. Syst. **52**(1), 108–118 (2011). https://doi.org/10.1016/j.dss.2011.05.008
Comes, T., Warnier, M., Feil, W., Van de Walle, B.: Critical airport infrastructure disaster resilience: a framework and simulation model for rapid adaptation. J. Manag. Eng. **36**(5), 04020059 (2020). https://doi.org/10.1061/(ASCE)ME.1943-5479.0000798
Comfort, L.K., Boin, A., Demchak, C.C.: Designing Resilience - Preparing for Extreme Events. University of Pittsburgh Press (2010)
Deutsches Institut für Normung e.V.: Analysetechniken für die Funktionsfähigkeit von Systemen – Verfahren für die Fehlzustandsart- und -auswirkungsanalyse (FMEA). European Union: Deutsches Institut für Normung e.V., Berlin (2006)
Doorn, N.: How can resilient infrastructures contribute to social justice? Preface to the special issue of sustainable and resilient infrastructure on resilience infrastructures and social justice. Sustain. Resilient Infrastruct. **4**(3), 1–4 (2019). https://doi.org/10.1080/23789689.2019.1574515
Eberhardt, O.: Risikobeurteilung mit FMEA. Renningen Expert Verlag (2012)
EPRI (Electric Power Research Institute): Cyber security metrics for the electric sector, 1 September 2017. Palo Alto (2017). https://publicdownload.epri.com/PublicDownload.svc/product=000000003002011685/type=Product
Faber, M.H., Miraglia, S., Qin, J., Stewart, M.G.: Bridging resilience and sustainability - decision analysis for design and management of infrastructure systems. Sustain. Resilient Infrastruct. **5**(1–2), 102–124 (2020). https://doi.org/10.1080/23789689.2017.1417348
Fiksel, J.: Designing resilient, sustainable systems. Environ. Sci. Technol. **37**(23), 5330–5339 (2003)
Fletcher, D., Sarkar, M.: Psychological resilience. Eur. Psychol. **18**(1), 12–23 (2013). https://doi.org/10.1027/1016-9040/a000124
Folke, C.: Resilience: the emergence of a perspective for social–ecological systems analyses. Glob. Environ. Change **16**, 253–267 (2006)
Folke, C., Carpenter, C.S.R., Walker, B., Scheffer, M., Chapin, T., Rockström, J.: Resilience thinking: integrating resilience, adaptability and transformability. Ecol. Soc. **15**(4), 20 (2010). https://doi.org/10.5751/ES-03610-150420
Folke, C., Biggs, R., Norström, A.V., Reyers, B., Rockström, J.: Social-ecological resilience and biosphere-based sustainability science. Ecol. Soc. **21**(3) (2016). https://www.jstor.org/stable/10.2307/26269981
Funtowicz, S., Ravetz, J.: Science for the post-normal age. Futures **26**(6), 568–582 (1993)
Gallopin, G.C.: Linkages between vulnerability, resilience and adaptive capacity. Glob. Environ. Change **16** (2006)
GIZ (Deutsche Gesellschaft für Internationale Zusammenarbeit): Assessing and monitoring climate resilience. In: From Theoretical Considerations to Practically Applicable Tools – A Discussion Paper (2014). https://www.adaptationcommunity.net/?wpfb_dl=233
von Gleich, A., Giese, B.: Resilient systems as a biomimetic guiding concept. In: Ruth, M., Gößling-Reisemann, S. (eds.) Handbook on Resilience of Socio-Technical Systems, pp 89–99, Edward Elgar Publishing, Cheltenham, Northampton (2019)
von Gleich, A., Gößling-Reisemann, S., Stührmann, S., Woizeschke, P., Lutz-Kunisch, B.: Resilienz als Leitkonzept. Vulnerabilität als analytische Kategorie. In: Fichter, K., von Gleich, A., Pfriem, R., Siebenhüner, B. (eds.) Theoretische Grundlagen für erfolgreiche Klimaanpassungsstrategien, pp. 13–49. Projektkonsortium ‚nordwest2050', Delmenhorst (2010)
Gleißner W (2017) Risikoüberwachung, Controlling und die Organisation des Risikomanagements. In: Gleißner W (2017) Risikomanagement, München Vahlen

Gluckman, P., Bardsley, A.: Uncertain but inevitable: the expert-policy-political nexus and high-impact risks. In: Tū, K. (ed.) The Centre for Informed Futures, University of Auckland (2021). https://informedfutures.org/high-impact-risks/

Gluckman, P., Bardsley, A., Kaiser, M.: Brokerage at the science-policy interface: from conceptual framework to practical guidance. Humanit. Soc. Sci. Commun. **8**(84), 1–10 (2021). https://doi.org/10.1057/s41599-021-00756-3

Gößling-Reisemann, S., Wachsmuth, J., Stührmann, S., von Gleich, A.: Climate change and structural vulnerability of a metropolitan energy system. The case of Bremen-Oldenburg in Northwest Germany. J. Ind. Ecol. **17**(6), 850 (2013)

Haimes, Y.Y.: On the definition of resilience in systems. Risk Anal. **29**(4), 498–501 (2009)

Hendriks, C.: Policy design without democracy? Making democratic sense of transition management. Policy. Sci. **42**(4), 341–361 (2009)

Henrickson, L., McKelvey, B.: Foundations of "new" social science: institutional legitimacy from philosophy, complexity science, postmodernism, and agent-based modelling. PNAS **99**(3), 7288–7295 (2002)

Hills, T., Pramova, E., Neufeldt, H., Ericksen, P., Thornton, P., Noble, A., Weight, E., Campbell, B., McCartney, M.: A monitoring instrument for resilience. In: CCAFS Working Paper no. 96. CGIAR Research Program on Climate Change, Agriculture and Food Security (CCAF), Copenhagen, Denmark (2015). http://hdl.handle.net/10568/56757

Hoffman, R.R., Hancock, P.A.: Measuring resilience. Hum. Factors **59**(4), 564–581 (2017). https://doi.org/10.1177/0018720816686248

Holling, C.S.: Resilience and stability of ecological systems. Annu. Rev. Ecol. Syst. **4**, 1–23 (1973). https://doi.org/10.1146/annurev.es.04.110173.000245

Holling, C.S.: Engineering resilience versus ecological resilience. In: Schulze (ed.) Engineering Within Ecological Constraints, vol. 31, p. 32. National Academy of Engineering (1996)

Hollnagel, E., Pariès, J., Woods, D.D., Wreathall, J. (eds.): Resilience Engineering in Practice. A Guidebook. Ashgate, Farnham UK (2011)

Hollnagel, E.: FRAM: the functional resonance analysis method: modelling complex socio-technical systems. Ashgate (2012). https://doi.org/10.1201/9781315255071

Hollnagel, E.: Resilience engineering (2016). https://erikhollnagel.com/ideas/resilience-engineering.html

Hutter, G.: Organizing social resilience in the context of natural hazards: a research note. Nat. Hazards **67**, 47–60 (2013). https://doi.org/10.1007/s11069-010-9705-4

Hynes, W., Trump, B., Love, P., Linkov, I.: Bouncing forward: a resilience approach to dealing with COVID-19 and future systemic shocks. Environ. Syst. Decis. **40**, 174–184 (2020). https://doi.org/10.1007/s10669-020-09776-x

IRGC (International Risk Governance Council): White paper on risk governance: towards an integrative framework. Geneva (2005)

IRGC (International Risk Governance Council): Risk governance: towards an integrative approach. White Paper (2017). https://irgc.org/wp-content/uploads/2018/09/IRGC_WP_No_1_Risk_Governance__reprinted_version_3.pdf

IRGC (International Risk Governance Council): Guidelines for the governance of systemic risks (2018). http://infoscience.epfl.ch/record/257279

Jackson, S., Ferris, T.L.J.: Resilience principles for engineered systems. Syst. Eng. **16**(2), 152–164 (2012)

Jackson, S., Ferris, T.L.J.: Designing resilient systems. In: Linkow, I., Palma-Oliveira, J.M. (eds.) Resilience and Risk, pp. 121–144. Springer, Heidelberg (2017)

Kaiser M, Chen ATY, Gluckman P (2021) Should policy makers trust composite indices? A commentary on the pitfalls of inappropriate indices for policy formation. Health Res Policy Sys 19(40). https://doi.org/10.1186/s12961-021-00702-4

Kaiser, M., Gluckman, P.: Looking at the future of transdisciplinary research. International Science Council. https://doi.org/10.24948/2023.05

Kaplan, R.S., Norton, D.P.: The Balanced Scorecard: Translating Strategy Into Action. Harvard Business Press (1996)
Keck, M., Sakdapolrak, P.: What Is social resilience? Lessons learned and ways forward. Erdkunde **67**(1), 5–19 (2013). https://doi.org/10.3112/erdkunde.2013.01.02
Klinke, A., Renn, O.: The coming age of risk governance. Risk Anal. **41**(3), 544–557 (2021). https://doi.org/10.1111/risa.13383
Kröger, W.: Securing the operation of socially critical systems from an engineering perspective: new challenges, enhanced tools and novel concepts. Eur. J. Secur. Res. **2**, 39–55 (2017). https://doi.org/10.1007/s41125-017-0013-9
Kunz, M., Mühr, B., Kunz-Plapp, T., Daniell, J.E., Khazai, B., Wenzel, F., Vannieuwenhuyse, M., Comes, T., Elmer, F., Schröter, K., Fohringer, J., Münzberg, T., Lucas, C., Zschau, J.: Investigation of superstorm Sandy 2012 in a multi-disciplinary approach. Nat. Hazards Earth Syst. Sci. **13**(10), 2579–2598 (2013). https://doi.org/10.5194/nhess-13-2579-2013
Letwin, O.: Apocalypse How? London Atlantic Books (2020)
Levin, S., Xepapadeas, T., Crépin, A., Norberg, J., de Zeeuw, A., Folke, C., Hughes, T., Arrow, K., Barrett, S., Daily, G., Ehrlich, P., Kautsky, N., Mäler, K., Polasky, S., Troell, M., Vincent, J., Walker, B.: Social-ecological systems as complex adaptive systems: modeling and policy implications. Environ. Dev. Econ. **18**(2), 111–132 (2013)
Linkov, I., Trump, B., Trump, J., Panda, A.: Resilience stress testing for critical infrastructure. Int. J. Disaster Risk Reduct. **82**(6), 103323 (2022). https://doi.org/10.1016/j.ijdrr.2022.103323
Liu, H., Renn, O.: Polycrisis and systemic risk: assessment, governance, and communication. Int. J. Disaster Risk Sci. **16**(4), 526–549 (2025). https://doi.org/10.1007/s13753-025-00636-3
Longstaff, P.H., Armstrong, N.J., Perrin, K., Parker, W.M., Hidek, M.A.: Building resilient communities: a preliminary framework for assessment. Homel. Secur. Aff. V **I**(3), 1–23 (2010)
Lorenz, D.F.: The diversity of resilience: contributions from a social science perspective. Nat. Hazards **67**(1), 7–24 (2013). https://doi.org/10.1007/s11069-010-9654-y
Lucas, K., Renn, O., Jaeger, C., Yang, S.: Systemic risks: a homomorphic approach on the basis of complexity science. Int. J. Disaster Risk Sci. **9**(3), 292–305 (2018)
Lucas, K., Renn, O.: Systemic risks: Intersections between science and society with policy implications for sustainability. Int. J. Environ. Sci. & Nat. Resour. **23**(2), 44–47 (2020). https://doi.org/10.19080/IJESNR.2020.23.556107
Maturana, H., Varela, F.: Autopoiesis and Cognition: The Realization of the Living. D. Reidel, Boston (1980). ISBN 978-90-277-1016-1
Ma, X., Chen, X.: Scenario analysis on the macroeconomic impact of COVID-19: a computable general equilibrium approach. Emerg. Mark. Financ. Trade **58**(1), 102–115 (2022). https://doi.org/10.1080/1540496X.2021.1987213
McKelvey, B.: Complexity science as order creation science: new theory new method. Emerg. Complex. Organ. **6**(4), 137–157 (2004)
McPhail, C., Maier, H.R., Kwakkel, J.H., Giuliani, M., Castelletti, A., Westra, S.: Robustness metrics: how are they calculated, when should they be used and why do they give different results? Earth Future **6**(2), 169–191 (2018). https://doi.org/10.1002/2017EF000649
Meyer R, Kunreuther H (2017) The ostrich paradox: Why we underprepare for disasters. University of Pennsylvania Press
Merz, M., Hiete, M., Comes, T., Schultmann, F.: A composite indicator model to assess natural disaster risks in industry on a spatial level. J. Risk Res. **16**(9), 1077–1099 (2013)
Molyneaux, L., Brown, C., Wagner, L., Foster, J.: Measuring resilience in energy systems: insights from a range of disciplines. Renew. Sustain. Energy Rev. **59**, 1068–1079 (2016)
Nationalakademien Leopoldina, Acatech and Union der deutschen Akademien der Wissenschaften: Das Energiesystem resilient gestalten. Maßnahmen für eine gesichterte Versorgung. Report. Nationale Akademien, Munich (2017)
Ross, R., Graubart, R., Bodeau, D., McQuaid, R.: Systems Security Engineering - Cyber Resiliency Considerations for the Engineering of Trustworthy Secure Systems, vol. 2, pp. 800–160. National

Institute of Standards and Technology (NIST), The MITRE Corporation, Draft NIST Special Publication (2018)
Naudé, W., Santos-Paulino, A.U., McGillivray, M.: Measuring vulnerability: an overview and introduction. Oxf. Dev. Stud. **37**(3), 183–191 (2009). https://doi.org/10.1080/13600810903085792URL:doi:10.1080/13600810903085792
Nespeca, V., Comes, T., Meesters, K., Brazier, F.: Towards coordinated self-organization: an actor-centered framework for the design of disaster management information systems. Int. J. Disaster Risk Reduct. **51**, 101887 (2020). https://doi.org/10.1016/j.ijdrr.2020.101887
Nicola, P.: Efficiency and Equity in Welfare Economics. Springer, Berlin (2013)
Norris, F.H., Stevens, S.P., Pfefferbaum, B., Wyche, K.E., Pfefferbaum, R.L.: Community resilience as a metaphor, theory, set of capabilities, and strategy for disaster readiness. Am. J. Community Psychol. **41**, 127–150 (2008)
Ouyang, M.: Review on modeling and simulation of interdependent critical infrastructure systems. Reliab. Eng. Syst. Saf. **121**, 43–60 (2014). https://doi.org/10.1016/j.ress.2013.06.040
Palazzo, E.: From water sensitive to floodable: defining adaptive urban design for water resilient cities. J. Urban Des. **24**(1), 137–157 (2019). https://doi.org/10.1080/13574809.2018.1511972
Park, J., Seager, T.P., Rao, P.S.C., Convertino, M., Linkov, I.: Integrating risk and resilience approaches to catastrophe management in engineering systems. Risk Anal. **33**(3), 356–366 (2013). https://doi.org/10.1111/j.1539-6924.2012.01885.x
Peletier, M.A., Van Santen, R.A., Steur, E.: Complexity Science. An Introduction. World Scientific Publishing, London and Singapore (2018)
Quagliariello, M. (ed.): Stress-Testing the Banking System: Methodologies and Applications. Cambridge University Press, Cambridge (UK) (2009)
Radtke, J., Holstenkamp, L., Barnes, J., Renn, O.: Concepts, formats, and methods of participation: theory and practice. In: Holstenkamp, L., Radtke, J. (eds.) Handbuch Energiewende und Partizipation pp. 21–42. Springer, Wiesbaden (2018)
Reid, R., Botterill, L.: The multiple meanings of 'resilience': an overview of the literature. Aust. J. Public Adm. **72**(1), 31–40 (2013)
Renn, O.: Risk Governance. Coping with Uncertainty in a Complex World. Eartscan, London (2008)
Renn, O.: Stakeholder and public involvement in risk governance. Int. J. Disaster Risk Sci. **6**(1), 8–20 (2015)
Renn, O.: Risk governance: concept and application to systemic risk. In: Kasperson, R.E. (ed.) Risk Conundrums. Solving Unsolvable Problems, pp. 243–259. Routledge, London and New York (2017)
Renn, O., Lucas, K.: Systemic risk: the threat to societal diversity and coherence. Risk Anal. **42**(9), 1921–1934 (2021). https://onlinelibrary.wiley.com/doi/10.1111/risa.13654
Renn, O., Jaeger, A., Deuschle, J., Weimer-Jehle, W.: A normative-functional concept of sustainability and its indicators. Int. J. Glob. Environ. Issues **9**(4), 291–317 (2009)
Renn, O., Schweizer, P.: Inclusive risk governance: concepts and application to environmental policy making. Environ. Policy Gov. **19**(3), 174–185 (2009)
Renn, O., Schweizer, P.: Inclusive governance for energy policy making: conceptual foundations, applications, and lessons learned. In: Renn, O., Ulmer, F., Deckert, A. (eds.) The Role of Public Participation in Energy Transitions, pp. 39–79. Elsevier, Academic Press, London (2020)
Ricci, N., Rhodes, D.H., Ross, A.M.: Evolvability-related options in military systems of systems. In: Conference on Systems Engineering Research (CSER 2014) Redondo Beach, CA (2014)
Rinaldi, S.M., Peerenboom, J.P., Kelly, T.K.: Identifying, understanding, and analyzing critical infrastructure interdependencies. IEEE Control. Syst. Mag. **21**(6), 11–25 (2001). https://doi.org/10.1109/37.969131
Robertson, M.: Sustainability. Principles and Practice, Second edn. Routledge, New York (2017)
Rockström, J., Norström, A.V., Matthews, N., et al.: Shaping a resilient future in response to COVID-19. Nat. Sustain. (2023). https://doi.org/10.1038/s41893-023-01105-9
Roege, P.E., Collier, Z.A., Mancillas, J., McDonagh, J.A., Linkov, I.: Metrics for energy resilience. Energy Policy **72**, 249–256 (2014)

Rose, A.: Economic resilience to natural and man-made disasters: multidisciplinary origins and contextual dimensions. Environ. Hazards **7**(4), 383–398 (2007)
Rowe, G., Frewer, L.J.: Public participation methods: a framework for evaluation. Sci. Technol. Hum. Values **25**(1), 3–29 (2000)
Saja, A.M.S., Goonetilleke, A., Ziyath, A.M.: A critical review of social resilience assessment frameworks in disaster management. Int. J. Disaster Risk Reduct. **35**, 101096 (2019). https://doi.org/10.1016/j.ijdrr.2019.101096
SAPEA (Science Advice for Policy by European Academies): Strategic crisis management in the European Union. Berlin (2022). https://doi.org/10.26356/crisismanagement
Savaget, P., Geissdoerfer, M., Kharazzi, A., Evans, S.: The theoretical foundations of sociotechnical systems change for sustainability: a systematic literature review. J. Clean. Prod. **206**, 878–892 (2019)
Sharpe, S., Lenton, T.M.: Upward-scaling tipping cascades to meet climate goals: plausible grounds for hope. Clim. Policy **21**(4), 421–433 (2021)
Shen, L., Cassottana, B., Heinimann, H.R., Tangs, L.C.: Large-scale systems resilience: a survey and unifying framework. Qual. Reliab. Eng. Int. **36**(4), 1386–1401 (2020). https://doi.org/10.1002/qre.2634
Schmidt, J.C.: Instabilität in Natur und Wissenschaft. Eine Wissenschaftsphilosophie der nachmodernen Physik. De Gruyter, Berlin, New York (2008). https://doi.org/10.1515/9783110209693.
Schnuelle, C., Kisjes, K., Stuehrmann, T., Thier, P., Nikolic, I., von Gleich, A., Goessling-Reisemann, S.: From niche to market - an agent-based modeling approach for the economic uptake of electro-fuels (power-to-fuel) in the German energy system. Energies **13**(20), 5522 (2020). https://doi.org/10.3390/en13205522
Skelcher, C.: Jurisdictional integrity, polycentrism, and the design of democratic governance. Governance **18**(1), 89–110 (2005)
Stirling, A.: Multicriteria diversity analysis. Energy Policy **38**(4), 1622–1634 (2010). https://doi.org/10.1016/j.enpol.2009.02.023
Stockholm Resilience Centre: What is resilience? (2015). https://www.stockholmresilience.org/research/research-news/2015-02-19-what-is-resilience.html
van Mierlo, B., Beers, P.J.: Understanding and governing learning in sustainability transitions: a review. Environ. Innov. Soc. Trans. **34**, 255–269 (2020)
van der Leeuw, S.: Social Sustainability, Past and Future. Undoing Unintended Consequences for the Earth's Survival. Cambridge University Press, Cambridge, MA (2020)
Vasconcellos Oliveira, R.: Back to the future: the potential of intergenerational justice for the achievement of the sustainable development goals. Sustainability **10**(2), 427–443 (2018)
Voss, J.-P., Smith, A., Grin, J.: Designing long-term policy; rethinking transition management. Policy. Sci. **42**, 275–302 (2009)
Washington, H., Taylor, B., Kopnina, H., Cryer, P., Piccolo, J.J.: Why ecocentrism is the key pathway to sustainability. Ecol. Citiz. **1**(1), 35–41 (2017). https://openaccess.leidenuniv.nl/bitstream/handle/1887/50284/WashingtonetalWhyecocentrismisthekeypathwaytosustainability2017.pdf?sequence=1
Wong, C.M.L.: From risk management to risk governance. In: Wong, C.M.L. (ed.) Energy, Risk and Governance, pp. 199–248. Palgrave Macmillan, Cham (2018)
Xu, L.: An integrated framework for resilience research: a systematic review based on citation network analysis. Sustain. Sci. **13**, 235–254 (2018)

Chapter 3
Biomimetic Resilience—Performance Claim, Concept and Design Principles

Arnim von Gleich and Bernd Giese

Abstract The purpose of this paper is to explore the concept of biomimetic resilience and to elaborate the differences to resilience concepts based on established risk management. First, we briefly reconsider what biomimetics is all about and examine two important sources of biomimetic resilience design principles, evolution and ecosystem theory. Then, being aware that biomimetics is not an easy undertaking, we discuss restrictions for the transfer from natural role models to socio-techno-economic implementations. Finally, we discuss the claim of biomimetic resilience to offer a solution path to prepare for real surprises (unknown unknowns) in the sense of a far-reaching precautionary principle.

Keywords Biomimetic resilience · Risk management · Design principles · Unknown unknowns · Precaution

3.1 Introduction

Recently, there has been an intensifying debate about nature-based solutions. The European Commission for example published a report on 'Nature-based solutions—State of the art in EU-funded projects' and defined them as "inspired and supported by nature, which are cost-effective, simultaneously provide environmental, social and economic benefits and help build resilience" (European Commission 2020: 3). The term 'nature-based' is just one example of a variety of concepts of 'learning from nature', which range from 'nature inspired' over 'biomimicry' to 'bionic' or 'biomimetic'. This text focuses on biomimetics, which seems to be the most ambitious concept. If one takes 'learning from nature' seriously, beyond mere associations, it is necessary to follow the requirements of the 'biomimetic processes of

A. von Gleich (✉)
Technological Design and Development, Production Engineering, University of Bremen, Bremen, Germany
e-mail: gleich@uni-bremen.de

B. Giese
Institute of Safety and Risk Sciences (ISR), BOKU University, Vienna, Austria

D. van Doren et al. (eds.), *Towards Energy System Resilience*, Ethics of Science and Technology Assessment 55, https://doi.org/10.1007/978-3-032-03187-7_3

abstraction' (von Gleich 2007; von Gleich et al. 2010). That means, that the biological template (Dicks 2023) must be scientifically understood and explained (Vincent et al. 2006: 474, Nachtigall 2010, Krohs 2021). How exactly, based on what kind of physical, chemical or biological systemic principles, do the organisms produce and use the interesting effect? Then, reflections about the conditions of transferability are necessary, in terms of functional relationships, general conditions, and, not least, ethics. Human concepts of risk and failure for instance, especially the values they embody, are not at all identical with the rules in nature (which are sometimes extremely cruel, especially regarding singular creatures). The loss of an individual is more or less unimportant for populations in relation to the survival of the population or species. And finally, our socio-techno-economic (system-)possibilities must allow for a successful implementation. Many biomimetic projects have failed in their implementation, not least due to their inherent complexity.

We have to admit, that we do not learn from nature, we usually learn from biology. This essentially means learning from scientifically understood processes and results of evolution. That means learning from the constitutions, capabilities and behaviour of organisms, which can take place on different levels. At least three of them can be distinguished[1]:

(1) *Learning from the results of evolution.* This level has a long tradition and still dominates in biomimetic research. Well-known examples deal with the transfer of form-function-relationships: stork wing to airplane wing, burdocks to Velcro fastener, gecko´s foot to adhesion tapes (based on van der Waals forces), lotus leaves as template for self-cleaning surfaces etc. (Nachtigall 2002).
(2) *Learning from the process of evolution.* The process of evolution can be interpreted as a certain type of multi-dimensional optimization over an extremely long time period. Well-known examples of biomimetic optimization methods are genetic algorithms, ant colony algorithms, topological optimization, soft kill option, and evolution strategies (Darwish 2018; Mattheck 1998; Rechenberg 1994; Schwefel 1995; Branke et al. 2008).
(3) *Learning from principles of success in evolution.* These principles are not as prevalent as the previous solutions.[2] Well known examples are concepts like diversity, modularization, self-organization and homeostasis.

[1] Similar approaches to distinct levels of 'learning from nature' can be found in Borsari (2017) and Krohs (2021). Hayes et al. just distinguish form-, process- and system-level Hayes et al. (2020: 7–8).

[2] Some principles of success in evolution and possible applications are discussed in the field of applied evolutionary biology, although still very closely related to agricultural, medical and biotechnological topics, e.g. Hendry et al. (2011), Lankau et al. (2011). Other principles are discussed in ecosystem theory, e.g. Levin (2005). The identification of 'Life's principles' as a basis for biomimicry to support a sustainable and resilient infrastructure is a similar attempt: 1. Evolve to survive, 2. Adapt to changing conditions, 3. Be locally attuned, 4. Integrate development with growth, 5. Be resource efficient, 6. Use life friendly chemistry. Its subprinciples have a large overlap with the principles presented in Table 3.1. They include diversity, self-renewal, integrity, variation, redundancy, decentralization, cooperation, readily available resources, feedback loops, self-organization, modularity, multi-functionality, nested components and recycling, Benyus (1997), Biomimicry 3.8 (2013), Helmrich et al. (2020).

As stated in the EU Commission report cited above, there is a widespread prior understanding, that learning from nature will lead to more sustainable, ecology friendly and safer solutions. This pre-understanding may be called an implicit 'biomimetic promise' (von Gleich 2007; von Gleich et al. 2010). Grounds for this promise can be identified on every level, but its justification is quite questionable. It must be assessed on a case-by-case basis. Anyway, contributions for building more resilient systems may be identified on each of the three levels. Design principles that enhance resilience and their foundations can especially be extracted from evolutionary theory and ecosystem theory.

3.2 Evolution Theory as Source of Biomimetic Design Principles

Two approaches are important here. The first approach is about the ability to evolve at all, about the preconditions of evolvability by producing variation and abundance, e.g. about mutation and recombination, about diversity and variability, combined with genetic stability (stasis). The second approach follows the question of the preconditions of success under selection pressure.

Design principles of evolvability and evolutionary success that may contribute to building more resilient systems include in particular the aspects listed in Tables 3.1 and 3.2.

There are trade-offs between some of these design principles. In the economic discourse on competitiveness for instance as well as in the sustainability discourse efficiency plays an important role. But it is also well known in economic theory that the capacity for innovation is just as important as costs (Porter 1980). Also, in the sustainability discourse efficiency plays a dominant role. Advances in efficiency are seen as 'low hanging fruits', and as an essential strategy against resource depletion. But slack in systems, redundancy and up to a certain level even abundance is essential for adaptation, improvisation, innovation and evolvability.

Of particular interest for the debate on resilient systems is the fragile balance between stability and instability, as a basis for evolvability. Thus, several design principles pose an important challenge for the design of resilient systems. In particular (productive) instability is a prerequisite for evolvability and stands at the same time in complete contrast to common ideas about the construction of 'safe systems'. The same applies to dynamic stability in contrast to solidly founded stability. It appliesto self-organization, to homoeostasis (dynamic equilibrium) and to socio-techno-economic implementations of autopoiesis and emergence (the unexpected). But interestingly another aspect appears in this context. Especially at the beginning of the debate, resilience was often misunderstood as being structural conservative (Olsson et al. 2015). The reply was that resilience is about sustaining services not structures. Systems can be completely redesigned to maintain their services. But evolution theory seems to indicate, that this view may be too radical. In evolution

Table 3.1 Design principles of **evolvability** that may contribute to building more resilient systems

Design principle	References	Description
Diversity (variety, balance, disparity)	Stirling (1998)	In a paper about technology policy (Stirling 1998) distinguishes three basic properties of diversity: variety, balance and disparity. "Variety refers to the number of categories into which the quantity in question can be partitioned (for instance: the number of functionally redundant—but morphologically or operationally distinct—technological options sustained in parallel by a market)" (p. 38). "Balance refers to the pattern in the apportionment of that quantity across the relevant categories (i. e.: the 'market shares' of each of the technological options). This might most simply be seen as 'something analogous to variance'—a set of positive fractions which sum to unity. For a particular system of given variety, the more equal are the fractions, the more even is the balance, the greater is the diversity" (p. 38). "Disparity refers to the nature and degree to which the categories themselves are different from each other" (p. 39)
Modularity/Near decomposability	Baldwin and Clark (2000), Gass and Bolker (2003)	Molecules, cells, organs and individuals can be seen as modules. "Modularity has become a central and remarkably useful concept in evolutionary developmental biology, offering an explanation of how independent, interacting units make possible developmental events and evolutionary changes. These modules exist at several different levels of organization, from genes to signal transduction pathways to cell populations" (Gass and Hall 2007: 349) Near decomposability is "...a fundamental property that appears to be shared by all multicelled organisms. Such organisms consist of a hierarchy of components, such that, at any level of the hierarchy, the rates of interaction within components at that level are much higher than the rates of interaction between different components" (Simon 2002 abstract)
Homoeostasis (homoeodynamics)/ steady state/nested networks/ feed-back-control	von Bertalanffy (1968)	Homoeostasis of an open system is based on internal regulation of a steady state by feedback control

(continued)

Table 3.1 (continued)

Design principle	References	Description
Stable evolutionary kernels/Cohesion/ Canalization	Scharloo (1991)	"Canalization and developmental stability refer to the tendency of developmental processes to follow particular trajectories, despite external or internal perturbation" (Hallgrimson et al. 2002 abstract). Canalization is also called evolutionary robustness
Alternative stable states	Beisner et al. (2003)	Different possible combinations of state variables exist, representing locally stable equilibrium points (Beisner et al. 2003)
Emergent self-organization/ Autopoiesis	Maturana and Varela (1980)	The system produces and maintains itself
Multidimensional (multi-objective) optimization in dynamic environments	Schwefel (1995), Branke et al. (2008)	Multidimensional or multi-objective optimization is intensively discussed in systems theory (Zakharova and Minashina 2015), often with reference to biomimetic approaches like evolutionary, genetic or swarm intelligence algorithms (Olariu and Zomaya 2005). One-sided optimizations, while neglecting other circumstances that are essential for survival, are ultimately doomed to failure in evolution
Productive instability	Schmidt (2007)	Instability (e.g. due to chaos, chance or non-linear thermodynamics) as an enabler of change
Gradual and rapid change/Punctuated equilibrium	Eldredge and Gould (1972)	Gradualism, a slight change over long time periods, is one of the basic assumptions of Darwinian evolution. But fossil records seem to show rapid changes in phases called punctuated equilibrium (Eldredge and Gould 1972).There are "phases in which rare and minor external events are able to trigger major changes" by affecting the so-called dominant auto-catalytic sets (Dovrolis and Streelman 2010: 75 ff)

theory there is an important role of internal stability (stable evolutionary kernels, canalization).[3] Without information storage in genes, multiple variations would come to nothing.

Such balances between stability and productive instability, between variation and conservation might be one of the most difficult tasks of a socio-techno-economic realization of biomimetic resilient systems. Approaches to implement multidimensional optimization and self-healing or self-repair are somewhat more familiar, at least in the information technology (IT) sector (Amin 2001), but also to some extent in materials. Dynamic regulation (balance of positive and negative feed-back-loops),

[3] And of course, there is a conservative and even restrictive element in the process of evolution (path dependency). Next steps can only be realized on the basis of the underlying 'body plan'.

Table 3.2 Design principles of **evolutionary success** that may contribute to building more resilient systems

Design principle	References	Description
Resource opportunism	Levins (1962, 1963)	The use of what is proximately available, in particular, solar energy and other low exergy sources, materials from the immediate environment like water, CO_2, limestone or waste, as well as a broad range and variety of alternative resources in case that particular resources should become unavailable
Resource efficiency/ Abundance/Circular economy	Hodapp et al. (2019), Loreau (1994)	In ecosystems and in evolution, the principle of resource efficiency is important but it is not entirely dominant. Efficiency is important when there is scarcity, which is not the case at any time and everywhere. Furthermore, evolution in many cases works with diversity and abundance. And combined with abundance there is a different paradigm regarding the realization of precision. We are faced with the phenomenon of not trying to produce the right thing from the start, but to produce a variety and then select the right one. This appears not only at population level, but also at molecular level. The relationship between resilience and efficiency does not allow for a one-sided dominance of either goal; they must be balanced. They are, along with several others, principles used in a multidimensional optimization process
Redundancy (functional/structural)/ Degeneracy	Biggs et al. (2020)	If one source for a specific capacity fails, redundant structural or functional sources can take over (the same capacity is provided by identical elements) "Degeneracy is the ability of elements that are structurally different to perform the same function or yield the same output. Unlike redundancy, which occurs when the same function is performed by identical elements, degeneracy, which involves structurally different elements, may yield the same or different functions depending on the context in which it is expressed." (Edelman and Gally 2001: 13,763)
Buffer/Storage	Spohn et al. (2021)	Reserves are important in times of scarcity
Multifunctionality	Manning et al. (2018)	Multifunctionality of components and their interplay is important with regard to the functions and services of a system

(continued)

Table 3.2 (continued)

Design principle	References	Description
Hierarchy	Eldredge and Salthe (1984)	This can be approximately understood as the opposite of homogeneity. We find different structures on different system (or organismic) levels. Hierarchical structuring is particularly typical for biogenic materials, (cf. Fratzl 2002). It is a principle of success in the design of multifunctional materials such as wood, bone, mother-of-pearl or spider silk which plays an important role in uniting seemingly incompatible technical properties. Comparable themes are discussed in systems and control theory as 'layering' or 'layered architecture'. "Layering is a ubiquitous and essential architectural feature in complex networks of all types" (Doyle and Csete 2011: 15,624)
Regeneration/ Self-healing	Birnbaum and Sánchez Alvarado (2008)	The capabilities for regeneration are almost unbelievable in the realm of organisms. However, they decrease significantly in higher organisms (Birnbaum and Sánchez Alvarado 2008)

hierarchical design, systems of systems (nested networks) and modularity are highly topical issues in automation technology and network design. All this indicates, that information and communication technologies are currently closest to the technical means for designing biomimetic resilient systems.

And finally, an important role in the establishment of ecosystem and organismal resilience during evolution was played by the fact that organisms and ecosystems have few options and opportunities to control their environment. They essentially have to adapt. Such an orientation towards 'adaptation' as opposed to 'control' represents an additional paradigm shift that is necessary on the way to design biomimetic socio-techno-economic systems. This paradigm shift from the 'domineering' attempt of control to 'humble' adaptation is just as much a challenge for the biomimetic design of resilient systems as productive instability.

3.3 Ecosystems Theory as a Source of Biomimetic Design Principles for Resilient Systems

Holling developed his concept of resilience within ecosystems theory, at least not explicitly with reference to evolution theory (Holling 1973). In his tradition, we evaluated publications from ecosystem theory with a focus on resilience enhancing principles.

An insightful understanding of systemic resilience is expressed by Woods, who compares resilience to graceful extensibility: "Graceful extensibility is the opposite of brittleness, where brittleness leads to a sudden collapse or failure when events push the system up to and beyond its boundaries of being able to handle changing environments, disturbances and variations. As the opposite of brittleness, graceful extensibility is the ability of a system to extend its capacity to adapt when surprise events challenge its boundaries" (Woods 2018: 433).

In a review paper Oliver et al. (2015) discussed basic principles and functions of resilience in ecosystems at different levels, from organisms (physiology and behaviour) over species, communities and populations, to networks and interaction with landscapes. The underlying understanding of resilience as a guiding concept focused on capabilities of resistance and recovery, that are basic for maintaining ecosystem functions and services for society. "In this case the most relevant definition of resilience is the degree to which an ecosystem function can resist or recover rapidly from environmental perturbations, thereby maintaining function above a socially acceptable level" (Oliver et al. 2015: 674). Interestingly, the authors focus not only on the species or ecosystems themselves, but also on their environment, which can have significant influence on the resilience of systems.

According to Oliver et al. (2015) the following capabilities and principles are of particular significance with regard to resilience of ecosystems in the sense of their capacity to resist and recover:

On the level of species:

- Low sensitivity towards change
- High rate of reproduction (population growth)
- Adaptive phenotypic plasticity
- High genetic variability and dispersion
- No growth suppression in the case of low population density.

On the level of the communities of species:

- Low correlation between the traits affected by change and those traits which are important for ecosystem functions
- High functional redundancy (combined with differing responses to environmental perturbations)
- Highly connected nested networks dominated by generalized interactions (in contrast to networks dominated by strong specialized interactions).

On the level of landscape (ecosystem context):

- High level of heterogeneity in the local environment
- High landscape level functional connectivity
- Potential for the existence of alternative stable states
- Spacious habitat which generally promises resource wealth.

Table 3.3 Biomimetic design principles for the design of resilient systems derived from ecosystem theory

Design principle	References
High reproduction rate	Oliver et al. (2015)
Phenotypic plasticity	Oliver et al. (2015)
Variability/Dispersion/	Oliver et al. (2015)
Spaciousness	Oliver et al. (2015)
Capacity for innovation	Molyneaux et al. (2016)
Functional diversity	Molyneaux et al. (2016)
Unspecialized interactions	Oliver et al. (2015)
Nested networks/Cellularity	Oliver et al. (2015)
Landscape-level connectivity	Oliver et al. (2015)
Adaptive circle/Panarchy	Holling et al. (2002)
Potential for alternative stable states	Oliver et al. (2015)

Of special interest is their view on networks: "In general highly connected nested networks dominated by generalized interactions are less susceptible to cascading extinction effects and provide more resistant ecosystem functions, in contrast to networks dominated by strong specialized interactions" (Oliver et al. 2015: 678). Molyneaux et al. (2016) go beyond an understanding of resilience as sheer resistance and recovery by noting that the capacity for innovation is a key function in the ability to adapt. In reference to Allen und Holling (2010) they explain that innovation is necessary for an ecosystem to sustain a state of dynamic equilibrium (homoeostasis) under changing environmental conditions. The resulting variety and functional diversity of the species in an ecosystem contribute to its integrity. This is seen as essential due to the number of possible combinations but also to the possibilities for compensation of functions of the species among each other due to their overlapping range of functions (Peterson et al. 1998 quoted in Molyneaux et al. 2016). Biodiversity is considered necessary for the adaptation of an ecosystem under changing environmental conditions although, in the case of stable environmental conditions, even a small degree of diversity is sufficient to maintain necessary functions (Oliver et al. 2015). Holling identified repetitive dynamic development phases in certain ecosystems at different levels of space and time: (1) growth or exploitation, (2) conservation, (3) collapse or release (omega) and (4) reorganization (alpha). In this way he points out that not only growth and conservation are essential in (some) ecosystem dynamics but also collapse and reorganization (Holling et al. 2002). This reminds us again of productive instability, here as a phase of development bordered by phases of relative stability. With the term "panarchy" Holling and Gunderson emphasize the stabilizing function of nested hierarchies (Gunderson and Holling 2002).

From the characteristics discussed above, a number of principles can be derived that are relevant to the design of resilient systems. The principles listed in Table 3.3 represent the characteristics needed to increase the resilience of systems rather than being an expression of resilient behaviour already.

If we compare the design elements derived from evolution theory and ecosystems theory, we can recognize in the evolutionary approach a bouquet of fundamental innovations for risk reduction like redundancy, storage, hierarchy, modularity, multifunctionality, homoeostasis and self-healing. And we can see important drivers for change like variability, self-organization, autopoiesis and productive instability. Ecosystem theory provides additional elements that relate to system design or architecture, such as nested networks, panarchy, or generalized interactions that enable loose coupling. It is also interesting to note that both sources emphasize the balance of stability and instability as a principle of success. Not to forget the qualities that the environment of the systems can contribute to resilience such as heterogeneity, spatial connectivity and spaciousness.

3.4 Restrictions on the Biomimetic Transfer

All this may be seen as quite interesting considerations but far from the design of concrete socio-techno-economic systems. And there are in fact several constraints for an implementation of biomimetic design principles. Restrictions exist in at least five subject areas. And conflicting goals that need to be balanced play a major role there.

The first restriction is ambivalent. Nature does not know societal actors. Actor based design principles cannot be obtained from biomimetics. But societal actors are extremely important in implementing complex biomimetic design principles aimed at building more resilient systems. Actors (institutions) bring intentionality into the system. They are able to realize capabilities like anticipation, self-organization, homeostasis, and autopoiesis but also monitoring, conflict resolution governance and participation.

Main restrictions on the transfer lay in the field of socio-techno-economic feasibility. Implementation of design principles like redundancy, dispersion and buffer/storage are relatively easy to accomplish by technical means. They therefore dominate actual discourses on the design of resilient systems. More difficult to implement are modularity, multifunctionality and regeneration/self-healing. And still more difficult to realize are multi-objective optimization, adaptability, alternative stable states, hierarchical design, nested (intermeshed) networks with tight and/or loose coupling and dynamic stability (feedback control). But there are socio-technical approaches at least in the field of IT to support implementation. Particularly far-reaching problems of technical realizations exist in view of the capabilities of complex systems like self-organization (emergence, autopoiesis), balanced efficiency and abundance (precision by selection) as well as balanced stability and instability (homeostasis and productive instability). Some contributions can be made by system technologies and (bio)nanotechnologies (e.g., self-organization of micelles, self-healing of polymers, 3D-printing). But the most important contributions must come from social actors and institutions. They are able to balance conflicting goals and to realize self-organization, homeostasis and autopoiesis, assisted by technical means, not least from IT and Artificial Intelligence.

Three more restrictions result from paradigms in risk management, in the economic and sustainability discourse and in mathematic-experimental laboratory science.

A short review of literature about risk management in the fields of supply chain management, IT and energy systems revealed a bundle of more or less already mentioned design principles like resource availability (multiple sources, low exergy sources, free reserves), redundancy (n-1 criterion, substitutability, power to-x), backbone structures, diversity (geographic dispersion), storage, repair, drift correction, efficiency, modularity (standardized interfaces), loose coupling, interacting nodes, feedback-loops (load management), multi-layer protection, balanced centralization/decentralization and fail-safe supplemented by safe-to-fail design[4] (Brand et al. 2017: 98–109). And there are a lot of principles based on actors and institutions like sensing, monitoring, anticipation, learning, informed decisions, assuming responsibility, new business models and products, polycentric governance (subsidiarity), task forces and emergency drill. This gives the quite justified impression that the possibilities of designing resilient systems depend much more on actors and institutions than on technical means. Nevertheless, it is remarkable that important design principles about dealing with complex systems are missing in established technological risk management. These include in particular: a dynamic understanding of stability (equilibrium, homeostasis, steady state, multiple stable states, balance of stability and instability), the importance of autopoiesis or at least productive instability, self-organization and evolvability, and non-linearity or tipping points as sources of real surprises. Particularly instability as a design principle fundamentally contradicts the prevailing security paradigm. Established technology risk management seems to especially shy away from ‘productive instability’, which is seen as the antithesis of ‘safe function’. But instability, adaptability, evolvability (as the contrary of e.g., ‘lock in’ and ‘path dependency’) are essential for improvisation, innovation and the evolution of resilient systems. All these missing principles are about dealing with complex systems and evolvability. It seems that established technology related risk management is not jet on the level of systems theory and the handling of complex systems. Nevertheless, this impression changes when leaving the field of technological risk management and looking at the risk discourse on climate change or biodiversity (Simpson et al. 2021; Nicholson et al. 2009; Brooks et al. 2016).

---

[4] Fail-safe design means the system is protected against failure. Safe-fail design means, if the system fails, it will fall into another stable state (Ahern 2011).

Efficiency as a dominant paradigm in the economic and the sustainability discourse also constrains the implementation of some biomimetic design principles. Abundance e.g. plays an important role for evolvability. Producing a broad spectrum of diversity without a quick reference to functionality and usefulness is a great challenge in socio-techno-economic systems as well as the production of precision by selection. Such projects are expensive regarding financial as well as physical resources. At a first glance, producing diversity without quick reference to usefulness, seems to be far from realization. But society affords itself exactly this in two areas, in the funding of basic research and of culture.[5] So, the balance between efficiency and abundance is another complex task in the design of resilient systems and their environment. As a first step it will be important to accept more slack (more free resources) within socio-technical systems.

An additional major challenge, besides the balance of instability and stability as well as of abundance and efficiency, is the transition from the dominant (top down) control paradigm to adaptation as an alternative way to deal with (anyway hard to control) complex systems. The completely different relevance of complexity, instability and adaptation between evolution or ecosystem theory and established risk management has a specific background, whose roots reach back into the fundamentals of the natural and engineering sciences. Organisms and ecosystems have little influence on their environment in terms of complexity reduction, reduction of change dynamics, or fixation and/or exclusion of so-called disturbance factors. They are forced to survive in unreduced complexity. This clearly contradicts to the dominant paradigm in the mathematic-experimental laboratory sciences, which in experiments (and based on them in the design of technical systems) strive to keep the surrounding conditions stable and disturbing factors away and excluded. This is without doubt a problematic form of complexity reduction, often called reductionism. It is neither oriented on appropriateness nor on practicability/applicability but on a platonic-mechanistic world view (strict mathematical laws behind phenomena). And these theoretical and practical abstractions in models and laboratory experiments have consequences. When the stability and exclusion of so-called disturbing factors can no longer be guaranteed in the socio-techno-economic real-world implementation, not foreseeable side effects are expectable (von Gleich 1989). Against this background, far-reaching theory related paradigm shifts are necessary for the design and implementation of biomimetic resilient systems, paradigm shifts which are in full swing in complexity and systems theory, in biology and physics (Schmidt 2007) but not yet in engineering (besides electronics) and not yet in risk management. Especially evolution theory is an interesting type of science, that is obliged to deal with unreduced complexity. As a result, there is no way to make predictions. And systems theories also represent an important progress, because there are at least intense debates about the adequacy of models and the legitimacy of abstractions.

[5] Whereby the opposition against utilitarian claims is much higher in culture than in research.

3.5 Preparing for Unknown Unknowns

Actual debates about necessary extensions of precaution merely refer to black swans, extremely improbable events with particularly high impact potential (Taleb 2007). The uncertainty in these cases lies in the extremely low probability of occurrence, but there are at least valid ideas about confounding factors and their possible consequences. Another way to expand precaution is through modelling, based on knowledge of at least conceivable incidents. But such notions of possible incidents are as limited as the number of simulation runs. Efforts for a precautionary principle that also includes preparation for genuine surprises must go much further. It is not about dealing with uncertainties; it is about dealing with ignorance. Unknown unknowns in contrast to known unknowns refer to knowledge about things or events, we do not know that we do not know them (Attenberg et al. 2015).

The concept of biomimetic resilience claims to be able to alleviate the knowledge gap, to show a path that makes it possible to prepare for complete surprises (von Gleich and Giese 2019). But how can we generate knowledge about unknown unknowns? The answer is clear, we can't. But there is another way to prepare for real surprises, we can expand preparedness, we can refer to evolution, to the processes and results of millions of years of evolutionary optimization, to the production of an abundance of variants and subsequent selection of solutions. The focal point is here, evolutionary optimization is misunderstood, if we reduce it to a mere quantitative extension of stress tests over millions of years. Evolutionary optimization is not just learning from (good or bad) experience. Those elements, success factors and principles from which we want to learn in the biomimetic approach to resilience emerge through random variation and not through learning. This opens a cornucopia of diversity, a huge realm of the unthinkable and unexpectable. It is above all in this way that a preparation for complete surprises, appears possible. With focus on random variation, and in the second step selection of the appropriate, we must deny that the millions of years of evolutionary optimization are nothing else than 'learning', only that the learning process was unusually long. In this way we reject a—quasi-Lamarckian—misunderstanding of evolution. In the Darwinian evolution theory, new traits for survival are neither the product of bad or good experience, nor the result of just selection. New solutions are the product of a completely undirected chaotic generation of diversity through mutation and recombination, the production of abundance without influence of usefulness. Only in a second step selection picks out what contributes to success of survival. And only then begins the 'learning' for organisms (including epigenetics), populations and ecosystems, which admittedly also plays an important role in adaptation and evolution.

At its core, the biomimetic concept of resilience leaves behind the stimulus-responses paradigm. Designing a system that is prepared for surprises cannot be solely based on learning from (bad) experience in the past, from already experienced or imaginable perturbations. Following the evolutionary approach is not the generation and inheritance of designs that are a reaction to innumerable perturbation events. The creational 'motor of evolution' is far more innovative. It is the random generation of innumerable possible designs by mutation and recombination.[6] In this way absolutely new designs become possible. Even extremely counterintuitive solutions are generated, that would never have come about as experienced, learned or 'planned'.[7] And if these new designs, prove themselves over millions of years, if they have survived countless constellations and surprises, the probability is high that they also will be able to withstand surprises that are still to come. However, complete safety cannot be achieved in this way either. Surprises and constellations that have not yet occurred in millions of years, and which are able to even overcome the most creative resilient design, are still conceivable.

And we have to admit that learning from (good or bad) experience is also an effective risk reduction strategy. It is also common in organisms and ecological systems. But the results of this learning are generally not stored in the genome.[8] Learning may be stored in the memory of individuals or populations or even in the immune system. And experiences can be transferred to coming generations in various ways. But there remains a difference. This knowledge is limited to experience and if it is not used or stored it fades away. And most importantly it does not have the creative potential of completely new solutions.

3.6 Conclusion

Biomimetics claim to be able to open a way to prepare for real surprises. This way refers to millions of years of evolutionary optimization. Central to this is not just selection, the innumerable amount of stress tests in millions of years. Central is the outflanking of the stimulus–response approach, the creation of an overwhelming abundance of traits without reference to useability. Selection then picks out, what is fitting for survival. But biomimetics cannot provide actor-based solutions. So, a second source of design principles for resilient systems is technological risk management as well as risk management dealing with climate change and biodiversity that

---

[6] Interestingly some organisms have the ability to quantitatively increase the production of mutations under stress and even to increase the production of 'useful' phenotypic variations for adaptation now or in the future, which is an important aspect of evolvability, cf. Wagner (2007), Silver and de Weck (2007), Pigliucci (2008), Wagner (2011), Payne and Wagner (2019).

[7] E.g. a rough surface of shark skin to reduce drag, or a rough leaf surface with the ability to clean itself (lotus-effect).

[8] Epigenetics are a special case and another interesting element of evolvability.

leaves behind the constraints of the technology approach. Resilience as a risk reduction strategy must draw from both sources from biomimetics and from established risk management based on experience (Goessling-Reisemann and Thier 2019).

There are important restrictions for the development and implementation of biomimetic design principles, first in the economic and sustainability discourse regarding the importance of efficiency, second in risk management regarding the problematic attitude towards stability and control. Instability is seen as the opposite of safety. But instability is essential for adaptation and evolvability. Most complex systems cannot be completely controlled. They can be managed and stabilized by certain parameters. And the capability to balance stability and instability as well as the attention to non-linearities play a major role in resilient systems. And last but not least, the theoretical (models) and practical abstractions (laboratory experiments) in specific natural and engineering sciences must be taken into account.

Complexity reduction is essential in any type of science. But there are differences between certain paradigms and theories. To mention here are mathematic-mechanistic laboratory science, systems theory and evolution theory. The selection of models and the quality and quantity of theoretical and practicable abstractions play a major role in view of the appropriateness of their insights and in view of the application of their results. For a precaution-oriented design, it is recommended to resort to theories or paradigms that reflect intensively on the appropriateness of models and experiments and the corresponding abstractions.

Biomimetics based on evolutionary diversity on one side and risk management based on experience and models on the other provide important elements for the design of an ideal resilient system and its supporting surroundings. A selection of these design elements is shown in Fig. 3.1. This compilation can be used in resilience monitoring to check whether the planned design step contributes to resilience requirements and how close the actual resilient system comes to the ideal, or what is still missing on the way there.

Such a guiding principle is important for the development of resilient systems. Some of these solutions already exist, others can be developed in short time and still others will only become apparent in the future. In general, a two-step development is necessary in order to adequately deal with complexity and to implement biomimetic resilience: First in the mindset (paradigms, system theory, aversion to instability) and second in socio-techno-economic design steps in direction of the ideal resilient system based not least on actors (institutions) as well as system and nano technologies as well as IT and AI. Both approaches help to prepare for surprises and support the ability to adequately deal with self-organization and autopoiesis, to produce regeneration and to balance stability and instability (homeostasis) as well as efficiency and abundance (a high level of variability).

Fig. 3.1 Design principles and implementation measures (in red) for an ideal biomimetic resilient system (own representation based on sources mentioned above)

References

Ahern, J.: From fail-safe to safe-to-fail: sustainability and resilience in the new urban world. Landsc. Urban Plan. **100**(4), 341–343 (2011). https://doi.org/10.1016/j.landurbplan.2011.02.021

Allen, C.R., Holling, C.S.: Novelty, adaptive capacity, and resilience. Ecol. Soc. 15(3) (2010)

Amin, M.: Toward self-healing energy infrastructure systems. IEEE Comput. Appl. Power **14**(1), 20–28 (2001). https://doi.org/10.1109/67.893351

Attenberg, J., Ipeirotis, P., Provost, F.: Beat the machine: challenging humans to find a predictive model's unknown unknowns. J. Data Inf. Qual. 6(1):1:1–1:17 (2015)

Baldwin, C.Y., Clark, K.B.: Design Rules: The Power of Modularity. The MIT Press, Cambridge, MA (2000)

Beisner, M., Haydon, D.T., Cuddington, K.: Alternative stable states in ecology. Front. Ecol. Environ. **1**(7), 376–382 (2003). https://doi.org/10.1890/1540-9295(2003)001[0376:ASSIE]2.0.CO;2

Benyus, J.M.: Biomimicry: Innovation Inspired by Nature. Harper Collins Publishers Inc., New York (1997)

von Bertalanffy L.: General System Theory. Foundations, Developments, Applications. George Braziller, New York (1968)

Biggs, C.R., Yeager, L.A., Bolser, D.G., Bonsell, C., Dichiera, A.M., Hou, Z., Keyser, S.R., Khursigara, A.J., Lu, K., Muth, A.F., Negrete, B., Jr., Erisman, B.E.: Does functional redundancy affect ecological stability and resilience? A review and meta-analysis. Ecosphere **11**(7) (2020). https://doi.org/10.1002/ecs2.3184

Biomimicry 3.8: Life's principles: biomimicry design lens (2013). https://biomimicry.net/the-buzz/resources/designlens-lifes-principles/

Birnbaum, K.D., Sánchez Alvarado, A.: Slicing across kingdoms: regeneration in plants and animals. Cell **132**, 697–710 (2008). https://doi.org/10.1016/j.cell.2008.01.040

Borsari, A.: Human mimicry and imitation: the case of biomimetics. Aisthesis Pratiche Linguaggi E Saperi Dell'estetico **10**, 51–61 (2017)

Brand, U., Giese, B., von Gleich, A., Heinbach, K., Petschow, U., Schnülle, C., Stührmann, S., Stührmann, T., Thier, P., Wachsmuth, J., Wigger, H.: Auf dem Weg zu Resilienten Energiesystemen! Resiliente Gestaltung der Energiesysteme am Beispiel der Transformationsoptionen „EE-Methan-System" und „Regionale Selbstversorgung" Schlussbericht des BMBF-Projekts RESYSTRA (FKZ:01UN1219A-B), Bremen, Berlin (2017). https://doi.org/10.26092/elib/1465

Branke, J., Deb, K., Miettinen, K., Slowinski, R.: Multiobjective Optimization: Interactive and Evolutionary Approaches. Springer (2008). ISBN 978-3-540-88907-6

Brooks, T., Akçakaya, H., Burgess, N., et al.: Analysing biodiversity and conservation knowledge products to support regional environmental assessments. Sci. Data **3**, 160007 (2016). https://doi.org/10.1038/sdata.2016.7

Darwish, A.: Bio-inspired computing: algorithms review, deep analysis, and the scope of applications. Future Comput. Inform. J. **3**(2), 231–246 (2018)

Dicks, H.: Biomimetic Epistemology. First View Dez 2023, pp. 1–21. Cambridge University Press (2023). https://doi.org/10.1017/psa.2023.173

Dovrolis, C., Streelman, J.T.: Evolvable network architectures: what can we learn from biology? ACM SIGCOMM Comput. Commun. Rev. (2010). https://doi.org/10.1145/1764873.1764886

Doyle, J.C., Csete, M.: Architecture, constraints and behavior. PNAS **108**, 15624–15630 (2011)

Edelman, G.M., Gally, J.A.: Degeneracy and complexity in biological systems. PNAS **98**(24), 13763–13768 (2001). https://doi.org/10.1073/pnas.231499798

Eldredge, N., Gould, S.J.: Punctuated equilibria: an alternative to phyletic gradualism. In: Schopf, T.J.M. (ed.) Models in Paleobiology, pp. 82–115. Freeman Cooper, San Francisco, CA (1972)

Eldredge, N., Salthe, S.N.: Hierarchy and evolution. Oxf. Surv. Evol. Biol. **1**, 182–206 (1984)

European Commission, Directorate-General for Research and Innovation, Bulkeley, H., Naumann, S., Vojinovic, Z., et al.: Nature-based solutions – State of the art in EU-funded projects (Freitas,

T., Vandewoestijne, S., Wild, T. (eds.)). Publications Office of the European Union (2020). https://data.europa.eu/doi/10.2777/236007

Fratzl, P.: Von Knochen, Holz und Zähnen: Durch gezielte Strukturierung bringt die Natur Materialien mit erstaunlicher Festigkeit oder Elastizität hervor – Materialwissenschaftler können von ihnen lernen. Phys. J. **1**(5):1617–9439/02/0505 (2002)

Gass, G.L., Bolker, J.A.: Modularity. In: Hall, B.K., Olson, W.M. (eds.) Keywords and Concepts in Evolutionary Developmental Biology, pp. 260–267. Harvard University Press, Cambridge (2003)

Gass, G.L., Hall, B.K.: Collectivity in context: modularity, cell sociology, and the neural crest. Biol. Theory **2**, 349–359 (2007). https://doi.org/10.1162/biot.2007.2.4.349

von Gleich, A.: Der wissenschaftliche Umgang mit der Natur. Über die Vielfalt harter und sanfter Naturwissenschaften. Campus Verlag, Frankfurt/M (1989)

von Gleich, A.: Das Bionische Versprechen. Ökologisches Wirtschaften **3**, 21–23 (2007)

von Gleich, A., Pade, C., Petschow, U., Pissarskoi, E.: Potentials and Trends in Biomimetics. Springer, Heidelberg (2010)

von Gleich, A., Giese, B.: Resilient systems as a biomimetic guiding concept. In: Ruth, M., Goessling-Reisemann, S. (eds.) Handbook on Resilience of Socio-Technical Systems, pp. 89–99. Edward Elgar Publishing, Cheltenham (2019)

Goessling-Reisemann, S., Thier, P.: On the difference between risk management and resilience management for critical infrastructures. In: Ruth, M., Goessling-Reisemann, S. (eds.) Handbook on Resilience of Socio-Technical Systems, pp. 117–135. Edward Elgar Publishing, Cheltenham (2019)

Gunderson, L.H., Holling, C.S. (eds.): Panarchy: Understanding Transformations in Human and Natural Systems. Island Press, Washington, DC (2002)

Hallgrimsson, B., Willmore, K., Hall, B.K.: Canalization, developmental stability, and morphological integration in primate limbs. Am. J. Phys. Anthropol. **2002** (Suppl 35), 131–158 (2002)

Hayes, S., Deshaa, C., Baumeister, D.: Learning from nature – biomimicry innovation to support infrastructure sustainability and resilience. Technol. Forecast. Soc. Change. **161**, 120287 (2020). https://doi.org/10.1016/j.techfore.2020.120287

Helmrich, A.M., Chester, M.V., Hayes, S., Markolf, S.A., Desha, C., Grimm, N.B.: Using biomimicry to support resilient infrastructure design. Earth's Future **8**, e2020EF001653 (2020). https://doi.org/10.1029/2020EF001653

Hendry, A.P., Kinnison, M.T., Heino, M., Day, T., Smith, T.B., Fitt, G., Bergstrom, C.T., Oakeshott, J., Jørgensen, P.S., Zalucki, M.P., Gilchrist, G., Southerton, S., Sih, A., Strauss, S., Denison, R.F., Carroll, S.P.: Evolutionary principles and their practical application. Evol. Appl. **4**(2), 159–183 (2011). https://doi.org/10.1111/j.1752-4571.2010.00165.x

Hodapp, D., Hillebrand, H., Striebel, M.: "Unifying" the concept of resource use efficiency in ecology. Front. Ecol. Evol. **6**, 233 (2019). https://doi.org/10.3389/fevo.2018.00233

Holling, C.S.: Resilience and stability of ecological systems. Annu. Rev. Ecol. Syst. **4**(1), 1–23 (1973)

Holling, C.S., Gunderson, L.H., Ludwig, D.: In quest of a theory of adaptive change. In: Gunderson, L.H., Holling, C.S. (eds.) Panarchy: Understanding Transformations in Human and Natural Systems. Island Press, Washington D.C. (2002)

Krohs, U.: The epistemology of biomimetics: the role of models and of morphogenetic principles. Perspect. Sci. **29**(5), 583–601 (2021). https://doi.org/10.1162/posc_a_00385

Lankau, R.A., Jørgensen, P.S., Harris, D.J., Sih, A.: Incorporating evolutionary principles into environmental management and policy. Evol. Appl. **4**, 315–325 (2011)

Levin, S.A.: Self-organization and the emergence of complexity in ecological systems. Bioscience **55**(12), 1075–1079 (2005). https://doi.org/10.1641/0006-3568(2005)055[1075:SATEOC]2.0.CO;2

Levins, R.: Theory of fitness in a heterogeneous environment. I. The fitness set and adaptive function. Am. Nat. **96**, 361–373 (1962)

Levins, R.: Theory of fitness in a heterogeneous environment. II. Developmental flexibility and niche selection. Am. Nat. **97**, 75–90 (1963)
Loreau, M.: Material cycling and the stability of ecosystems. Am. Nat. **143**(3), 508–513 (1994). https://doi.org/10.1086/285616
Manning, P., van der Plas, F., Soliveres, S., et al.: (2018) Redefining ecosystem multifunctionality. Nat. Ecol. Evol. **2**, 427–436 (2018). https://doi.org/10.1038/s41559-017-0461-7
Mattheck, C.: Design in Nature - Learning from Trees. Springer, Heidelberg New York (1998)
Maturana, H., Varela, F.: Autopoiesis and Cognition: The Realization of the Living. D. Reidel, Boston (1980). ISBN 978-90-277-1016-1
Molyneaux, L., Brown, C., Wagner, L., Foster, J.: Measuring resilience in energy systems: insights from a range of disciplines. Renew. Sustain. Energy Rev. **59**, 1068–1079 (2016). https://doi.org/10.1016/j.rser.2016.01.063
Nachtigall, W.: Bionik - Grundlagen und Beispiele für Ingenieure und Naturwissenschaftler. Springer, Berlin (2002). https://doi.org/10.1007/978-3-642-18996-8
Nachtigall, W.: Bionik als Wissenschaft. Erkennen – Abstrahieren – Umsetzen. Springer, Berlin (2010)
Nicholson, E., Keith, D.A., Wilcove, D.S.: Assessing the threat status of ecological communities. Conserv. Biol. **23**, 259–274 (2009). https://doi.org/10.1111/j.1523-1739.2008.01158
Olsson, L., Jerneck, A., Thoren, H., Persson, J., O'Byrne, D.: Why resilience is unappealing to social science: theoretical and empirical investigations of the scientific use of resilience. Sci. Adv. **1**(4) (2015). https://doi.org/10.1126/sciadv.1400217
Olariu, S., Zomaya, A.Y.: Handbook of Bioinspired Algorithms and Applications. Chapman & Hall/CRC (2005)
Oliver, T.H., Heard, M.S., Isaac, N.J.B., Roy, D., Procter, D., Eigenbrod, F., Freckleton, R., Hector, A., Orme, C.D.L., Petchey, O.L., Proença, V., Raffaelli, D., Suttle, K., Mace, G.M., Martín-López, B., Woodcock, B.A., Bullock, J.M.: Biodiversity and resilience of ecosystem functions. Trends Ecol. Evol. **30**(11), 673–684 (2015)
Payne, J.L., Wagner, A.: The causes of evolvability and their evolution. Nat. Rev. Genet. **20**, 24–38 (2019). https://doi.org/10.1038/s41576-018-0069-z
Peterson, G., Allen, C.R., Holling, C.S.: Ecological resilience, biodiversity, and scale. Ecosystems **1**(1), 6–18 (1998). https://doi.org/10.1007/s100219900002
Pigliucci, M.: Is evolvability evolvable? Nat. Rev. Genet. **9**(1), 75–82 (2008). https://doi.org/10.1038/nrg2278.PMID18059367
Porter, M.E.: Competitive Strategy: Techniques for Analyzing Industries and Competitors - With a New Introduction. The Free Press, New York (1980). ISBN 0-684-84148-7
Rechenberg, I.: Evolutionsstrategie '94. Frommann-Holzbook Verlag, Stuttgart-Bad Cannstatt (1994)
Schmidt, J.C.: Instabilität in Natur und Wissenschaft: Eine Wissenschaftsphilosophie der nachmodernen Physik. De Gruyter, Berlin (2007)
Scharloo, W.: Canalization – genetic and developmental aspects. Annu. Rev. Ecol. Syst. **22**, 65–93 (1991). https://doi.org/10.1146/annurev.es.22.110191.000433
Schwefel, H.-P.: Evolution and Optimum Seeking. Wiley, New York (1995)
Simon, H.A.: Near decomposability and the speed of evolution. Ind. Corp. Chang. **11**(3), 587–599 (2002). https://doi.org/10.1093/icc/11.3.587
Simpson, N.P., et al.: A framework for complex climate change risk assessment. One Earth **4**, 489–501 (2021)
Silver, M.R., de Weck, O.L.: Time-expanded decision networks: a framework for designing evolvable complex systems. Syst. Eng. **10**(2), 167–188 (2007)
Spohn, M., Aburto, F., Ehlers, T.A., et al.: Terrestrial ecosystems buffer inputs through storage and recycling of elements. Biogeochemistry **156**, 351–373 (2021). https://doi.org/10.1007/s10533-021-00848-x
Stirling, A.: On the economics and analysis of diversity. In: SPRU Electronic Working Papers Nr. 28 University of Sussex (1998)

Taleb, N.N.: The Black Swan: The Impact of the Highly Improbable. Random House, New York (2007). ISBN 978-1400063512

Vincent, J.F.V., Bogatyreva, O.A., Bogatyrev, N.R., Bowyer, A., Pahl, A.: Biomimetics: its practice and theory. J. R. Soc. Interface. **3**, 471–482 (2006)

Wagner, A.: Robustness and Evolvability in Living Systems. Princeton Studies in Complexity. Princeton University Press (2007)

Wagner, A.: The Origins of Evolutionary Innovations: A Theory of Transformative Change in Living Systems. Oxford University Press (2011). ISBN 978-0-19-969259-0

Woods, D.: The theory of graceful extensibility: basic rules that govern adaptive systems. Environ. Syst. Decis. **38**(5), 433–457 (2018). https://doi.org/10.1007/s10669-018-9708-3

Zakharova, E.M., Minashina, I.K.: Review of multidimensional optimization methods. J. Commun. Technol. Electron. **60**, 625–636 (2015). https://doi.org/10.1134/S1064226915060194

Chapter 4
System and Context

Wolfgang Weimer-Jehle and Sigrid Prehofer

Abstract Systems are always embedded in environments with which they interact and which influence their behaviour—sometimes to a limited extent, but often in a formative way. This also applies to the resilience behaviour of energy systems, which means that the resilience of an energy system can never be solely a property of the system itself and its architecture, but also arises or fails as a product of the interaction between system and context. The chapter describes the analytical gaps that result from disregarding the system-context problem and some approaches to dealing with the system-context problem that can be observed in practice. The context-scenario approach is then presented in more detail and its application in the ReMoDigital project is outlined.

Keywords Resilience · Energy systems · Context sensitivity · Context scenarios · Cross-impact balances

4.1 Introduction

Resilience is a property that is always thought of in the context of systems: it is "systems" whose performance suffers more or less deeply after a shock and then recovers quickly, slowly, imperfectly or not at all. With this constitutive reference to the concept of systems, however, the concept of resilience also incurs some theoretically and practically significant complications of the concept of systems.

Systems (except the system "cosmos" as a whole) are always parts of a larger reality, of a larger system. And apart from highly idealized systems, this embedding is also expressed in the exchange of effects (i.e., energy, information, momentum, forces, mass, ...) between the system and its environment, the "context". The focusing of an observer or analyst on a section of reality, which he or she perceives as a "system", is thus approximately always a reductionistic act.

W. Weimer-Jehle (✉) · S. Prehofer
Center for Interdisciplinary Risk and Innovation Studies, University of Stuttgart (ZIRIUS), Stuttgart, Germany
e-mail: wolfgang.weimer-jehle@cross-impact.de

D. van Doren et al. (eds.), *Towards Energy System Resilience*, Ethics of Science and Technology Assessment 55, https://doi.org/10.1007/978-3-032-03187-7_4

It must be admitted, however, that this reductionistic act is unavoidable, for the world as a whole is neither observable nor understandable. We can only approach the understanding of the world and its phenomena piecemeal, which is why its mental division into smaller "systems" is a hypothetical, but pragmatically reasonable approach. Nevertheless, it remains necessary for the observer and analyst to remain aware of the artificiality of this division into system and context, to remain mindful that this artificiality will lead to distortions in system perception, and to make reasonable efforts to mitigate these distortions.

This "original sin" of systems analysis also manifests itself in the study of energy systems embedded in its natural and societal context, regardless of which properties of an energy system are the focus of our attention, be it its costs, its environmental impacts, or specifically its resilience properties.

Warnings that we are generally doing too little to take account of this system-context-problem have been around in manifold disguises for a long time in future-oriented systems analysis in and outside energy research. Already the old master of French futures research, Michel Godet, urged his guild colleagues in his 1983 essay "Reducing the Blunders in Forecasting" to supplement their technically sophisticated system calculations with a reflection of the uncertainties of the embedded context. The same period also saw the first attempts in energy system analysis, still disdained by funding agencies, to combine energy scenarios with "prescenarios" (Jochem, n.d.). The idea reached broad attention in energy, climate, and environmental research not before the development and application of the "story-and-simulation" concept in the context of global emissions scenario building for the IPCC (Nakićenović et al. 2000; Alcamo 2008). Since then, the system-context-problem has also gained increasing attention in energy system analysis (for a summary, see e.g. Weimer-Jehle et al. 2020).

This paper is divided into five parts. First, the system-context problem is grounded in the concept of systems, a brief typology of dealing with this problem in systems analysis is presented, and its "ubiquity" in a wide variety of thematic contexts is illustrated (Sect. 4.2.3). The concept of context scenarios, as a powerful approach of dealing with this problem, is then presented in more depth (Sect. 4.3), and how the system context problem manifests itself in the application of the resilience concept to energy systems is discussed (Sect. 4.4). Section 4.5 describes the application of the context scenario approach in the ReMoDigital project and reflections and conclusions are presented in Sect. 4.6.

4.2 System Boundaries

The concept of resilience is widely understood as a property of systems and is inseparably linked to the concept of systems. This is already expressed in the common definitions of resilience, which always interpret resilience as a phenomenon in systems. For example, von Renn et al. write in Chap. 2 of the ability "…to respond adequately when the system is under stress" and at the same time refer to the origins in ecosystem research (Holling 1973), in which resilience is described as "…the resistance of natural ecosystems to cope with stressors".

The list of examples could be continued. Obviously, the concept of resilience can only be thought of and applied in relation to the concept of systems, and this has the important consequence that basic characteristics of the system concept must have their effects on the concept of resilience and on the operationalization, measurement and interpretation of resilience. A central and defining property of systems, as soon as we dispense with the consideration of highly idealized systems, is their openness.

4.2.1 System Boundaries

A pile of sand is not a system because it remains a pile of sand after you remove part of it. A pile of gears, winding springs and screws is also not a system, but a clock is. With examples like these, many texts on systems research and systems analysis approach the question of what a system is. The examples suggest what systems constitute (Bossel 2004):

- system purpose (from the perspective of the observer)
- system elements and interactions between them
- system integrity (loss of identity in case of division).

In addition to the system elements and their interactions, the essential constituents of a system also include the system boundary (Imboden and Koch 2003), because apart from the cosmos as a whole, every conceivable system represents only a section of a larger whole. In closed systems, the system boundary would not be traversed by effects of the system or by effects of the system environment on the system; in open systems, this is different (Fig. 4.1).

However, large-scale systems of practical importance are almost always open, and essentially open. This is also very distinctly true of energy systems of any kind and at any level. They are embedded in society, which develops and builds their technical components, provides operating material, supplies operating personnel, consumes their products, decides on regulations, absorbs their waste, risks and side effects. In parallel, energy systems are also embedded in their natural environment, which affects them and provides—or does not provide—climatic and other operating conditions within design limits.

The idea that the boundaries of a system result from the thing itself is a myth. Rather, how one draws the boundaries of a system is strongly influenced by the perspective of the observer and, to some extent, by pragmatic considerations, and is guided, among other things, by the criteria listed in Table 4.1.

Often enough, these motives for choosing the system boundary are in tension with each other. When trying to understand and manage a complex system, ignoring this tension may already be the first step to failure (Dörner 1997).

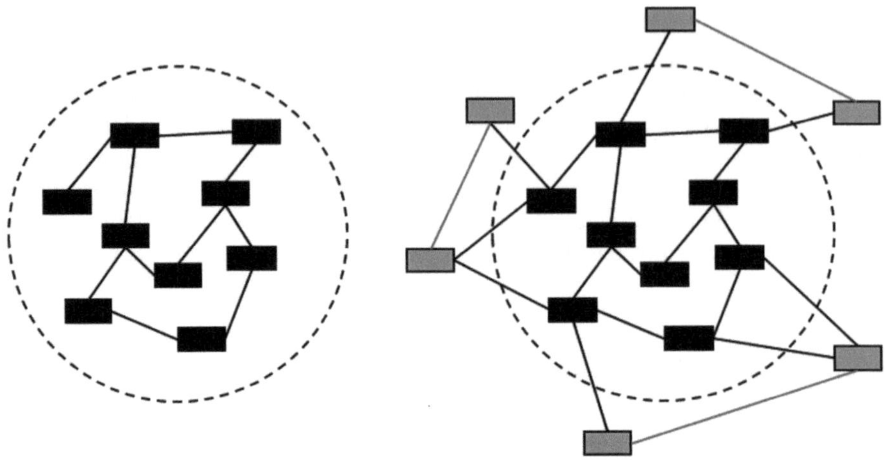

Fig. 4.1 Closed (left) and open systems (right)

Table 4.1 Criteria for determining system boundaries (selection)

Criterion	Guiding question
System purpose	Which elements directly serve the *perceived* system purpose?
Analysis purpose	What do I need to consider in order to gain the desired insights about the system?
Designated analysis methods and data availability	How do I have to define the system boundary, so that I can examine the system with the designated method toolbox and available data?

4.2.2 Dealing with System Boundaries in System Analysis

The inevitable challenge of addressing the question of system boundaries in the study of systems leads to very different practices, which are condensed here into four prototypes (Fig. 4.2).

The way that is easiest to carry out, but does least justice to the system-analytical challenges of the system boundaries, is to ignore the world outside the system boundary. The model only represents the section of the world that we have recognized as part of the system, and for the rest it is assumed (as a working fiction) that the world outside the system boundaries exerts no relevant influences on the system. This approach would be appropriate for the rare special case of a (sufficiently) closed system. However, it becomes all the more inappropriate, the less the investigated system corresponds to the concept of a closed system.

The opposite strategy to this is to extend the system boundaries until all relevant factors are included in the system definition, so that the subsequent ignoring of the outwardly shifted system boundaries seems justified (Fig. 4.3). This strategy seems

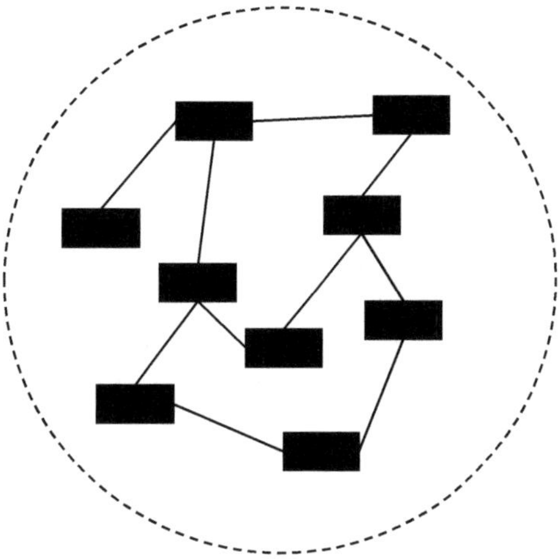

Fig. 4.2 Dealing with system boundaries. Strategy I: ignore

particularly comprehensive, but has the disadvantage that it would often require the models to swell to a size that is difficult to handle. Most importantly, it has the serious disadvantage that the enlarged system often cannot be treated with the specialized methods that would have been available for a high-performance analysis of the core system. Comprehensive system coverage must then be "paid for" with a loss of analysis quality, which is in many cases not an option.

A very practicable and frequently chosen alternative is to recognize the sources of influence outside the system boundaries as external factors, i.e. not to integrate them into the system, but to regard them as framework conditions of the system, which are not scrutinized in a system-analytical way, but for which fixed assumptions are made (Fig. 4.4). Fixed assumptions do not necessarily mean that they are defined as constant in time. Fixed time series can also be assumed.

This approach is resembling a common test procedure for electronic modules. Their intrinsic function is difficult to assess once they have been inserted into the extensive electronic system for which they are intended, because it is then no longer possible to separate which behaviour is caused by the module itself and which behaviour is merely a reflex of the actions of the surrounding system. In order to investigate their function, the modules are therefore first placed in a "testbed" in which they receive standardized impulses from an artificial environment and their reactions to these impulses are observed.

The same is true for this form of system analysis: In order to avoid the complexity of the interplay between the internal and external parts of the system, the core system receives "standardized impulses" from an "artificial environment". It is then assumed that the population development proceeds in this way and the economic development in that way, and so on. The uncertainty of the assumptions made and the question whether the combination of assumptions for the different framework factors (which in general form again a complex interdependent system) are consistent in themselves, and whether the assumed context conditions can withstand possible reactions from the system, these questions are left aside. Nevertheless, the approach is excellently suited to develop an understanding of the system under investigation. However, statements about how things might develop once the core system is taken out of its testbed and "inserted" into its interacting context can only be derived from it to a limited extent.

A recently developed approach meets this system analytic challenge better than the testbed, but at the same time, unlike in Fig. 4.3, leaves the analysis of the core system in the hands of specialized models. It has become known as story-and-simulation (Alcamo 2008; O'Mahony et al. 2013) or context scenarios (Weimer-Jehle et al. 2016), depending on its design. In it, the core system is not exposed to a single set of framing assumptions as in Fig. 4.4, but to a set of very different framing assumptions, the context scenarios (Fig. 4.5). In the creation of the context scenarios, the discernible interdependencies between the context parameters and the feedback effects from the core system on the context parameters are taken into account, and this is done with methods that no longer have to follow the requirements of the core system, but are optimized for the challenges of context analysis. In response, one obtains integrated pictures of context and system that reflect the uncertainty and complexity of context and also the mutual influence of context and system. This approach is discussed in more detail in Sect. 4.3. An intermediate form between the context scenario approach and the testbed approach is sensitivity analysis, in which one varies individual assumptions to investigate the response of the system, but one does not subject the system to fundamentally different configurations of assumptions.

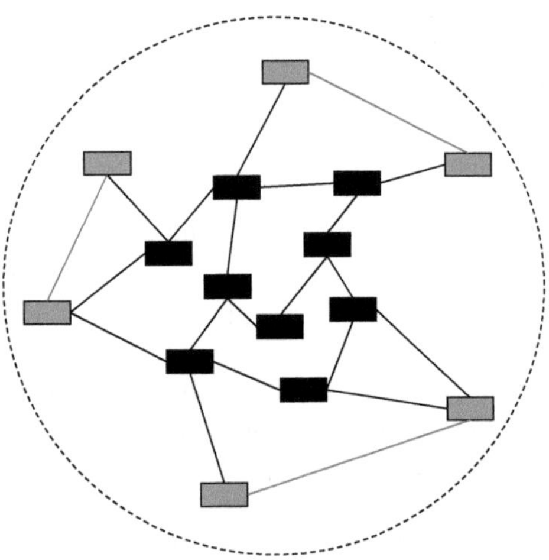

Fig. 4.3 Dealing with system boundaries. Strategy II: absorb

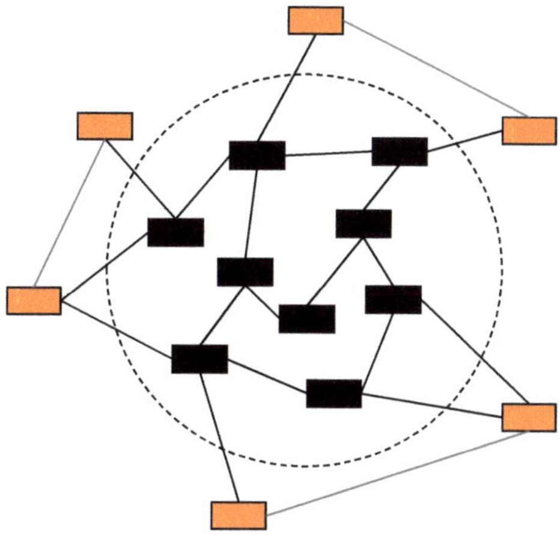

Fig. 4.4 Dealing with system boundaries. Strategy III: the testbed

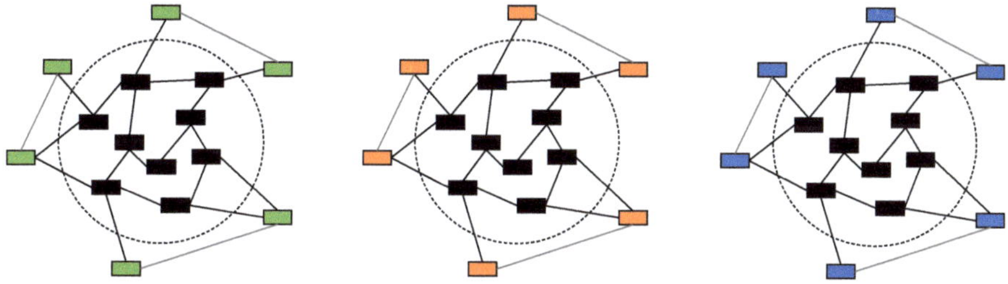

Fig. 4.5 Dealing with system boundaries. Strategy IV: context scenarios

4.2.3 Ubiquity of the System-Context Problem

4.2.3.1 Example: Energy Systems Analysis

As mentioned, the system context problem manifests itself clearly in the analysis of energy systems. In general, energy models consider the energy flows through the energy system from primary energy through the conversion and distribution of final energy and its transformation into energy services through application technologies, plus the associated costs and mass flows, including emissions. All things that parameterize these processes, such as future demand developments with the social drivers behind them, technology characteristics, political limit settings, factor costs, and the like, are fed in as assumptions to run the model (acatech et al. 2016). This "cloud" of assumptions forms the context of the energy system under study: itself not part of the system and rarely the object of in-depth reflection in energy system analyses, but a complex system in its own right and of determinative power for the results of model calculations (Prognos 2011).

In addition to the societal contexts described above, spatial contexts also affect national energy systems, such as connections to partner countries' energy systems or to international markets.

4.2.3.2 Example: Management Sciences

Another example is management science. A standard task here is to assess the prospects of success of a long-term strategy in its dependence on uncertain future conditions. For example, is it worthwhile for a company to open up a new market for its product in another country by setting up local production facilities, or to achieve this instead through other strategies, such as local cooperations? The answer will depend on many environmental conditions that are beyond the company's control, such as the market position of local competitors, consumer expectations, the availability of local human resources and upstream suppliers, regulatory conditions and much more. And all this has to be considered not only for the present, but for decades into the future. In the management science, it is suggested to develop a series of environmental scenarios and, for each combination of environment scenario and strategy option, to examine how the strategy would perform in this environment (Fink et al. 2002) (Fig. 4.7).

4.2.3.3 Example: REACH

REACH (Registration, Evaluation, Authorisation and Restriction of Chemicals) is an EU chemicals regulation that ties the placing of chemicals on the market to prior registration (EU 2006). A prerequisite for registration is a comprehensive assessment of the impact and risk aspects of the chemical, including a "socio-economic analysis" (SEA) that looks at the health, economic and social consequences of chemical use (BAuA 2010). It quickly becomes apparent that the impact assessment depends on the context conditions in which the impacts take place. For example, a loss of jobs in a particular occupational group due to the use of a new chemical is to be assessed completely differently in terms of its social consequences and also in terms of its economic effect if the occupational group concerned is in high demand on the labour market and those affected therefore have good prospects of finding an equivalent new job and, at the same time, personnel shortage-related obstacles to growth are also removed for the new employer, or if none of this is the case. As a way of capturing the assessment uncertainty associated with context sensitivity, REACH guidelines recommend the use of (context) scenarios, and at the same time warn against basing the analysis on arbitrary or inconsistent scenarios (BAuA 2010:21).

As the examples show, it would be misleading to think that the system-context problem is an academic problem of a certain discipline. Rather, it results from the common circumstances of any kind of system reflection. Accordingly, system analysts of the most different disciplinary perspectives, dealing with the most different tasks from science and practice find themselves confronted with this problem, according to the citations of their publications mostly without knowing about each other, and develop their own, always similar solutions.

4.3 Context Scenarios

As argued in Sect. 2.2, context scenarios are a powerful tool for dealing with the context-system problem in the analysis of open systems. In order to fulfil the expectations that can be placed on them in principle, they must meet certain quality requirements in practice. When describing the context-system problem in the case of the REACH Directive, the requirement that (context) scenarios must not be arbitrary and must not be inconsistent has already been cited in Sect. 4.2.3. This means:

- Context scenarios must be diverse enough as a scenario set to at least roughly represent the possibility space for the context—as far as we can know today. It would be arbitrary to choose some scenarios that illuminate only one part of the possibility space and leave other parts unrepresented
- Contextual scenarios must be inherently logical, free of contradictions, and represent what we can know today about the system's determining forces. Otherwise, they are not credible hypotheses about possible futures, because they are inconsistent in themselves or inconsistent with existing knowledge

The creation of context scenarios is a preparatory, but not an incidental task in systems analysis; rather, its result significantly shapes the outcome of the subsequent analysis of the core system. This step requires own research, own methods and, being a process of knowledge integration, (Prehofer et al. 2021) it requires generally distinct interdisciplinary competences. In practice, two very different approaches have emerged for their construction.

4.3.1 Intuitive Scenario Methods

Some practitioners prefer scenario processes to be as unstructured and discussion-based as possible in order to leave as much room as possible for intuition and free thinking. In the Intuitive Logics process ("IL", Wilson 1998), experts are asked to discuss the problem in scenario workshops and then to bundle their ideas about possible futures into some scenarios. Often, the special method variant of "Scenario Axis" is used (van 't Klooster and van Asselt 2006), in which one first agrees on two megatrends that are likely to be the main determinants of the context system under consideration, formulates two polar alternatives for each of these trends, and then fills the resulting four fields with one scenario each (possibly with subvariants) based on discussion (Fig. 4.8).

Intuition-based methods are regularly used in story-and-simulation (SaS) approaches (Alcamo 2008) to solve the context-system problem. The earliest prominent use of the SaS approach to develop storylines for the IPCC emissions scenarios (Nakićenović et al. 2000) was also done this way. Subsequently, many energy system analyses have also relied on this approach (e.g., Mander et al. 2008, O'Mahony et al. 2013, WEC 2016, and others).[1]

A great strength of the IL approach is the use of the intuitive complexity competences of the human brain, the accessibility for the participation of participants inexperienced in the scenario method, and the avoidance of time expenditures for complicated "methodical apparatuses". On the other hand, the method offers no guarantee that the future space is treated in an exhaustive way and that all available system knowledge is systematically incorporated into the scenarios. It is also hardly possible in this way to overcome existing thought patterns and preconceived notions of the future and to arrive at a genuine post-structuralist deconstruction (Scheele et al. 2018; Schweizer et al. 2018).

4.3.2 Semi-formalised Scenario Methods

For these reasons, recently there has been a growing interest in semi-formalised scenario methods as alternatives to intuitive methods. They are expected to provide more comprehensive, consistent, and objective context analysis. Frequently used for this purpose is the cross-impact balances (CIB) method (Weimer-Jehle 2006, 2023), in some cases also the consistency matrix method (Rhyne 1974; von Reibnitz 1987; Johansen 2018). Since the CIB method was used in the ReMoDigital project (see Sect. 4.5), it will be described here as a typical example of a semi-formalised scenario method.

In a CIB context analysis (Weimer-Jehle et al. 2016), usually 10–20 essential context factors that interact with the system are listed first. For each context factor, typically 2–5 alternative, clearly distinguishable future developments are then formulated. In the next step, expert assessments about the question which of the collected future developments would promote or hinder other developments are collected in a "cross-impact matrix". This is usually done on a 7-point integer scale from strongly hindering (−3) to strongly promoting (+3). In this way, the context system is represented by a discrete-valued influence network (Fig. 4.9).

Software-based, all combinatorial network states are then examined, the promoting and hindering influences between the network nodes are balanced, and the Nash equilibria of the network are determined. These represent self-stabilizing trend configurations and are interpreted as consistent scenarios. As a rule, this yields a set of very different context system futures, each with its own internal logic and plausibility.

[1] For more examples, see, e.g., Weimer-Jehle et al. (2020).

In the literature, CIB-generated scenarios are considered to have higher consistency, completeness and objectivity than IL-generated scenarios (Girod et al. 2009; Kemp-Benedict 2012; Lloyd and Schweizer 2014; Carlsen et al. 2017; Schweizer 2020). However, compared to IL, CIB requires higher methodological skills and also a higher time investment, so that both methodological options can be said to have its meaningful range of application.

Examples of CIB-based context analysis for energy systems include Wachsmuth (2015), Ruth et al. (2015), Pregger et al. (2020), and Mier et al. (2023). Examples of using the consistency matrix method for this task include Droste-Franke et al. (2020) and Burger et al. (2022).

4.4 The System-Context Problem in Resilience Analysis and Assessment

In view of the above, the question arises as to whether and how the system-context problem also arises in resilience research—in general and specifically in resilience considerations of energy systems. It seems obvious that the resilience concept, once applied to energy systems, "inherits" the marked context sensitivities of energy systems mentioned in Part 2. Nevertheless, this will be explicitly underpinned again in the second part of this Section. First, however, we want to ask ourselves whether the system-context problem, due to its generic nature, is already manifested in the resilience concept itself, quite independent of the topic.

There is indeed some evidence for this. Figure 4.10 shows a representation of the prototypical reactions of a system to an external shock according to Kröger (2017). Depending on the resilience of the system, the same shock can lead to collapse, limited recovery of system performance, its full recovery, or even an improved system. However, the path the system actually takes through the crisis is determined not only by the characteristics of the system itself, but also by its contextual conditions.

Even the immediate shock effect on the system is not given by the system properties alone, but also, for example, by the operating conditions into which society has led the system. A system may have had high resilience under the conditions for which it was originally designed, but lose some of that resilience when society has moved to demand higher output from the system without making the necessary system improvements. An example would be an electrical grid that is increasingly burdened by higher and more fluctuating amounts of electricity over time, while on the other hand investment restraint or social conflicts prevent adequate retrofitting. If the system is already in a tense state in the new "normal case", external shocks can plunge it into deep dysfunction all the more easily.

Conversely, the damaging effect of the temporary performance reduction of a system *on the context* is also influenced by the context in which the system is located and how vulnerable it is. Consider a chemical plant that causes hazardous emissions for some time due to an incident, i.e., whose containment performance is reduced for some time due to shock. How effectively the authorities coordinate the emergency measures, how many civil protection forces are available and how well trained they are for the emergency, how level-headed and informed the local residents do their part to control the hazardous situation and minimize its effects, all this cannot be attributed to the technical plant. It is not only the local context (the responsible authorities, civil protection forces, and the local residents), but also the broader societal context, which, through regulations, information, awareness-raising, stimuli and others, determines whether the local context is prepared or not.

At the end of the shock effect, the question arises as to the performance level at which the system stabilizes again. Ideally, the lessons learned from the event will enable a higher and more stable performance than in the initial system. Whether this learning effect can occur is again a context-dependent question. Whether, for example, the local socio-economic system benefits from learning effects after a flood disaster such as the heavy precipitation event in Western and Central Europe in 2021 and thus subsequently result in a more resilient system, e.g. through changes in development planning, improved organization of disaster control or the breaking up of local political structures that hinder resilience, depends not only on the local systems. Rather, the question is also whether the larger socio-political framework is engaged in this task, or whether society as a whole and at the local level is distracted, for example, by other priorities, by a political instrumentalization of the event, or by other concurrent major crises.

4.4.1 Context Sensitivity in Energy Systems

As indicated in Fig. 4.6, energy systems are subject to numerous, sometimes obvious, sometimes subtle contextual influences due to their embeddedness in the society that builds, maintains and uses them. In a review of 16 energy scenario studies that had explicitly addressed this embeddedness in terms of societal "storylines" as a starting point for their energy system analysis, Weimer-Jehle et al. (2020) found a wide range of identified social, economic, political, and cultural contextual influences (Table 4.2).

The examples suggest that the contextual conditions, in addition to the system properties themselves, can have a considerable influence on how resiliently energy systems behave under shock. A more detailed insight into this dependency is provided in Sect. 4.5, which describes the application of these basic considerations in the ReMoDigital project.

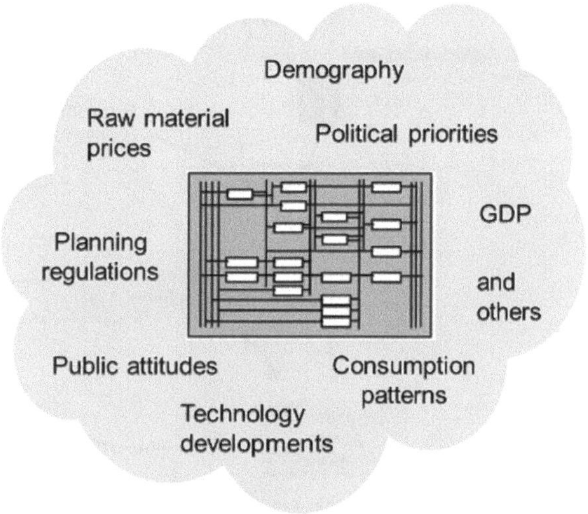

Fig. 4.6 System and context in energy systems analysis

Fig. 4.7 The use of "Context Scenarios" in the management sciences: Assessing the performance of strategy options under different environmental assumptions. Own depiction based on Fink et al. (2002)

	Environmental scenario I	II	III	IV
Strategy option A	++	++	+	-
Strategy option B	+	o	++	o
Strategy option C	--	--	-	++
Strategy option D	++	o	-	--
Strategy option E	o	-	--	--

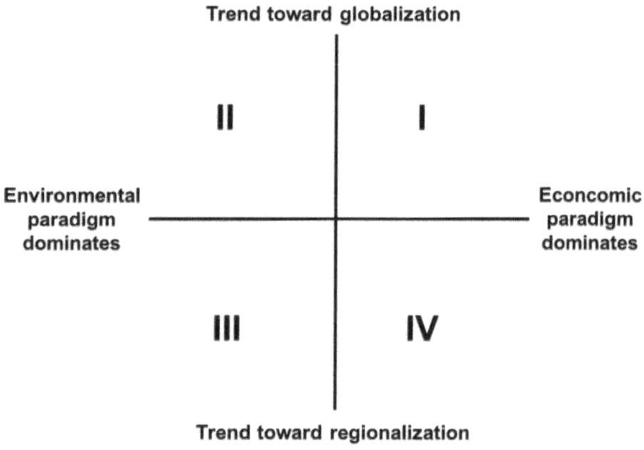

Fig. 4.8 The scenario-axes method for generating context (and other) scenarios. The example is inspired by Nakićenović et al. 2000

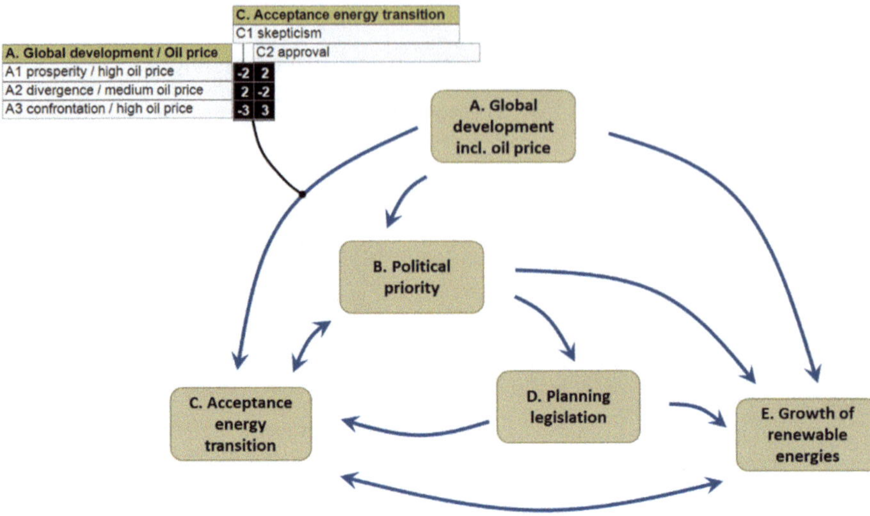

Fig. 4.9 Example of a simple CIB network of energy system context factors. Each arrow is substantiated by a detailed impact assessment as shown for the arrow A → C

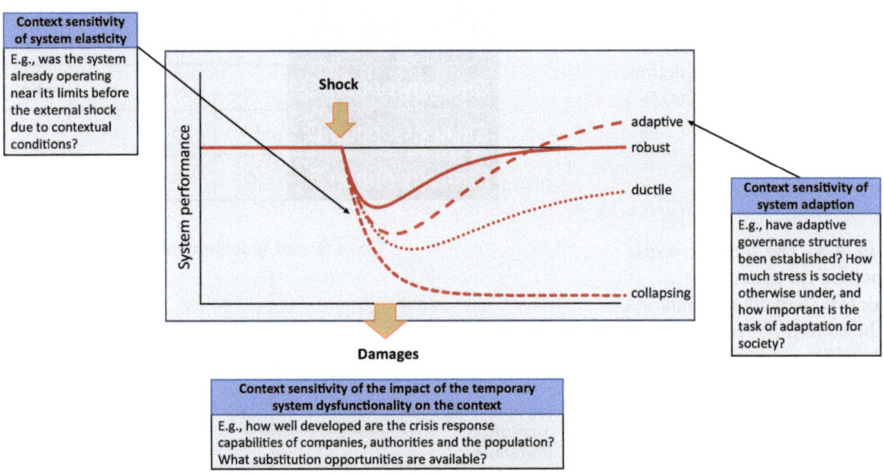

Fig. 4.10 Examples of context sensitivities in system response behaviors after an external shock. *Source* Kröger (2017) with modifications (boxes in blue and grey)

Table 4.2 Societal context influences on energy systems as found in 16 energy scenario studies analysed by Weimer-Jehle et al. (2020)

International	Environment (local and global)	Politics	Economy	Social	Culture
Climate protection policies	Climate change impacts	Influence of government	GDP growth	Migration	Value shift
Global development (general and conflicts	Climate change visibility	Overall development of the (UK) union	Technological innovation drivers, innovation ability/ dynamics	Demographic change	Public sustainability values
EU agricultural policy	Weather extremes/ Heat waves	Competences of regions	Technology development and diffusion	Scarcity-driven riots	Energy security concerns
EU funds/budget	Water scarcity	Spirit of international relations	Role of service economy/ Tertiarization	Disposal income and wealth distribution	Cultural identity
EU energy/ transport/ chemicals/trade policy	Resource scarcities	Governance pattern and quality	Post-crisis performance	Consumption patterns	Scope of liberalism
Global cooperation, attitudes, consciousness, responsibility towards environment		Centralized versus decentralized governance	Sector roles	Role of civil society	Lifestyle
Global versus regional focus		Political system and state configuration	Role of market versus state	Societal well-being	Social norm changes
Sovereignty/ identity		Spatial planning	Infrastructure aspects	Institutional changes	Relationship between economy, society and energy
Diversity/ tensions between regions/cultures		Subsidies	Business strategies	Key actors and actor roles	Role of religions

(continued)

Table 4.2 (continued)

International	Environment (local and global)	Politics	Economy	Social	Culture
Techno-economic paradigm		Electricity market regulations	Landscape and regime uncertainties (Multi-Level Perspective	Key multi-level patterns	Environmental concern
Globalization		Policy stability	Economic specialization	Learning processes	Public attitudes about the energy transition
Role of religions		Infrastructure planning legislation	Financial stability/ sustainability	Technology user practices	
Rule setting procedures		Intervention mode preferences	Strategic leadership and pro-activity of the economic agents	Institutional capacity building	
Level of free trade		Political design model	Digital and physical connectivity	Ability to create social capital	
Oil/fuel prices		Political priorities	Goods traffic	Evolution of the social cohesion model	
Resource trade		Markets versus state directive policy	Willingness to invest	Typology and role of the cities	
Interest rates			Knowledge integration	Generational uncertainty	
EU integration			Asset replacement	Evolution of the education and training system, education goals and social divides in education	
Transnational trade flows			Degree of centralization	Consumer participation and behavior	

(continued)

Table 4.2 (continued)

International	Environment (local and global)	Politics	Economy	Social	Culture
Transnational energy projects			Employment market	Regional diversification	
International integration of electricity grids			Centralized versus decentralized energy structures	Welfare state	
Success of global economic transition to sustainability				Technology acceptance	
Productivity growth				Media discourse	
Change of power blocks				Knowledge generation	

Source Own illustration

4.5 Application in the ReMoDigital Project

4.5.1 Goals of the Context Scenario Concept Application in ReMoDigital

The stress testing toolkit developed in ReMoDigital consists of four quantitative models, a "digitalization model", six context scenarios, 2 natural and 3 digital stress cases, and a set of (quantitative) resilience indicators. The four models, which each cover different parts of the energy and transport system and affect multiple levels, form the core system. The stationary energy system is considered, which is effective nationally, as well as the distribution grid level, which is effective regionally. A national transport model and a charging model for electric vehicles complete the models in the toolkit.

The context scenarios play both a moderating and an independent role in the toolkit. The context scenarios have a moderating effect between the models in the sense that they provide common framework conditions for all models and thus ensure that the model inputs are consistent at different levels and subsystems with different needs. The harmonization of framework conditions, which enables a comparative and complementary analysis of the subsystems in the first place, is also the basis on which (simulated) monitoring of (simulated) system processes can take place. In addition, the context scenarios also fulfil an independent role in the stress test toolkit. On the one hand, this role is filled by the diversity of possible developments, in that they can provide orientation from today's perspective as to which transformation paths could possibly lead to less resilient energy and transport systems in the future in the face of (certain) stress cases. On the other hand, the concept of framework twins, i.e., of two context scenarios that differ from one another solely in terms of the transformation of digitization in the energy system, offers the possibility, in the sense of different "measurement points in time" as in real-world monitoring, of having the stress case hit the energy system at two different points in time—once at a point in time when the energy system is still less digitized, and once when it is already more digitized.

4.5.2 *Process Overview*

4.5.2.1 Identifying Relevant Context Factors

Context scenarios are developed in several steps. In the first step, we identified relevant context factors in ReMoDigital through literature research and five expert workshops with project partners and three expert workshops with the project group and stakeholders. Relevant factors from the context of the core system are defined in ReMoDigital in two ways. From the perspective of the core systems (energy, network and transport models), they represent important and uncertain framework assumptions as well as other indirect influences on them. From the perspective of stress events that may occur, societal conditions are illuminated that reveal vulnerabilities or resiliencies for the transforming or transformed digitized energy and transportation system. In this way, interfaces to the core system can be established on the one hand, and the resilience of the socio-technical system can be analysed and interpreted more comprehensively on the other. This collection of factors was sorted, summarized and from the resulting list 22 factors were selected according to different criteria in cooperation with the project partners and defined in short essays with plausible future alternatives. The context analysis integrates developments at international, European and national level from the fields of economy, technology (transport, electricity, ICT), politics and culture (examples of selected descriptors are shown in Table 4.3). The full list and process will be presented in Prehofer and Weimer-Jehle (2026).

A further result of this process step is the collection of potential stress cases and the subsequent selection of two natural (combined) stress cases as well as three digital stress cases, which affect the energy system either top-down or bottom-up.

4.5.2.2 Assessing Context Interdependencies

In the next step, the interactions of the socio-technical system factors are evaluated. For this purpose, we created a cross-impact matrix in which we evaluated the influences on a factor in (mostly) 2-way interviews of experts (project partners, colleagues

Table 4.3 Examples of context descriptors and their future variants in ReMoDigital

B European Development	**M Cybersecurity**
B1 EU Rennaissance	M1 Low
B2 Nobody Cares	M2 Moderate
B3 EU Under Threat	M3 High
D Innovation capacity of the economy	**P Governance of energy transition**
D1 Declining	P1 Muddling Through
D2 Stable	P2 Central governance
D3 Increasing	P3 Locally anchored—collectively extended
E Technological development in transport	**R Trust in institutions**
E1 Electric world	R1 Declining
E2 Hydrogen world	R2 Steady
E3 Synthetic fuel world	R3 Increasing
E4 Mix world	**V Value orientation**
J Development of grid infrastructure	V1 Competitive mode
J1 Expansion in line with demand	V2 Speed differences
J2 Slower expansion	V3 The bonus system
	V4 Ecological rationalization

from the participating institutes and project group members) according to CIB evaluation logic in promoting and inhibiting influences and documented the respective argumentation. Of the 22 factors, the two international factors were interpreted as autonomous (i.e., not influenced by the other factors) and the remaining 20 factors were subjected to the evaluations. A total of 25 experts were involved in the survey and evaluation.

4.5.2.3 Calculating and Assessing the Context Scenarios

Finally, the completed matrix could be evaluated by the balance algorithm and resulted in a total of 125 fully consistent scenarios from a number of about 52 billion possible configurations. From these, six scenarios were selected together with the project partners on the basis of further analyses and criteria for joint processing in all subprojects (for an example, see Fig. 4.11).

The six context scenarios are divided into three scenario twins which, apart from the technical digitization factors, do not differ in the framework conditions, i.e. both a more and a less digitized energy system are consistent under certain context conditions. This makes it possible to analyse the effects of a stress case not only in differently transformed contexts, but also in differently digitized energy systems with the same context. Thus, the effects of stress cases and the resilience of different transformed societies with their differently designed energy transition can be compared

	Twin scenario no. 1
A International Development:	A3 Fortress World
B European Development:	B1 EU Renaissance
C Development of forms of work and employment:	C2 Strongly increasing
D Innovation capacity of the economy:	D3 Increasing
E Technological development in transport:	E4 Mix world
F Development of electric charging infrastructure:	F1 Low investment
G Degree of decentralization in energy conversion:	G3 Building
H Internationalization of energy supply:	H2 Stronger European interconnection
I Expansion of conversion facilities for renewable energies:	I2 High expansion
J Development of grid infrastructure:	J1 Expansion in line with demand
K Technological plurality in the electricity sector:	K2 High
L Degree of automation in the energy transition:	L3 LV level | L4 Full automation
M Cybersecurity:	M2 Moderate | M3 High
N Islanding capability:	N3 LV level
O State design modes:	O1 Focus state control
P Governance of energy transition:	P2 Central governance
Q Governance elements in the energy sector:	Q3 Technology-unspecific economic instruments
R Trust in institutions:	R2 Steady
S Acceptance of new technologies and infrastructures:	S3 Slightly increasing
T Education:	T1 STEM / access restriction low
U Digital mission statements and lifestyles:	U2 OMline
V Value orientation:	V2 Speed differences

Fig. 4.11 Example of a selected twin scenario

and interpreted. In total, 30 different combinations are available for a comparative analysis in the stress test toolkit (3 context scenarios × 2 different, digital designs × 5 stress cases).

Figure 4.11 shows the compilation of the future variants of twin scenario 1, which is rather strongly digitized and politically rather centrally controlled. A look at this aspect in the alternating effects shows a network that strongly promotes each other (Fig. 4.12). The expert arguments here suggest the following storyline:

> European integration is moving forward. It is expanding further east and south, creating political, military and fiscal-political integration. With rising uncertainty at the global level, a strong EU takes on great importance. A strong EU would also create a stronger European interconnection in the electricity market, which in turn would promote central governance in Germany. The reason for this is, on the one hand, that EU law breaks national law, but above all, that a strongly internationally regulated energy supply also tends to need a counterpart at the national level. The focus on state control in the political design objectives also facilitates central governance. [...]. While the EU is moving closer together, inequalities and (military) conflicts are growing on the international level ("Fortress World"), e.g. in the course of the increasing demand for resources. The pressure on the energy transition in Germany is growing significantly as a result. [...]. All resources must be exploited, which is why Germany is relying on a technology mix for both transport and power generation. For this reason, power generation should also be implemented with decentralized solutions at the building level through, for example, PV on the roof, battery storage, heat pumps, vehicle-to-building etc. Broad STEM education among the population has, on the one hand, ensured that digitization has arrived among the population and that they are taking advantage of its benefits. On the other hand, it has reduced inhibitions about technological innovations and made it possible to operate more complex systems. Acceptance of externally operated technologies is also rising slightly overall. The more complex and decentralized the energy system becomes, the more

> important automation down to the lower voltage levels becomes. The increasing capacity for innovation in Germany will make various automation technologies available for the power system. More innovations reduce the cost of technological and digital products, which is why automation technologies down to the low-voltage level will also become affordable and come into use.
>
> (Excerpt from the storyline designed from factor essays and expert arguments).

As already analysed in Sect. 4.4, the resilience of the energy and transport system is not only explained by the system's own characteristics, but also by its contextual conditions. Since the resilience aspect was already the focus of the search for and definition of relevant societal factors, the selected context factors also reflect at various points, among other things, the three response phases of a system to an external shock. Table 4.4 shows examples of the context factors and the resilience reference identified by experts. On the one hand, this allows the system's reaction to a specific stress case to be anticipated directly in the model calculation; on the other hand, however, the resilience assessment of the model calculations can also be supplemented in the follow-up. For the scenario twin 1 briefly described above, for example, this could indicate a short response time in the recovery phase to a stress case, because clearer responsibilities can be anticipated with central governance than would be expected, for example, with the "muddling through" future variant.

4.5.3 *Current Steps*

Currently the modelers are analysing the 30 different combinations of framework conditions, digitization options and stress cases. This creates a portfolio for monitoring that enables an analysis and discussion of resilience in the digitized energy system. In addition, a publication is being prepared that will present the complete results from the context scenario and stress case development (Prehofer and Weimer-Jehle 2026).

4.6 Conclusion

The preceding parts show that the behaviour of real-world complex systems typically depends on the environmental conditions in which they are embedded. This is also, and especially, true for the energy system and its resilience behaviour. In other words, the resilience behaviour of a system (and the consequences of failure) is the result of an *interplay between its system architecture and its context sensitivity.*

For energy systems (or generally, for systems) in transformation, resilience behaviour can thus only be comprehensively assessed if the future of their context is also scrutinized. Since the context is uncertain and itself a complex system, this should be done on the basis of context scenarios substantiated by system analysis.

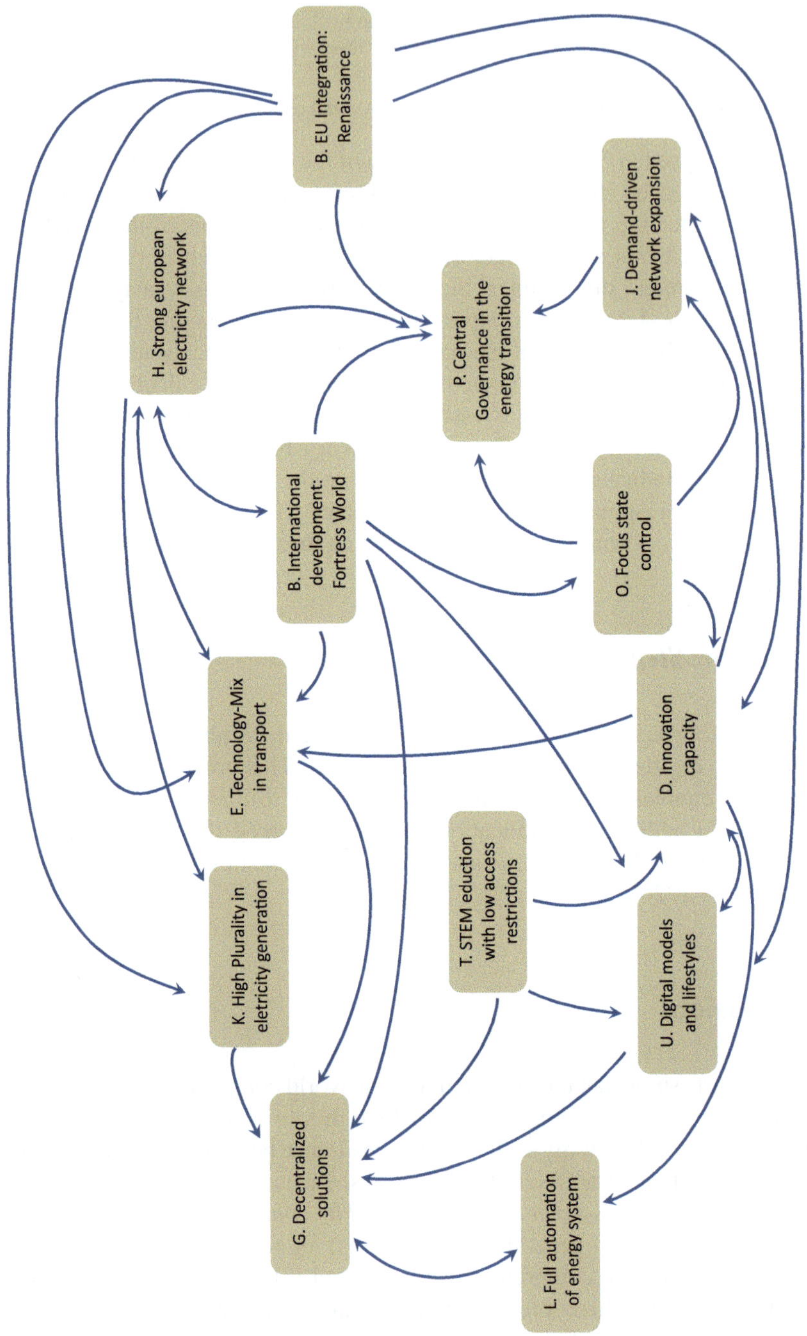

Fig. 4.12 (Incomplete) Section of the influence diagram for the high digitized variant of scenario twin 1

Table 4.4 Examples of context descriptors and connected technical system properties and expert assessments of their relevance for the reaction of the energy system after an external shock (cf. Fig. 4.10)

Phase I: system elasticity (degree of performance loss immediately after shock)	
B. European development	Climate agreement and smart integration of national supply networks (→ flexibility, cross-border lines as resilience enhancement), intra-European coordination to respond to cyber disruptions, innovation capabilities
K. Technological plurality in the electricity sector	Diversity of systems (promotes resilience)
R. Trust in institutions	General awareness of Hacker Attacks. Sensitivity to information campaigns (if trust is low, information campaigns have less effect)
S. Acceptance of new technologies and infrastructures	Acceptance of external control that would be required for system stability, e.g. smart meter, smart grid
Phase II: recovery	
M. Cybersecurity	Monitoring, rapid fault detection rather in the case of high cybersecurity. In the case of low cybersecurity, the gaps are large and people may not even be aware of the dangers. Lack of security mentality
P. Governance of energy transition	Are there clear responsibilities and processes in the event of a disaster? Have communication, cooperation, harmonization and understanding between systems and different levels been promoted?
Phase III: adaption (to the post-crisis situation)	
P. Governance of energy transition	With local governance, people at all levels are involved and may have the resources and expertise to respond more quickly. "Subsidiarity: polycentric governance, local capacities, culture of failure, reduction of complexity" (von Gleich, Renn). "Hierarchical Organization: balanced centralization-decentralization, multi-layer protection, local capacities" (von Gleich, Renn)
T. Education	General education (rather than technical education) promotes adaptivity. "Discussion skills need to be improved and you need to be able to think about different areas and not just stay in your own small area of expertise."

In the ReMoDigital project, a context scenario analysis was therefore carried out to make the context sensitivity of digitized energy systems tangible and accessible for model analyses. In expert workshops and interviews, context factors and their interdependencies were collected, their resilience effects interpreted, context scenarios developed using the CIB method, and these were used for model analyses of the transforming German energy system.

Resilience analyses of transforming systems based on context scenarios offer an analytical advance over classical analyses that capture the co-transformation of their environmental conditions only through assumptions or, at best, pointwise sensitivity analyses. Nevertheless, the limitations of the approach must also be kept in mind: The relevant contextual factors, their interdependencies and effects are captured through expert elicitation, and aspects may manifest themselves in the future that are unrecognized today due to lack of experience. The used analytical division of the overall system into core system and context is pragmatic and a step forward compared to previous analysis practices. A more ambitious vision, however, would be to also model all subsystems that make up the context in disciplinary detail and thus to move to multisystem modelling (which, however, raises many questions that are still unanswered today).

In the ReMoDigital project, the focus was on looking at the effect of stress events on the resilience of energy systems. This focus enabled several important findings. However, the concept of context scenarios also points to the question of how the reaction of the energy system to stress events then affects society again, and how it affects the framework conditions of the energy system for subsequent stress events. These considerations, however, require an iterative approach that was beyond the resources of the ReMoDigital project.

Similarly, the project required focusing primarily on the function of the context scenarios for harmonizing the model analyses. This limited the ability to address in any relevant detail the effects of context developments on the individual aspects of energy system resilience behaviour. Here, further research may contribute still deeper insights.

References

Acatech/Leopoldina/Akademienunion (eds.): Consulting with energy scenarios. Requirements for scientific policy advice (Series on Science-Based Policy Advice), acatech, Munich (2016)

Alcamo, J.: The SAS approach: combining qualitative and quantitative knowledge in environmental scenarios. In: Alcamo, J. (ed.) Environmental Futures: The Practice of Environmental Scenario Analysis, vol. 2, pp. 123–150. Elsevier, Amsterdam (2008)

BAuA: Die sozioökonomische Analyse. REACH-Info 7. Bundesanstalt für Arbeitsschutz und Arbeitsmedizin, REACH-CLP Helpdesk, Dortmund (2010)

Bossel, H.: Systeme – Dynamik – Simulation. Modellbildung, Analyse und Simulation komplexer Systeme, Norderstedt (2004)

Burger, P., Emmenegger, R., Sohre, A.: Energy futures as societal futures: co-evolutionary scenarios linking futures of society and decarbonized energy systems in Switzerland 2050. Futures **140**, 102955 (2022)

Carlsen, H., Richard, J.T., Klein, R.J.T., Wikman-Svahn, P.: Transparent scenario development. Nat. Clim. Chang. **7**, 613 (2017)

Dörner, D.: The Logic of Failure: Recognizing and Avoiding Error in Complex Situations. Basic Books, New York (1997)

Droste-Franke, B., Voge, M., Kanngießer, A.: Achieving transparency and robustness of regional energy scenarios by using morphological fields in inter- and transdisciplinary project groups. Energy Strat. Rev. **27**, 100430 (2020)

EU: Verordnung (EG) Nr. 1907/2006 des Europäischen Parlaments und des Rates vom 18. Dezember 2006 zur Registrierung, Bewertung, Zulassung und Beschränkung chemischer Stoffe (REACH) (2006). https://eur-lex.europa.eu/legal-content/DE/TXT/?uri=CELEX:32006R1907

Fink, A., Schlake, O., Siebe, A.: Erfolg durch Szenario-Management – Prinzip und Werkzeuge der strategischen Vorausschau. Campus, Frankfurt (2002)

Girod, B., Wiek, A., Mieg, H., et al.: The evolution of the IPCC's emissions scenarios. Environ Sci Policy **12**, 103–118 (2009)

Godet, M.: Reducing the blunders in forecasting. Futures **15**, 181–192 (1983)

Holling, C.S.: Resilience and stability of ecological systems. Annu. Rev. Ecol. Syst. **4**(1), 1–23 (1973)

Imboden, D.M., Koch, S.: Systemanalyse. Einführung in die mathematische Modellierung natürlicher Systeme. Springer, Berlin (2003)

Jochem, E.: Personal communication (n.d.)

Johansen, I.: Scenario modelling with morphological analysis. Technol. Forecast. Soc. Chang. **126**, 116–125 (2018)

Kemp-Benedict, E.: Telling better stories: strengthening the story in story and simulation. Environ. Res. Lett. **7**, 041004 (2012)

van 't Klooster, S.A., van Asselt, M.B.A.: Practising the scenario-axes technique. Futures **38**(1), 15–30 (2006)

Kröger, W.: Securing the operation of socially critical systems from an engineering perspective: new challenges, enhanced tools and novel concepts. Eur. J. Secur. Res. (2017)

Lloyd, E.A., Schweizer, V.J.: Objectivity and a comparison of methodological scenario approaches for climate change research. Synthese **191**, 2049–2088 (2014)

Mander, S.L., Bows, A., Anderson, K.L., Shackley, S., Agnolucci, P., Ekins, P.: The Tyndall decarbonisation scenarios. Part I: development of a backcasting methodology with stakeholder participation. Energy Policy **36**, 3754–3763 (2008)

O'Mahony, T., Zhou, P., Sweeney, J.: Integrated scenarios of energy-related CO2 emissions in Ireland: a multi-sectoral analysis to 2020. Ecol. Econ. **93**, 385–397 (2013)

Mier, M., Siala, K., Govorukha, K., Mayer P.: Collaboration, decarbonization, and distributional effects. Appl. Energy (2023)

Nakićenović, N., et al.: Special report on emissions scenarios. Report. Intergovernmental Panel on Climate Change (IPCC). Cambridge University Press, New York (2000)

Pregger, T., Naegler, T., Weimer-Jehle, W., Prehofer, S., Hauser, W.: Moving towards socio-technical scenarios of the German energy transition - lessons learned from integrated energy scenario building. Clim. Change **162**, 1743–1762 (2020)

Prehofer, S., Kosow, H., Naegler, T., Pregger, T., Vögele, S., Weimer-Jehle, W.: Linking qualitative scenarios with quantitative energy models. Knowledge integration in different methodological designs. Energy Sustain. Soc. **11**(25) (2021)

Prehofer and Weimer-Jehle: Qualitative scenarios in the resilience monitoring of the digitization of the energy transition in Germany 2050 (forthcoming) (2026)

Prognos: Analysis and comparison of relevant mid- and long-term energy scenarios for EU and their key underlying assumptions. Prognos AG, Bases (Switzerland) (2011)

von Reibnitz, U.: Scenario Techniques. McGraw-Hill, New-York (1987)

Rhyne, R.: Technological forecasting within alternative whole futures projections. Technol. Forecast. Soc. Chang. **6**, 133–162 (1974)

Ruth, M., Özgüna, O., Wachsmuth, J., Gößling-Reisemann, S.: Dynamics of energy transitions under changing socioeconomic, technological and climate conditions in Northwest Germany. Ecol. Econ. **111**, 29–47 (2015)

Scheele, R., Kearney, N.M., Kurniawan, J.H., Schweizer, V.J.: What scenarios are you missing? Poststructuralism for deconstructing and reconstructing organizational futures (Chap. 8). In: Krämer H., Wenzel M. (eds.) How Organizations Manage the Future - Theoretical Perspectives and Empirical Insights. Springer International Publishing (2018)

Schweizer, V.J.: Reflections on cross-impact balances, a systematic method constructing global socio-technical scenarios for climate change research. Clim. Change **162**, 1705–1722 (2020)

Schweizer, V., Scheele, R., Kosow, H.: Practical poststructuralism for confronting wicked problems. In: 9th International Congress on Environmental Modelling and Software. Fort Collins, Juni 2018 (2018)

Wachsmuth, J.: Cross-sectoral integration in regional adaptation to climate change via participatory scenario development. Clim. Change **132**, 387–400 (2015)

WEC: World energy scenarios 2016. World Energy Council (2016). https://www.worldenergy.org/publications/2016/world-energy-scenarios-2016-the-grand-transition/

Weimer-Jehle, W.: Cross-impact balances: a system-theoretical approach to cross-impact analysis. Technol. Forecast. Soc. Chang. **73**(4), 334–361 (2006)

Weimer-Jehle, W.: Cross-Impact Balances (CIB) for Scenario Analysis - Fundamentals and Implementation. Springer, Berlin, New-York (2023)

Weimer-Jehle, W., Buchgeister, J., Hauser, W., Kosow, H., Naegler, T., Poganietz, W., Pregger, T., Prehofer, S., von Recklinghausen, A., Schippl, J., Vögele, S.: Context scenarios and their usage for the construction of socio-technical energy scenarios. Energy **111**, 956–970 (2016)

Weimer-Jehle, W., Vögele, S., Hauser, W., Kosow, H., Poganietz, W.-R., Prehofer, S.: Socio-technical energy scenarios: state-of-the-art and CIB-based approaches. Clim. Change **162**(4), 1723–1741 (2020)

Wilson, I.: Mental maps of the future: an intuitive logics approach to scenario planning. In: Fahey, L., Randall, R.M. (eds.) Learning from the Future: Competitive Foresight Scenarios, pp. 81–108. John Wiley and Sons (1998)

Part III
Governance of Resilience

Chapter 5
Risk, Resilience and Precaution

Ortwin Renn

Abstract Almost all definitions of the precautionary principle emphasize the need for regulatory measures (actions) before it is established beyond doubt that an "intolerable" risk exists. Such measures are even more necessary the greater the depth of intervention is and the more irreversible the consequences may be. Risk mitigation in this case is not related to a reduction of risk, but rather to a policy that avoids irreversibility as much as possible. In the international literature, this approach is also popularly referred to as resilience strategy. In terms of assessment, it includes the need for conservative estimates in characterizing uncertainty and in terms of risk management it implies the need for ensuring opportunities for adapting to new situations, surprises and unexpected stress.

Keywords Ambiguity · Complexity · Precautionary principle · Risk assessment · Risk management · Uncertainty

5.1 Introduction

Conventional forms of risk assessment are based on the proposition that the effects of the risks to be assessed are known, both in terms of the maximum amount of harm and in terms of the probability distribution of potential damages. The severity and frequency of the consequences of an activity or event are used as a yardstick for assessing its acceptability.

But how should one assess situations where the effect itself is still uncertain? How can one assess hazards for which one cannot, or not yet, calculate the probability of their occurrence? How should risk managers respond to risks where detrimental effects are suspected, but at the same time there is a lack of data to estimate the amount of damage in any reliable way? In these situations, the precautionary principle comes into play (Elliott and Charnley 2002). The precautionary principle applies when there is high uncertainty in the risk estimate (Bennett 2000). How should we assess, manage

O. Renn (✉)
Research Institute for Sustainability/ at GFZ (RIFS), Potsdam, Germany
e-mail: ortwin.renn@rifs-potsdam.de

D. van Doren et al. (eds.), *Towards Energy System Resilience*, Ethics of Science and Technology Assessment 55, https://doi.org/10.1007/978-3-032-03187-7_5

and regulate risks where we know nothing or little about the possible consequences of an activity or event, or where there is uncertainty about the probability distribution?

This situation is present, for example, in the debate about the consequences of the energy transition. There is still great uncertainty about the risks of energy security as well as carbon pricing due to a radical transformation towards renewable energy sources within a short period of time. In addition, the various options of how to reach a carbon–neutral energy system are all associated with large uncertainties about economic, social and political impacts (Linzenich et al. 2020). Many analysts have only rough ideas of how to calculate or estimate the transition risks that society might face when implementing different policy options to reach the desired goal vis-vis changing geopolitical conditions (Chishti et al. 2023). In this situation, the precautionary principle comes into play: If a policy might lead to high uncertainty about potential impacts, it may be better to reconsider this policy, try to find less uncertain alternatives or at least start a rigorous monitoring program to assure that one can change the policy if major negative impacts are observed (Hynes et al. 2022). In the case of energy transition, the decision about appropriate policies also depends on a balance between taking the risk of failing to accomplish the goal of carbon-neutrality and the risk of taking actions that endanger the security of supply, potential blackout or inacceptable price hikes.

Institutionally, it makes sense not to leave the question of the right balance between too much and not enough caution in assessing the still uncertain consequences solely to the scientists or even the political and administrative authorities. After all, the question of proportionality cannot be answered unambiguously in scientific terms but requires a value-based balancing exercise. To this end, those who are affected by the risk of failing to reach carbon neutrality and the risk of facing major economic or social hardships or even disasters should be included in the decision-making process (Renn 2008: 193ff.). Of course, this is only possible if the institutional and procedural framework is in place before the policies are enacted. To this end, the German Risk Commission proposed the establishment of a risk council in its final report, which was submitted to the environment and health ministries in June 2003 (Risikokommission 2003). This council's task is to ensure a precautionary risk assessment and to guarantee an implementation of the precautionary principle over all potential risks and policy domains, including climate change, energy transitions or other transformation risks towards sustainability.

The following sections are intended first to explain the precautionary principle in the context of risk assessment and risk evaluation, secondly to explain the risk management strategies developed by the International Risk Governance Council and thirdly to reflect on the implementation of the principle in the current debate on precaution and resilience.

5.2 Precaution in the National and International Context

In German law, a distinction is made between risk, danger, precaution and residual risk (Rehbinder 1997). If damage is to be expected with a high probability, the German law defines this situation as a danger (Gefahr). In this case, danger must be averted, even if the potential benefits are high. If, however, the probability is lower (which is not numerically specified), the risk can be regulated from a precautionary point of view. In this case it is up to the regulatory bodies to initiate a regulation, but it is not mandatory to reduce or minimize the risk. There is room for discretion within the general frame of more specific regulations within each domain (such as air pollution or accident prevention). Often the ALARA principle (as low as reasonably archivable) or the status of technology and science (regulation to the degree that it is technical feasible or even scientifically possible) is used as a criterion for the degree of regulatory interventions (see Sect. 5.6 of this chapter). In the case of very small probabilities, one speaks of residual risks. Member of society are obliged to accept such residual risks—assuming there is a corresponding benefit of the activity that serves the common good.

In EU and international law, the precautionary principle is essentially related to the problem of uncertainty. Although it has been defined and interpreted in a variety of ways in the literature, the formulation from the Rio Declaration remains the most used and cited version of this principle (cited from Paterson 2005):

> In order to protect the environment, the precautionary approach shall be widely applied by States according to their capabilities. Where there are threats of serious or irreversible damage, lack of full scientific certainty shall not be used as a reason for postponing cost-effective measures to prevent environmental degradation (Rio Declaration 1992, Principle 15).

However, there remain serious ambiguities and queries about the nature and appropriate role of the precautionary principle in the context of EU law and regulation (Morris 2000; Majone 2002; Löfstedt 2004; O'Riordan 2014: 17). For a better understanding of the debate and its implication for policymaking, it is helpful to distinguish three positions or mental framings of the precautionary principle (Resnik 2003). Within the first frame of *stochastic reasoning,* risk assessment and management rely on the best scientific estimates of probabilities and associated damages to judge the ability to tolerate the risk and design protective measures that are cost-effective, proportional to the threat, and fair to the affected population. In these terms, precaution might best be interpreted as being conservative when making judgements and erring on the side of caution when calculating exposure or determining safety factors (see Sect. 5.4 on risk assessment). In addition, as Martin Peterson points out "the precautionary principle can be interpreted as an analogous epistemic principle which prescribes that it is always more desirable to avoid false negatives than false positives when it comes to assessing risks." (Peterson 2007, 306).

Within the frame of "*better safe than sorry*", risk is seen from the perspective of pervasive uncertainty, ambiguity and in particular ignorance. Precautious risk management therefore aims to ensure prudent decisions in situations where there is a high incertitude about probabilities, outcomes or both, and a high vulnerability of the population at risk. Policy and regulatory instruments of precaution include minimization requirements, diversification of risk agents, containment in time and space, and close monitoring (Klinke and Renn 2021). It is not—as many critics believe—a call to ban all potentially hazardous activities. As Andrew Stirling states "[p]recaution does not automatically entail bans and phase-outs but calls instead for deliberate and comprehensive attention to contending policy or technology pathways" (Stirling 2007: 311).

A third frame of deliberation has been advocated as *"discursive approach to risk"*. This approach can be regarded as an alternative or an addition to purely analytical procedures of both assessing and managing risks. Here, the task of risk management is to involve stakeholders and an interested public in designing risk management strategies based on the knowledge and value system of each stakeholder (Renn 2015; Renn and Schweizer 2020). Both probabilistic risk assessment as well as precautionary strategies can be complemented by such a discourse among the major actors, but the approach has been advocated either as an independent path to handling risk democratically or, more often, as a policy-oriented implementation of the precautionary approach (van den Daele 2000).

In the recent past, advocates of each side have launched a fierce debate over the legitimacy of their approaches. One side argues that precautionary strategies ignore scientific results and lead to arbitrary or inconsistent decisions (Sunstein 2005; Peterson 2007). The precautionary proposition that it is better to be safe than sorry could therefore be interpreted as a mandate to ban everything that might result in negative side effects. Such a rule would logically apply to any substance or human activity and would lead to total arbitrariness (Majone 2002). Accordingly, the principle has been labelled as ill-defined and absolutist, and is considered to lead to increased risk-taking, to be an ideology, to be unscientific or to marginalize the role of science (Sandin et al. 2002). Some analysts claim that using the precautionary principle runs the risk that science might be held "hostage to interest group politics" (Charnley and Elliott 2002); others contest that policymakers could abuse the precautionary principle to protect economic interests and to impede world trade (Majone 2002).

Conversely, advocates of the precautionary approach argue that it does not automatically mean an immediate ban but rather a step-by-step diffusion of risky activities or technologies until more knowledge and experience is available (Fisher 2001; Stirling 2003). They have accused their critics of ignoring the uncertainty and ambiguity of most hazardous situations and relying on data that often turns out to be insufficient for making robust judgements (van Asselt and Vos 2006). They also claim that risk assessment techniques pretend to meet a high standard of scientific validity, rigor and reliability that cannot be sustained under scrutiny. Too many unpleasant surprises during the past few decades would therefore justify a more cautious and humble approach to claiming knowledge about possible risks (Gee et al. 2001; Harremoes et al. 2002).

So, there is still some confusion about the exact meaning of precaution and the precautionary principle. However, regardless how the principle is articulated, there is a consensus among most scholars that precaution allows risk managers to initiate regulatory actions even if the scientific evidence about risks is not yet conclusive (Renn 2007). But how much evidence is evidence enough to trigger such actions?

Despite the intensity of attention that the principle has received in the policy arenas, there remains a large number of serious ambiguities and queries concerning the nature and appropriate role of the precautionary principle in governance. These are addressed—if not resolved—in a burgeoning academic and more policy-oriented literature (Reviews in Fisher 2002; O'Riordan et al. 2001; O'Riordan, 2014; Renn and Schweizer 2023). In order to understand the principle in the context of a larger array of risk management strategies it is necessary to look more deeply into the fabrics of uncertainty characterization. This will be done in the next section.

5.3 Key Variables: Complexity, Uncertainty and Ambiguity

Risks are *mental 'constructions'*. They are not real phenomena but originate in the human mind (IRGC 2005; Renn 2008, 4). Actors, however, creatively arrange and reassemble signals that they get from the 'real world' providing structure and guidance to an ongoing process of creating representative images of the real world. The status of risk as a mental construct has major implications on how risk is conceptualized and managed. Unlike trees or houses, one cannot scan the environment, identify the objects of interest, and count them. Risks are created and selected by human actors. What counts as a risk to someone may be an act of God to someone else or even an opportunity for a third party. Since risks represent mental constructs, the quality of their explanatory power depends on the accuracy and validity of their (real) predictions. Unlike some other scientific constructs, validating the results of risk assessments is particularly difficult because, in theory, one would need to wait indefinitely to prove that the probabilities assigned to a specific outcome were correctly assessed. If the number of predicted events is frequent and the causal chain obvious (as is the case with car accidents), validation is relatively simple and straightforward. If, however, the assessment focuses on risks where cause-effect relationships are difficult to discern, effects are rare and difficult to interpret and variations in both causes and effects are obscuring the results (as is often the case with ecological risks), the validation of the assessment results becomes a major problem. In such instances, assessment procedures are needed to characterise the existing knowledge with respect to complexity, remaining uncertainties and ambiguities. What do these terms mean (Renn et al. 2011; Klinke and Renn 2021)?

- ***Complexity*** refers to the difficulty of identifying and quantifying causal links between a multitude of potential causal agents and specific observed effects. The nature of this difficulty may be traced back to interactive effects among these agents (synergism and antagonisms), long delay periods between cause and effect, inter-individual variation, intervening variables and multiple feedback loops leading to non-linear causal structures with dynamic fluctuations and often distinct tipping points or tipping corridors. There is no doubt that the transition risks to reach the goal of carbon neutrality in 2045 met all the requirements and criteria for complex causal and functional relationships.

- ***Uncertainty*** is different from complexity but often results from an incomplete or inadequate reduction of complexity in modelling cause-effect chains. Whether the world is inherently uncertain is a philosophical question that we will not pursue here. It is essential to acknowledge in the context of risk assessment that human knowledge is always incomplete and selective and thus contingent on uncertain assumptions, assertions and predictions. It is obvious that the modelled probability distributions within a numerical relational system can only represent an approximation of the empirical relational system with which to understand and predict uncertain events. It therefore seems prudent to include other, additional, aspects of uncertainty. Although there is no consensus in the literature on the best means of disaggregating uncertainties, the following categories appear to be an appropriate means of distinguishing the key components of uncertainty (Klinke and Renn 2021):

  - unrecognized or only estimated variability in the endpoints (targets) of risk analyses, especially inter-individual sensitivity to equal exposure (for example, exposure to particulate matters from car exhausts);
  - the sum of systematic and random measurement errors and extrapolations in the interpretation of statistical distributions of harmful impacts (for example, the confidence intervals with respect to distribution of hardships among citizens when CO pricing is in place);
  - genuinely stochastic processes where there is indeterminacy (complex pathways to cause a blackout in Germany);
  - remaining ignorance, and excluding events outside of the system boundaries of consideration that may impact the energy situation (such as wars, international conflicts, rising populism tec.)

The first two components of uncertainty qualify as epistemic uncertainty and therefore can be reduced by improving the existing knowledge and by advancing the present modelling tools. The last three components are genuine uncertainty components and thus can be characterised to some extent using scientific approaches but cannot be reduced to numeric confidence intervals.

- *(Interpretative and normative) ambiguity* is the last term in this context (Renn and Klinke 2016) Whereas uncertainty refers to a lack of clarity over the scientific or technical basis for decision making, (interpretative and normative) ambiguity is a result of divergent or contested perspectives on the justification, severity or wider 'meanings' associated with a given threat. In relation to risk it is understood as 'giving rise to several meaningful and legitimate interpretations of accepted risk assessment results. It can be divided into ***interpretative ambiguity*** (different interpretations of an identical assessment result: e.g. as an adverse or non-adverse effect) and ***normative ambiguity*** (different concepts of what can be regarded as tolerable referring e.g. to ethics, quality of life parameters, distribution of risks and benefits, etc.). A condition of ambiguity emerges where the problem lies in agreeing on the appropriate values, priorities, assumptions, or boundaries to be applied to the definition of possible outcomes.

5.4 Precaution in Risk Assessment

Because of the complexity, uncertainty and ambiguity about the range and intensity of how different risk targets (people, animals, buildings, or biotopes) are affected and due to the occurrence of systematic and random measurement errors, precautionary risk assessment implies the need for conservative assessment of risks, i.e. to be on the safe side of the distribution. Thus, safety factors from 2 up to 100,000 are used to include variability. These factors set the threshold for the safety margin that is deemed acceptable. Extrapolation is often based on conservative assumptions, i.e., rather cautious values in selecting the parameters. For example, linear relationships in dose. Response functions are often assumed when extrapolating from high to low doses, although a quadro-linear or even logarithmic dose–response curve would be more appropriate given the empirical data. At the same time, however, it is clear that no probabilistic risk assessment can include the worst case and the most sensitive individual. One can argue about the breadth of the spectrum (about 99% of individuals), but not about the need for a pooled i.e., group-based approach (Renn 2007; Renn and Schweizer 2023).

What does this mean for precaution? If precautionary thinking is the starting point, then risk assessments should err on the side of caution rather than on the side of “daring" to the extent that it is reasonable to do so (Renn 2002). At the same time, the determination of reasonable assumptions for prudent courses of action is not itself predetermined with scientific precision and always requires a “value judgment" in the sense of weighing between too much and too little prudence. This is because maximum caution is incompatible with probability theory; a justifiable degree of caution must always be defined, otherwise one would in principle have to categorically prohibit everything that could involve any risk. But no one can seriously advocate this. The justifiable degree of caution in risk assessment is usually referred to as convention in the literature (Akademie der Wissenschaften zu Berlin 1992: 342ff.). First, this includes the definition of what is called damage and what is included in the conceptualization of the risk under consideration (adverse effect), It also depends on the choice of the reference variable (such as expected damage per unit of time or per Kilowatt-hour, or per unit of gross national product, etc.). Secondly, the precautionary approach suggests that even unlikely scenarios (with low probability) need to be considered as the potential damage associated with this event may be particularly high.

In the estimation of risks pertaining to the energy transition, the main cause for uncertainties and ambiguities lies in the lack of knowledge about side effects of the respective policy measures such as carbon pricing, phasing-out of fossil fuels, setting standards for maximum CO_2 emissions etc. Only during the implementation of these measures, there will be realistic chance to determine the degree of risk and in addition the potential side effects on economy, equity, social acceptance and political governance. Unemployment, bankruptcies, dramatic loss of income, blackouts, lack of competitiveness are only a few of the potential side effects that need to be considered. At this point, regulators would need the best estimates about these side effects from a variety of different disciplines. Furthermore, the disciplines themselves need to review their methodologies and approaches to assess risks under the present circumstances of high uncertainty.

From a normative perspective, it is suggested that these revised methods and assessment procedures should be found and justified within the scientific system itself. Within the respective scientific communities, such precautionary approaches can best be negotiated in the discourse of the disciplines concerned (including social, economic and cultural studies), because this requires appropriate expertise and experience.

5.5 Precaution in Risk Management and Governance

Almost all definitions of the precautionary principle emphasize the need for regulatory measures (actions) before it is established beyond doubt that an "intolerable" risk exists (see overview in: Sand 2000; Gilbert 2019). Such measures are even more necessary if the depth of intervention is high and if irreversible consequences may be expected. Risk mitigation in this case is not related to a reduction of risk, but rather to a change of situation that avoid irreversibility as much as possible. In the international literature, this approach is also popularly referred to as resilience strategy (Linkov et al. 2014a, b). In Sect. 5.7 we will return to that strategy when we discuss precaution and resilience. Regulatory measures aim to make the risk control system robust and less vulnerable.

The imperative of precaution against as yet unknown effects of risks and threats such as pandemics is brought to bear in two different ways (Akademie der Wissenschaften zu Berlin 1992, p. 370ff.):

- *The* "ALARA principle" (as low as reasonably achievable): according to this principle, any exposure must be reduced as far as possible, within the limit of economic feasibility.
- *The state of the art in technology and science:* under this principle, every available technique must be used to reduce risk. In the case of energy transition, any technology that would provide less risks to the economy or the consumer of energy should be preferred over another technology.

The application of both principles leads to suboptimal solutions from a purely rational point of view because the costs and burdens to reduce risks and the benefits that are expected by the reduction methods are not systematically compared.

In the more recent discussion, innovative forms of risk regulation based on the precautionary principle have been developed in addition to the classical elements of minimizing exposure and the state of the art. Foremost among these are the so-called containment methods, according to which risk measures, such as carbon pricing, are introduced step by step in controlled environments until it is reasonably clear whether the measure is associated with tolerable side effects (Chen 2020). Limiting measures in space and introducing measures step by step in time avoids potentially irreversible outcomes or at least limits their effects when detected early. New liability and insurance models are also under discussion, which are intended to motivate risk generators to identify possible risks at an early stage and to monitor them continuously. In this context, public subsidies for insurance premiums can also be provided if this serves the common good.

The possibility of using the properties of the risk itself as a criterion for applying the precautionary principle in cases of high uncertainty about effects that are not yet known is also particularly interesting. In its 1998 annual report, the German Advisory Council on Global Change proposed using the criteria of ubiquity, persistence and irreversibility as substitute variables for the assessment of risks; namely, in cases where sufficiently precise knowledge about effects is not yet available (Wissenschaftlicher Beirat der Bundesregierung 1999: 53ff.). In an EU-funded project, a team led by chemist Ulrich Müller-Herold developed a quantitatively measurable filter system that enables pre-selection according to the precautionary principle based on certain substance properties (hazards) (Müller-Herold et al. 2005).

Institutionally, it makes sense not to leave the question of the right balance between too much and not enough precaution in assessing the still uncertain consequences solely to scientists or even to the authorities. Since certain groups—if too much precaution is taken, it is the beneficiaries of the risk; if too little precaution is taken, it is the victims of the risk—must bear the consequential costs of this balancing act, these groups should also be involved in determining the extent of precaution (Klinke and Renn 2014). To this end, a group of experts of the then Academy of Sciences in Berlin had proposed the establishment of a Risk and Environment Council in analogy to the model of the German Science Council (Akademie der Wissenschaften zu Berlin 1992: 42ff. as well as 475ff.). The first instance (essentially scientists) should ensure precautionary risk assessment, the second instance should ensure implementation of the precautionary principle in the sense of a rationally justifiable compromise that can be agreed to by the parties bearing the costs. This proposal was taken up by the ad hoc Federal Risk Commission, which advised the Federal Government from 2001 to 2003. It was recommended that the federal government establish a risk council to take on this task (Risikokommission 2003). Unfortunately, this recommendation was not followed up in the course of the political discussion. In contrast, other countries, such as the Netherlands, have taken up this idea and developed it methodically in the sense of a comparative "national risk assessment".

5.6 The IRGC Approach to Precautionary Risk Management

In accordance with the three major components "complexity, uncertainty, and ambiguity", The International Risk Governance Council designed a risk management taxonomy depending on the degree or intensity of complexity, uncertainty and ambiguity (Renn 2008, 173ff.). For example, if a risk is characterized by high complexity, low uncertainty and low ambiguity, the main emphasis will be on scientific modelling and consensual methods for performing a most accurate assessment. If a risk is characterized by high uncertainty the main emphasis will be on strengthening resilience of the system to cope even with surprises (Bloesch et al. 2015). In a situation of high ambiguity, a more discursive approach with stakeholder participation is asked for to find an acceptable way to reconcile conflicting values and world views (Renn 2015). More specifically, risk handling can pursue one of four strategies or a combination thereof. These strategies are summarized in Table 5.1 and described in the following paragraphs (Renn 2008: 182–183).

- *Simple risk problems:* This class of risk problems requires hardly any deviation from traditional decision making. Data is provided by statistical analysis; goals are determined by law or statutory requirements and the role of risk management is to ensure that all risk reduction measures are implemented and enforced. Traditional risk–risk comparisons (or risk–risk trade-offs), risk–benefit analysis and cost-effectiveness studies are the instruments of choice for finding the most appropriate risk reduction measures. Additionally, risk managers can rely on best practice and, in cases of low impact, on trial and error. Staff members of risk management agencies, industrial risk managers and experts in the field are the only groups that need to be involved for dealing with simple risks.
- *Assessment-based strategy:* This strategy includes the search for unintended consequences (CO_2 emission levels, pricing system for CO_2, performance standards for wind-energy systems, auction rules for certificates on greenhouse gas emissions, etc.). Within this second strategy of scientific risk analysis, risk management relies on the best scientific estimates of probabilities and potential consequences and uses expected values as main input to judge the tolerability of risk as well as to design risk reduction measures that are cost-effective, proportional to the threat and fair to the affected population (Linkov et al. 2014a, b). In this frame precaution may best be interpreted as being conservative in making risk judgments and choosing cautious assumptions when calculating exposure or determining safety factors (of 10, 100 or more) to cover inter-target variability. Since risk assessments for complex problems rely on sophisticated models and simulations, a broad representation of the various scientific communities including the social scientists, is required to assure a broad and inclusive handling of the risk.

Table 5.1 Risk characteristics and their implications for risk management (taken from IRGC 2005)

Knowledge characterisation	Management strategy	Appropriate instruments	Stakeholder participation
1. 'Simple' risk problems	Routine-based (tolerability/ acceptability judgement) (risk reduction)	a. Applying 'traditional' decision-making – Risk–benefit analysis – Risk–risk trade-offs – Trial and error – Technical standards – Economic incentives – Education, labelling, information – Voluntary agreements	Staff members and specialists in the field
2. Complexity-induced risk problems	Assessment-based strategy	a. Characterising the available evidence – Expert consensus seeking tools like: – Delphi or consensus conferencing – Meta analysis – Scenario construction, etc. – Results fed into routine operation	Wide range of experts and scientist, including social scientists
		a. Improving buffer capacity of risk target through – Additional safety factors – Redundancy and diversity in designing safety devices – Improving coping capacity – Establishing high reliability organisations	

(continued)

Table 5.1 (continued)

Knowledge characterisation	Management strategy	Appropriate instruments	Stakeholder participation
3. Uncertainty-induced risk problems	Precaution-based strategy	a. Using exposure characteristics such as persistence, ubiquity etc. as proxies for risk estimates Tools include – Containment – ALARA (as low as reasonably achievable) and ALARP (as low as reasonably possible) – BACT (best available control technology), etc.	Inclusion of major stakeholders such as industry and environmental groups
		a. Improving capability to cope with surprises – Diversity of means to accomplish desired benefits – Avoiding high vulnerability – Allowing for flexible responses – Preparedness for adaptation – Orientation towards resilience	
4. Ambiguity-induced risk problems	Discourse-based:	a. Application of conflict resolution methods for reaching consensus or tolerance for risk evaluation results and management option selection – Integration of stakeholder involvement in reaching closure – Emphasis on communication and social discourse	Wider public, including different social and cultural interests and values

- *Precaution-based strategy:* This strategy refers to risk reduction activities that are derived from the *application of the precautionary principle.* Within this third frame, the concept of risk is seen from the perspective of pervasive uncertainty and in particular ignorance and non-knowledge (Goldstein 1999; Bloesch et al. 2015). Precautious risk management entails to ensure prudent handling of decision options in situations of high uncertainty about causes and effects and of high vulnerability of the ecosystems under risk. Instruments of precaution include ALARA (as low as reasonably achievable), state of the art in technology and science, diversification of risk agents, containment in time and space and close monitoring (see Sect. 5.4). It is difficult to draw the line between too much precaution and not enough precaution in the face of uncertainty. How can one judge the severity of a situation when the potential damage and its probability are unknown or highly uncertain? In this dilemma, risk managers are well advised to include the main stakeholders in the evaluation process and ask them to find a consensus on the extra margin of safety in which they would be willing to invest in exchange for avoiding potentially catastrophic consequences. Furthermore, such a balancing procedure needs to include major reasons for additional concern if potential consequences may lead to a range of cascading effects and irreversible changes in the natural or social environment.
- *Discourse-based strategy:* This strategy includes *discursive processes* such as roundtables, consensus conferences, deliberative rulemaking, mediation or citizen panels. The task of risk management here is to organize in a structured and effective manner the involvement of stakeholders and interested public for designing risk management strategies based on each stakeholders' knowledge (epistemic community) and value system (Renn 2015). For managing these risks, the main groups representing different vision of future development should be consulted and, if possible, a consensus among the main actors accomplished.

5.7 Implications for Resilience

Taking a closer look at the risk literature demonstrated that the application of the precautionary principle does not conflict with the use of best available science and the need for regulatory clarity. Any regulatory regime based on precaution is faced with the question of how to make regulatory decision under uncertainty or even ignorance. It may be helpful in this respect to resort to a differentiation that Resnik (2003, 332) has proposed:

- decisions under certainty: the outcomes of different choices are known,
- decisions under risk: probabilities can be assigned to the outcomes of different choices,
- decisions under ignorance: it is not possible to assign probabilities to the outcomes of different choices.

A similar distinction has been made by Stirling (2003). One of the main conclusions has been that using precaution for the first two cases is neither necessary nor prudent given that regulation need to meet both objectives protection of public health and the environment and securing economic welfare. The legitimate realm of using precaution lies in the case of ignorance or other forms of remaining uncertainties (such as system boundaries or truly stochastic events). In order to avoid misunderstanding some analysts have proposed to avoid the term precaution and replace it with the more adequate term of *principle of insufficient reason* (see Peterson 2003: 71). This describes the situation where precaution is required.

In our view, the main purpose of precaution is to avoid irreversible decisions. Although highly critical about the use of the precautionary principle itself, policy analyst Giandomenico Majone concedes that it does have a role in risk analysis, namely where "losses (or utilities) are unbounded" and where it is "clearly impossible to calculate expected values", for example, when there is a threat of "serious and irreversible damage" (Majone 2002). In these cases, it is prudent to proceed in a way in which a decision can be reversed quickly if the outcomes or their probabilities turn out to be higher than expected (Klinke and Renn 2021).

The issue of irreversibility is directly related to the understanding of resilience as previously advocated.[1] It is focused on the ability and capacity of systems to resist shocks and to have the capability to deal and recover from threatening events (Carpenter et al. 2001; Rose 2007). This idea of resistance and recovery can also be applied to social and economic systems (Review in Norris et al. 2008; Adger et al. 2005) as explained in Chap. 2. Our approach to precautionary risk management and resilience has been inspired by Lorenz (2010), who distinguishes adaptive, coping and participative aspects of resilience. These three aspects correspond with the three management styles described above: risk-informed (corresponding to adaptive capability); precaution-based (corresponding to coping capability) and discourse-based (corresponding to participative capability).

5.8 Conclusion

Risk assessment and risk management according to the precautionary principle are instruments to protect humans and the environment against uncertain threats and risks. At the same time, however, trade-offs must always be made in order not to jeopardize cultural, economic and social goals and to uphold the principle of proportionality. Despite all risk precautions, society must refrain from the idea that science can predict all dangerous events and developments and rule them out through preventive action. At best, risk analyses are able to identify the possible threats and reveal weak points in the system. In addition, risk assessments help us to make better decisions in the trade-off between the side-effects of risk-taking (such as exposure in

[1] See Chap. 2 of this volume.

the case of a pandemic) and the side-effects of the protective measures introduced (economic, social, cultural consequences).

The precautionary principle can play an important role here. In the case of high uncertainty associated with the analysis of risks, precautionary thinking means to start from a rather conservative estimate of the risks (about 95% or 99% percentile on a normal distribution). Here, it is prudent to assume rather pessimistic, but by no means unrealistic, assumptions. For this purpose, conventions and rules are needed from the respective specialist science or from interdisciplinary committees. For risk management, the main issue is the careful balancing of measures according to the principle of proportionality, with accompanying measures such as containment, strict monitoring or exposure avoidance justified in cases of high uncertainty (European Commission 2000).

Sufficient capacity, diversification and flexibilization are three key means of making systems adaptable and resilient to recurrent surprises. This is especially true in preparatory planning (preparedness) before a crisis occurs. The health care system must be able to cope with particular stresses, there must be contingency plans that can be implemented consistently and quickly in the event of an emergency, and there must be anticipatory planning that enables coherent political action even under high time pressure. But even with the best precautions, societies will have to live with a certain degree of uncertainty and ambivalence.

What does this view of the problem suggest for the implementation of the precautionary principle? The more complex the consequences of a threat situation, the more important the instruments of precaution in the field of risk assessment. The more uncertain the consequences of possible countermeasures, such as continuous increase in carbon pricing over time, the more important precautionary risk management measures are. And the more controversial the prescribed risk reduction measures are, the more likely it is that a discourse will have to be held with the relevant shaping forces in society on the acceptability of the respective measures. However, this should be prepared before the negative impacts show up, because protracted discourses are not possible under time pressure.

Whether this early balancing of precautionary measures succeeds has an influence not only on the further development of risk analysis as a means of providing for the future. Rather, the opportunities and possibilities of a society are decided here: To what extent will it be able to act responsibly in times of rapid technical and ecological change on its own responsibility and with a view to the values of being human that are recognized as essential.

References

Adger, N., Hughes, T., Folke, C., Carpenter, S., Rockstrom, J.: Social-ecological resilience to coastal disasters. Science **309**, 1036 (2005)

Akademie der Wissenschaften zu Berlin: Umweltstandards. Grundlagen, Tatsachen und Bewertungen am Beispiel des Strahlenrisikos. De Gruyter, Berlin (1992)

Asselt, M., Vos, E.: The precautionary principle and the uncertainty paradox. J Risk Res. **9**, 313–336 (2006). https://doi.org/10.1080/13669870500175063

Bennett, P.G.: Applying the precautionary principle: a conceptual framework. In: Cottam, M.P., Harvey, D.W., Pape, R.P., Tait, J. (eds.) Foresight and Precaution, Bd. 1. Rotterdam & Brookfield: A. A. Balkema, pp. 223–227 (2000)

Bloesch, J., von Hauff, M., Mainzer, K., Mohan, V., Renn, O., Risse, V., Song, Y., Takeuchi, K., Wilderer, P.: Sustainable development integrated in the concept of resilience. Probl. Sustain. Dev. **10**(1), 7–14 (2015)

Carpenter, S., Walker, B., Anderies, J.M., Abel, N.: From metaphor to measurement: resilience of what to what? Ecosystems **4**(8), 765–781 (2001)

Charnley, G., Elliott, E.D.: Risk versus precaution: environmental law and public health protection. Environ. Law Report. **32**(2), 10363–10366 (2002)

Chen, W.: Promise and challenges in the development of COVID-19 vaccines. Hum. Vaccin. Immunother. **16**(11), 2604–2608 (2020)

Chishti, M. Z., Sinha, A., Zaman, U., Shahzad, U.: Exploring the dynamic connectedness among energy transition and its drivers: understanding the moderating role of global geopolitical risk. Energy Econ. **119**, Article 106570 (2023). https://doi.org/10.1016/j.eneco.2023.106570

Elliott, E.D., Charnley, G.: Risk versus precaution: environmental law and public health protection. Environ. Law Report. **32**, 10363–10366 (2002)

European Commission: Communication from the commission on the precautionary principle. COM(2000) 1. European Union, Brussels (2000)

Fisher, E.: Is the precautionary principle justiciable? 13th J. Environ. Law **3**, 315–334 (2001)

Fisher, E.: Precaution, precaution everywhere: developing a 'common understanding' of the precautionary principle in the European community. Maastricht J. Eur. Comp. Law **9**(1), 7–46 (2002)

Gee, D., Harremoes, P., Keys, J., MacGarvin, M., Stirling, A., Guedes Vaz, S., Wynne, B.: Late lesson from early warnings: the precautionary principle 1898–2000. European Environment Agency, Copenhagen (2001)

Gilbert, S.G.: Precautionary principle. In: Wexler, P. (ed.) Information Resources in Toxicology, pp. 489–494. Academic Press, New York, NY (2019)

Goldstein, B.D.: The precautionary principle and scientific research are not antithetical. Environ. Health Perspect. **107**, A594-595 (1999)

Harremoës, P., Gee, D., MacGarvin, M., Stirling, A., Keys, J., Wynne, B., Guedes Vaz, S. (eds.): The Precautionary Principle in the 20th Century: Late Lessons from Early Warnings. Earthscan, London (2002). https://doi.org/10.4324/9781315071985

Hynes, W., Trump, B.D., Kirman, A., Haldane, A., Linkov, I.: Systemic resilience in economics. Nat. Phys. **18**(4), 381–384 (2022)

IRGC: White paper on risk governance: towards an integrative framework. International Risk Governance Council, Geneva (2005)

Klinke, A., Renn, O.: Expertise and experience: a deliberative system of a functional division of labor for post-normal risk governance. Innov. Eur. J. Soc. Sci. Res. **27**(4), 442–465 (2014)

Klinke, A., Renn, O.: The coming age of risk governance. Risk Anal. **41**(3), 544–557 (2021). https://doi.org/10.1111/risa.13383

Linkov, I., Anklam, E., Collier, Z.A., DiMase, D., Renn, O.: Risk-based standards: integrating top–down and bottom–up approaches. Environ. Syst. Decis. **4** (2014). https://doi.org/10.1007/s10669-014-9488-3

Linkov, I., Bridges, T., Creutzig, F., et al.: Changing the resilience paradigm. Nat. Clim. Chang. **4**(6), 407–409 (2014b)

Linzenich, A., Zaunbrecher, B.S., Ziefle, M.: Risky transitions? Risk perceptions, public concerns, and energy infrastructure in Germany. Energy Res. Soc. Sci. **68**, 101554 (2020)

Löfstedt, R.E.: The swing of the pendulum in Europe: from precautionary principle to (regulatory) impact assessment. AEI-Brookings Joint Center for Regulatory Studies Working Paper 04–07. Kings College Publication, London (2004)

Lorenz, D.F.: The diversity of resilience: contributions from a social science perspective. In: Hutter, G., Kuhlicke, C., Glade, T., Felgentreff, C. (eds.) Natural Hazards. Special Volume: Resilience in Hazards Research and Planning – A Promising Concept?, pp. 1–18 (2010)
Majone, G.: What price safety? The precautionary principle and its policy implications. J. Common Mark. Stud. **40**(1), 89–109 (2002)
Morris, J.: Defining the precautionary principle. In: Morris, J. (ed.) Rethinking Risk and the Precautionary Principle, pp. 1–21. Butterworth Heinemann, London (2000)
Müller-Herold, U., Morosini, M., Schucht, O.: Choosing chemicals for precautionary regulation: a filter series approach. Environ. Sci. Technol. **39**(3), 683–691 (2005)
Norris, F.H., Stevens, S.P., Pfefferbaum, B., Wyche, K.E., Pfefferbaum, R.L.: Community resilience as a metaphor, theory, set of capabilities, and strategy for disaster readiness. Am. J. Community Psychol. **41**, 127–150 (2008). https://doi.org/10.1007/s10464-007-9156-6
O'Riordan, T., Cameron, J.C., Jordan, A. (eds.): Reinterpreting the Precautionary Principle. Cameron May, London (2001)
O'Riordan, T.: Environmental Science for Environmental Management. Routledge, London (2014)
Paterson, J.: Sustainable development, sustainable decisions and the precautionary principle. University of Aberdeen, Aberdeen (2005)
Peterson, M.: The precautionary principle should not be used as a basis for decision-making. EMBO Rep. **8**, 305–308 (2007)
Peterson, M.: Transformative decision rules. Erkenntnis **58**, 71–85 (2003)
Rehbinder, E.: Stoffrecht. In: Arbeitskreis für Umweltrecht (eds.) Grundzüge des Umweltrechtes, pp. 13/001–013/051. Erich Schmidt, Berlin (1997)
Renn, O.: Vorsorge als Prinzip: Besser in der Vorsicht irren als im Wagemut. GAIA – Ökologische Perspektiven in Natur-. Geistes- und Wirtschaftswissenschaften **1**, 44–46 (2002)
Renn, O.: Precaution and analysis: two sides of the same coin? EMBO Rep. **8**(4), 303–305 (2007)
Renn, O.: Risk Governance. Coping with Uncertainty in a Complex World. Earthscan, London (2008)
Renn, O.: Stakeholder and public involvement in risk governance. Int. J. Disaster Risk Sci. **6**(1), 8–20 (2015)
Renn, O., Klinke, A.: Complexity, uncertainty and ambiguity in inclusive risk governance. In: Andersen, T.J. (ed.) The Routledge Companion to Strategic Risk Management, pp. 13–30. Routledge, Milton Park and New York (2016)
Renn, O., Klinke, A., van Asselt, M.: Coping with complexity, uncertainty and ambiguity in risk governance: a synthesis. Ambio **40**(2), 231–246 (2011)
Renn, O., Schweizer, P.-J.: Inclusive governance for energy policy making: conceptual foundations, applications, and lessons learned. In: Renn, O., Ulmer, F., Deckert, A. (eds.) The Role of Public Participation in Energy Transitions, pp. 39–79. Elsevier Academic Press, London (2020)
Renn, O., Schweizer, P.-J.: Precautionary approach. In: Wallenhorst, N., Wulf, C. (eds.) Handbook of the Anthropocene, pp. 1629–1638. Springer, Cham (2023). https://doi.org/10.1007/978-3-031-25910-4_265
Resnik, D.: Is the precautionary principle unscientific? Stud. Hist. Philos. Biol. Biomed. Sci. **34**, 329–344 (2003)
Sand, P.: The precautionary principle: a European perspective. Hum. Ecol. Risk Assess. **6**(3), 445–458 (2000)
Sandin, P., Peterson, M., Hansson, S.O., Rudén, C., Juthé, A.: Five charges against the precautionary principle. J. Risk Res. **5**(4), 287–299 (2002)
Risikokommission: Abschlussbericht der Risikokommission „Neuordnung der Verfahren und Strukturen zur Risikobewertung und Standardsetzung im gesundheitlichen Umweltschutz der Bundesrepublik Deutschland“. München: Bundesanstalt für Strahlenschutz (2003). https://doris.bfs.de/jspui/handle/urn:nbn:de:0221-201004071338
Rose, A.: Economic resilience to natural and man-made disasters: multidisciplinary origins and contextual dimensions. Environ. Hazards **7**, 383–398 (2007)

Stirling, A.: Risk, uncertainty and precaution: some instrumental implications from the social sciences. In: Berkhout, F., Leach, M., Scoones, I. (eds.) Negotiating Change, pp. 184–212. Elgar, London (2003)
Stirling, A.: Risk, precaution and science: towards a more constructive policy debate. EMBO Rep. **8**, 309–315 (2007)
Sunstein, C.: Laws of Fear: Beyond the Precautionary Principle. Cambridge University Press, Cambridge (2005)
van den Daele, W.: The precautionary principle – political versus legal perspectives. In: Cottam, M.P., Harvey, D.W., Pape, R.P., Tait, J. (eds.) Foresight and Precaution, vol. 1, pp. 213–221. Bakema (2000)
Wissenschaftlicher Beirat der Bundesregierung „Globale Umweltveränderungen“: Welt im Wandel: Strategien zur Bewältigung globaler Umweltrisiken. Jahresgutachten 1998. Springer, Berlin (1999)

Chapter 6
Resilience and Values—A Conceptual Approach

Matthias Kaiser

Abstract How does human behaviour influence societal resilience in the face of sudden shocks? In the following chapter this question is approached from the angle of human/social values. It is argued that our values are providing us with important orientations and are defining our social identities. The diversity of social identities is, in turn, defining the responsive pattern with which a society meets a shock. Despite the recognition that human behaviour can either be dominated by a rational and reflective decision pattern or by an automated and/or group respectively cultural norm-driven decision pattern, the system of societal value landscapes provide a key to understand or possibly strengthen societal resilience. It is argued that the complexity of these systems needs to be studied with a multi-system approach, which also forms the basis for advanced models of social learning. With more sophisticated models of societal complexities and learning, improving societal resilience through active public engagement can be possible.

Keywords Societal resilience · Social identities · Value landscapes · Norm-driven decision making · Public engagement

6.1 Introduction

The years 2020 and 2021 have shown us how vulnerable our societies are to relatively sudden and unexpected disruptors and harms. The COVID-19 pandemic posed a rapidly evolving global threat to virtually most fundamental functions of social life. It affected the "hard infrastructure" (like air traffic, transport, hospitals, etc.) as well as the "soft infrastructure" (educational systems, cultural institutions, economy etc.). The flash floods in Germany, Netherlands and Belgium during the summer of 2021 posed a point-disruptor to the affected regions, with devastating forces to hard infrastructure (roads, bridges, railways, housing, electricity, etc.) and to a lesser extent the soft infrastructure (communication networks, schooling etc.). With hindsight one

M. Kaiser (✉)
University of Bergen—Centre for the Study of the Sciences and the Humanities, Bergen, Norway
e-mail: Matthias.kaiser@uib.no

D. van Doren et al. (eds.), *Towards Energy System Resilience*, Ethics of Science and Technology Assessment 55, https://doi.org/10.1007/978-3-032-03187-7_6

may question why our societies were not better prepared for events of this kind, given that for instance in regard to pandemics experts have claimed for a long time that the occurrence of a pandemic was highly probable and called for preparedness. With regard to the flash floods experts have pointed to the increased probability of extreme weather events caused by climate change, and they have pointed out how our conventional measures to mitigate flooding may be insufficient or even increasing the nature of problems.

Raising the question why available expertise was not better translated into policies touches upon the area of science-for-policy. It highlights the role of evidence-synthesis and evidence-brokerage as contributions to policy formation (cf. Gluckman et al. 2021; Gluckman 2014; SAPEA 2019). It also casts a critical light on how the scientific community often inadequately conceives of and prepares for crisis events (Kaiser et al. 2021). Even good cognitive tools like indicators and modelling often fall short of intended functions because they are used blindly without recognition of their systemic boundaries or are used unilaterally and not e.g. as a multi-model approach (cf. Page 2021). At the same time, it puts into a sharper focus the critical and positive contributions of transdisciplinary research (Kaiser and Gluckman 2023; Kaiser et al 2020; OECD 2020; Renn 2021) and a post-normal science approach (Funtowicz and Ravetz 1993, 2020). In the light of the above-mentioned events, it seems clear that we need to rethink our science-policy nexus and probably must improve it.

However, hindsight and backward-looking criticism appears pointless when the harm is done. What we need to do is looking forward with the intention to design better policies. While risk studies and the Precautionary Principle can help us to design strategies to hopefully avoid or at least mitigate possible future harms, resilience studies may prepare us for those harms and catastrophes that may occur despite our best efforts to avoid them. This is what we will look at now.

6.2 Social Identities in a Crisis Situation

One highly interesting observation of both the Covid-19 pandemic and the flash floods is related to individual and social responses to crisis. We could observe how social cohesion, social identities, trust in institutions, and individual values played a crucial role in responses to the crisis once it had occurred.

With respect to the Covid-19 pandemic one could observe a varied response to policies in the population in various countries. An article by Lam (2021) identified no less than 16 different identity types. Among them such behavioural types like: *deniers* (who did not believe there was a real threat to them, and wanted to go on with business as usual), *hoarders* (who were gripped by panic and hoarded all kind of goods like toilet paper in order to quell their insecurity), *blamers* (they needed to vent their frustration which often led to racism), *spreaders* (who want the virus to spread and herd immunity to develop), *worriers* (who stay informed and safe to manage their uncertainty and fear of viral threat), *realists* (who recognize the reality of the possible harm and adjust their behaviour accordingly), *contemplators* (who isolate and reflect on life and the world), *supporters* (who show their solidarity in support with others through e.g. clapping, songs etc.), *altruists* (who like the caremongers help the vulnerable), and some other types as well. "Such salient viral identities are fluidly emerging (and disappearing) in society in dynamic response to COVID-19" (Lam 2021, 3). What is perhaps important to note in our context is how social identities evolved which led to varied responses to crisis and policies.

From a policy perspective, it is assumedly most important to note how some of these social types tend to cooperate to mitigate the crisis while some others defect or reject policy measures. Obviously, the amount of trust to politics, science, and governmental agencies played a role.

With respect to the floods in 2021 one of the most encouraging observations was perhaps the surprising degree of *solidarity* that emerged in the population, both among those affected and those not directly affected. Individuals gave cloths and money and travelled to help through physical labour. Small companies went into the areas to help the victims without pay (https://www.dw.com/en/showing-solidarity-after-the-floods-people-helping-people/a-58612983). One may suspect that this amount of solidarity was spreading through close networks: "If others can help, I can help as well!". Given the picture of the populace which emerges for instance in elections and political debates, and which conveys the impression of deep splits and perhaps even of narrow self-centred interests, the massive solidarity shown in the immediate aftermath of the floods seemed surprising.

This raises the more general question how to account for cooperation and solidarity in a society. From a policy point of view, one would like to know how this can be brought about in order to strengthen social resilience in times of crisis. We shall come back to that. First, we shall look at values and what they do for our behaviour.

6.3 The Concept of Values

Values are here defined as (from the *Value-isobars* project)[1]: "*Values are reference points for evaluating something as positive or negative. Values are rationally and emotionally binding and they give long-term orientation and motivation for action.*" This conception of values opens a path for empirical research (cf. Brosch and Sander 2015), placing values between attitudes and preferences on one side and norms and principles on the other.[2] The noun "value" is what we can describe as the general feature of the active process of valuing something, a "valuation". The metaphor of a *value landscape* is meant to convey the following complex of insights: (a) values relate to each other in terms of proximity; (b) different values may exhibit

[1] The definition is taken from the EU project "Value-Isobars" 2009–2011, coordinated by Kaiser; cf.: https://cordis.europa.eu/project/id/230557.

[2] "A social norm is a prescribed guide for conduct or action which is generally complied with by the members of a society" (Ullmann-Margalit 1977, p. 12).

different intensity (defining peaks in the landscape); (c) each value acquires various meanings dependent on from where it is perceived (contextuality); and (d) values have an inertia but are malleable over time, particularly in interaction with belief states (Kaiser 2024). Our value landscapes are seen as the bedrock of our social identities. Everyone has multiple social identities, relating us to different groups and networks. Some of the values may be more salient in some network, while others may relate to other groups. They are the value-based bridges we (intend to) share with others.

Looking at value research there are some disappointing news.[3] First, there is no generally accepted or well proven theory of values. In spite of several attempts in social psychology, sociology, and anthropology to devise an all-encompassing global theory of values and to support it by empirical data, no such theory meets all critical standards of theory appraisal. Maslow (1943, 1959, 1970), Kluckhohn and Strodtbeck (1961), Rokeach (1973), Inglehart (1977, 1997) and Schwartz (1992, 2012, 2016) are classics that introduced the conceptual and theoretical foundations for some of these studies, each adopting different modes of structuring values. Kluckhohn and Strodtbeck (1961), Hills (2002) introduced an influential theory of basic human values that examined diverse cultural orientations to time, man and nature, relating to other people, and motive for behaving. They have identified three orientations in the interaction between man and nature: 1. Mastery, 2. Harmonious, and 3. Submissiveness (Hills 2002), but these have not been robustly studied empirically. Inglehart (1977, 1990, 1997) proposes a basic dualism between materialist versus post-materialist values, a framework which has entered the World Values Surveys. Shalom H. Schwartz proposed a different model based on the relationships of compatibility and incompatibility between values (cf. also Schwartz 2012; Lee et al. 2019), and on this basis has collected massive amounts of data. Schwartz operates with basic dimensions of self-transcendence vs. self-enhancement, and openness-to-change vs. conservation, which gives him ten values along these axes. Strong weight on one side of the axis implies lesser weight on the opposite side. All these approaches can be characterized as functionalist theories.

Spates (1983) identifies three fundamental problems in these functionalist theories of values. First, there is the problem of empirical support. Spates claims that very few scholars actually attempt empirical studies to test their theories. One notable exception is Kluckhohn and Strodtbeck (1961) on value orientations, and a few others are concerned with particular selective cases. Second, there is the problem of deductive imposition. As Spates puts it: "subscribing to functionalist value theory, the researcher knows what to look for beforehand and imposes certain categories of response upon the empirical situation … Observable reality is forced into accord with a preconceived model. Thus, it is little wonder that functionalist research often finds '*evidence*' for its concepts: such evidence is never given a chance not to appear" (Spates 1983, italics in original, p. 34). Third, there is the problem of abstraction. This issue concerns the immunization from falsification through increased abstraction.

[3] The following ideas about value-landscapes, a theory of values and the apparent lack of it are further developed in Kaiser (2024).

"With the distinction between values and norms …, values were said to be neither situation-specific or function-specific. Ironically, their 'power' in shaping social life became even greater. Values were at the peak of the cultural 'cybernetic hierarchy of control'; they 'controlled' norms, and norms 'controlled behaviour'' (Spates 1983, p. 35).

Our tentative conclusion is therefore that one needs to be open-minded and context-specific when doing research on values.

Second, there is no definite list of (describing) all possible values held by all people; all lists are in an important sense incomplete. On the other hand, some lists may be sufficient given a context and the people involved. The often-shared assumption that there is a comprehensive list of universal values fails in the face of cultural variety.

Third, values are elusive and hard to detect reliably. One always faces the danger of values-as-truisms ("Friede, Freude, Eierkuchen" or "Motherhood and Apple-Pie") (Maio and Olson 1998; Bernard et al 2003). Fischhoff (1991) commented upon the differences that exist among empirical value studies. He differentiates between three different paradigms: articulated values, basic values, and partial perspectives. His point is that supporting a theory of values by data will imply a previous choice between these paradigms.

Given these difficulties, what then would speak in favor of explicitly including values in our efforts to build up resilience? We claim that it is the fact that values almost always figure in our explanations (or theories) of action, no matter what model we apply. Ever since Aristotle, philosophers explained human action on the basis of three elements of the human mind, working conjointly: the *cognitive*, the *affective*, and the *conative*. The first supplied the relevant beliefs about how the world is, the second provided for the individual drive and emotional stimulation, and the third for the direction which we desire to fulfil in the long run through the concrete action (our values and views of a good life). Versions of this we detect in all models of human action, particularly those on rational action (e.g. cost–benefit models), when we pursue strategies to achieve certain ends. But also, when actors are viewed as mainly driven by cultural norms, heuristics, and expectations, by automatic impulses, rule-based behaviours, without overt rational contemplation of options, the actions result from chosen social identities (and their norms and behavioural patterns) which in turn reflect our values. Currently, most scholars assume therefore a 2-tier or 2-modes model of human behaviour (Strack and Deutsch 2004; Marteau 2017) combining these two modes; cf. Fig. 6.1. But—importantly—both modes of behaviour imply some bedrock of values.

It is noteworthy that learning can take place within both of these modes, though in different forms. While the rational approach is typically slow and deliberative in reaching considered conclusions, it is relatively fast in the learning, i.e. it corrects its strategies if events come out in an unexpected way. The automatic response and cultural push model (rule-based behaviour) on the other hand is quick in response—supplying useful heuristics—but is slow in learning when things go wrong (Page 2021). It is furthermore noteworthy that both behavioural frames respond also to

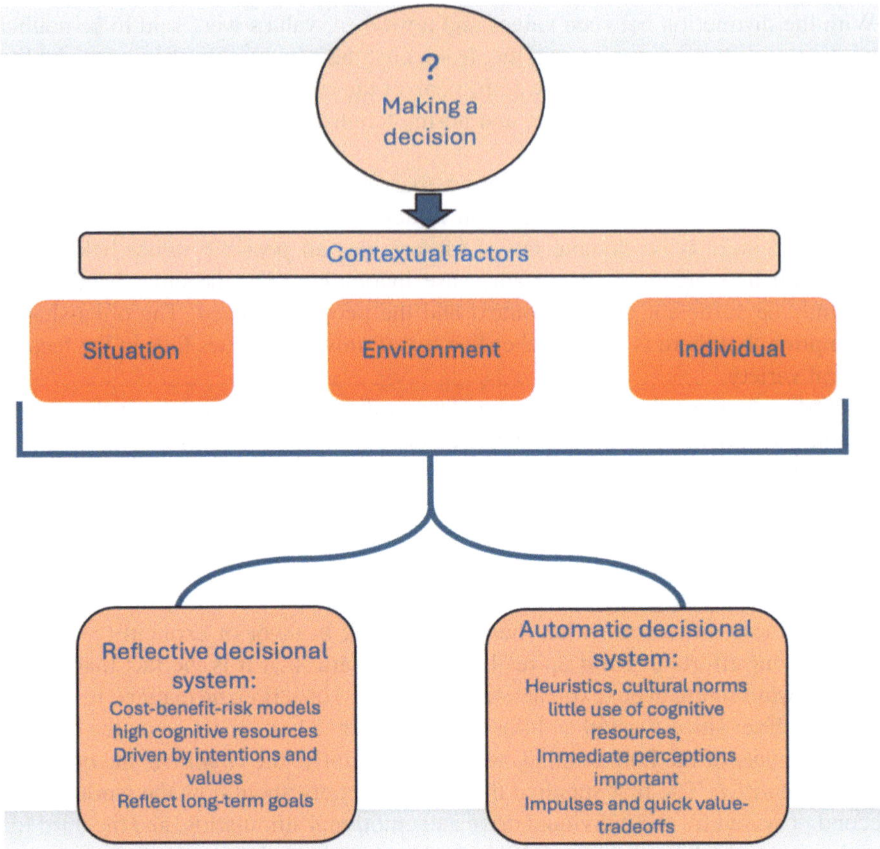

Fig. 6.1 Two modes of behaviour; figure inspired by Strack and Deutsch (2004), here drawn by the author

their immediate social surrounding and network, i.e. if behavioural changes take place in the network, then individual learning is inspired.

What is seldom happening is that specific values lead directly and consistently to specific behaviour or action. The motivation they provide is typically expressed post hoc, as rationalizations perhaps, while isolated values seldom have anticipatory value. That changes somewhat in the more specific cases when they are the backbone of a given social identity.

To sum up what we have said so far:

- In spite of the lack of a unified theory of values, there is an intuitive agreement among many scholars that values are an important aspect of understanding human behaviour.
- Values can be assessed empirically, though always embedded in specific contexts, and with due respect to uncertainty and individual and cultural variations.

- The complex set of values we adopt is the basis and bedrock for our social identities and cultural identification.
- Human behaviour can be depicted as either based on contemplative, rational deliberation, or on cultural or rule-based behaviour, following simple heuristics. Both are including values, though in a different manner.
- Learning can take place in both modes of human behaviour, in the rational mode as a quick adaptation to changed circumstance, and in the rule-based behaviour as a slow convergence of social identity to a changed environment.

6.4 Reminder on Specifics of Complex Systems

Systems in general were previously defined in this book.[4] Real world problems are typically based on highly complex systems. A complex system exhibits interactions and dependencies among its elements which make them hard to model because of nonlinear relationships, emergence of novel properties, adaptation or other mechanisms. Furthermore, they may interact with their environment and several of its systems in novel and surprising ways. Complex systems may have multiple stable states. Scale and scope are important when describing such systems.

This is also important when discussing resilience of e.g. digitized energy systems. Gunderson and Holling (2002) in the book *Panarchy* and particularly Holling et al. (2002) in this book discuss nested systems with interactions affecting scale and scope. The small scale may consist of e.g. individuals who react to a certain disturbance individually, but their reaction may provide feedback to institutions on a higher level. The small scale is often able to, or compelled to respond quickly, while the higher level needs a larger reaction time. The feedback between levels can typically either force change and innovation (revolt), or feedback with responses from earlier stressors (remembering).

Thus, this type of interaction introduces complexity between different scales of modelling the system. There is, of course, also the complexity that arises around scope and how we define the boundaries of the system. We may e.g. model the man–machine interaction as one big, albeit highly complex system, or we may split it up into two interacting systems, where the parts in each are modelled separately. It is not always desirable (or actually practically feasible) to aim for one big holistic model of the total system. The O'Neill conjecture is that for a given problem there is an optimal balance between model complexity and model error. In other words, adding more detail to a model does not always guarantee an increase in reliability and precision (cf. Kaiser et al. 2022 for more detail).

We also need to note that models will serve different purposes and functions. It is essential to realize that models as such are not true or false (Oreskes et al. 1994), not even in the sense of being closer to the real system of the world studied. Instead, they are better or worse in relation to the purpose they are designed for. We may

[4] See Chap. 2 of this book.

construct a model to understand and explain a certain effect or state of affairs. Or we may want a model in order to predict a future state of the system, or alternatively we may want a model to help us design a policy to deal with a desired change of the system. These different uses point us to trade-offs between model realism, precision and generality (Levins 1966).

The way out of the conundrum of the complexity of the world is then the application of several alternative and supplementary systems and to adjust them to our given purpose. This is the advantage of a multi-system/many-model approach (Page 2021).

To briefly sum up the previous observations:

- Real world problems appear in the form of open and complex systems.
- Complex systems display features of dependencies and emergent properties which limit their accuracy for the different functions of modelling.
- Scope and scale are always crucial characteristics of complex systems and interactions across scale and scope may need varied approaches.
- There is balance to be achieved between system complexity and system error.
- Models are to be designed for different uses and functions, and this needs to be mirrored in the design of the models.
- Given that no model is true of the world, we will have to settle for a many-model approach to deal with real world problems.

6.5 Learning and Preparing for Resilience

We have noted earlier[5] how resilience can be perceived along different trajectories. It can be seen as the structural elements in a system that are designed to provide the functioning of this system in times of stress and disruption. A technical system may include a fallback system or relate to other systems with similar functions, thus providing technical robustness. However, resilience becomes more complex when we include the human/societal response to a harm. In order to reach resilient adaptivity or evolvability of a given system, and in order to prepare for stressors by anticipation and precaution, social responses matter, and for a given problem they will often imply behavioural changes. These changes typically require some form of learning.

Aiming at preparing the ground for increased resilience of our energy systems (or any other complex system for that matter) will imply having a grip on how people learn to adapt to new challenges of this kind. In practice, we may want to do this by modelling how people learn and adapt.

There are basically two modes (different scales) we have to consider, namely *individual learning* versus *social learning*. Individual learning follows largely the rational mode of behaviour and is typically achieving its dynamic structure through *reinforcement* strategies. Note that this can apply to social learning as well if we view the result of societal outcomes as the result of the sum of individual processes.

[5] See Chap. 2 of this book.

Here a set of alternative options is given a weight, and they have connected to them a set of rewards. Economic incentives—like for buying/driving electric cars—can be an example of this. Over time the alternatives with the highest rewards will win out:

"In the **learning-the-best-alternative framework**, reinforcement learning with the aspiration level set equal to the average earned reward (eventually) almost always selects the best alternative" (Page 2021, p. 308).

In order to design a policy that appeals to rational individual learning, the policy-maker would have to identify available alternatives of the individual and to stimulate rewards for the societal best solutions.

However, things become more complicated if we look at some (other) forms of social learning. The idea behind these models is that individuals are strongly influenced by their social identities and behaviour within their networks. This is e.g. what we above have captured in the 2-tier system of behaviour and coined rule-based behaviour. Individuals learn from their own behaviour in conjunction with the behaviour of others in their network, they copy other behaviour, or they implement alternative heuristics. In its simplest form we get a replicator dynamic where the probability of a specific action comes as the result of *reward x popularity*. The latter factor is not necessarily holistic but is mostly to be drawn from the actor's network which defines his social identity.

One factor to be considered is that people can be motivated by diverse rewards. Do absolute or relative payoffs matter most? It is shown that people are not always aiming at maximizing their own gain, but often are optimizing it in relation to what the others gain. Thus, one may prefer a lesser reward if this implies to remain at the same or higher level in social hierarchy rather than gaining a higher reward while other groups advance further than oneself. International security politics where one aims at retaining one's power position relative to other states can be an example here (cf. Powell 1991). Interactions of this kind we normally cast in a game-theoretic form where actions depend on the individuals' expectations of the other actors' choices. Basically, this individual attitude seems to be related to our perception of social justice: the amount of welfare or reward is more dependent on how others are rewarded for similar behaviour than on the amount of happiness this benefit can incur in oneself. If this is cast in a game-theoretic framework combined with a replicator dynamic, one may generate a process where individual choices may converge towards generous benefits for all. Were one to only apply replicator dynamics alone, the process would lead to more actors choosing malevolent options, since every actor who chooses egoistically and malevolent to others earns higher payoffs than the others every time (cf. Page 2021 313ff).

It is a long-standing conflict between two basic paradigms: those who maintain that culture always dominates strategic choices versus those who maintain that strategy trumps culture. The former we typically find in organizational theory and the latter in economics. Our point here is to argue for the point that it all depends on contexts, and that we need to be open to both if we want to model human behaviour for resilience.

One thing we have not covered yet is the question how cooperation can evolve in a social structure, e.g. the kind of cooperation and solidarity we saw in the German floods in 2021. One way to approach this is by models-feedback and threshold-based behaviour. The most popular way to approach this issue is by way of so-called **agent-based models**. In those models we attempt to include behavioural diversity among people and place them in space and time. It is obvious that these models are both more granular but also more complex than what we typically do in systems theory. How far down to individual differences do we want to go? Here it is important to keep in mind what we said above about the O'Neill conjecture, the optimal balance between error and complexity, and keeping in mind the payoffs between realism, precision and generality. Obviously, we need to focus on a small selection of behavioural types. These behavioural types are connected to individual and social identity of the agents. For instance, we may just differentiate between risk-aversive and risk-seeking individuals. Or we may want to differentiate between conformity, defection and rebellion. In the example of the pandemic, we have seen how different identities eventually form which tend to fall into one of these boxes.

When modelling these behavioural types, we need to take account of thresholds and the reward feedback on these thresholds. Thresholds occur when a behavioural type is triggered or dampened by positive or negative feedback. Thresholds may for instance be the number of people who have adopted an alternative kind of behaviour, or they may be the tolerance level when too many or too few people adopt a behavioural kind. Heterogeneity and homogeneity in group behaviour depends on the diversity of feedback and the diversity of thresholds. We may have to model these differentiations differently, depending on what we know, what kind of data we have, about the population we study. Again, this refers us back to social identities and those again on the diversity of value landscapes.

Looking at cooperation from the other end, from the point of view of institutional arrangements, one will typically face the well-known *Prisoners' dilemma*, a situation where individual defection rather than cooperation emerges as the dominant strategy in a game-theoretic setting: self-interest leads to actions which are collectively worse. The challenge is to overcome this institutional arrangement in our social institutions, and it is shown that repetitions of the game can stabilize this situation towards more collective actions. This can among others be achieved through punishment or reputational gains. Ostrom (2000) adds to this the introduction of institutionalized mechanisms to resolve disputes. Again, such a process requires learning, institutional learning. If we conceive of our social institutions as the conjunction of hard-law and soft-law regulations, we can see the challenge of institutional reform. Institutional arrangements are on a higher scale than individual behaviour, and they often may face the need to revolt against lower scale behaviour, i.e. to implement innovative game-changers to force different behaviour to emerge. Response-time on higher scale is typically slower than on lower scales, and therefore when planning for increased resilience one needs to be anticipatory of the potential behavioural responses to crisis events. Cooperation is the one behavioural pattern which institutions want to bring about, rather than defection. All mechanisms of both hard- and soft-law regulations can be considered to achieve this. An implication of the famous *Campbell's Law* may be that in the end soft-law may be a more effective measure than hard-law. It states: "The more any quantitative social indicator is used for social decision-making, the more subject it will be to corruption pressures and the more apt it will be to distort and corrupt the social processes it is intended to monitor" (Campbell 1979, 85). Given that hard-law needs clear indicators for compliance and defection, the law would assume that people respond in ways which make it less effective in the long-run. Soft-law, on the other hand, appeals to people's morality and striving for socially accepted good practices. This is usually embedded in ethical guidelines or charters of institutional values, and this refers us again back to our starting point about the importance of values.

To sum up what we have said so far:

- When preparing for increased resilience to potential stressors to our crucial functions, social responses matter, and the potential responses may not always be adequate or optimal to face these stressors.
- Preparing for resilience implies therefore also to navigate learning processes that lead to increased resilience.
- Typically, individual learning models will differ from models aiming at social learning.
- Individual learning is most often depicted as the rational-actor framework, maximising benefits in isolation from others.
- Social learning processes are based on the assumption that individual learning is affected by processes, actions, and behaviour in one's social network.
- Both the rational actor approaches and the cultural heuristics approaches have deep roots in people's behaviour.
- Reinforcement and replicator dynamics can be dominant features in social learning.
- Agent-based models add behavioural types of behaviour to the description of the system.
- Some rule-based behaviours may change when certain thresholds in the network behaviour are reached.
- Cooperation may arise through repeated games, improving the institutional structure through punishment or (reputational) gains.
- Institutional mechanisms to avoid merely selfish behaviour may include hard-law and/or soft-law regulations, with the latter sometimes having a better chance of success in relation to individual compliance.

6.6 The Complexity of an Example

Let us just briefly touch upon an imagined example of a systemic short-term stressor in the energy and internet sector which hits a system at a point of high vulnerability, and halts some highly important digitized functions of the system. We will assume a regional scope and look at mobility and electricity supply as affected by an all-out hacker attack, one that is highly effective. (Alternatively, one may, of course, also assume that the event is caused by other stressors like extreme weather events, but our point is here that basic digitized nodes of all internet connections and crucial energy distribution are set out of function whatever the precise cause.) How realistic such a hazard can turn out to be may be another matter, but recent experiences have taught us that "shit happens" and that we will do well to be prepared also for worst-case events.

Let us spell out some features of such an imaginary. Assuming that the society will have moved substantially in the direction of digitized services in all sectors of life, that society has also moved towards renewable electricity supplies, many specific and essential functions of social life will be affected. *Work-life*: what was a fall-back solution in our work-life during the pandemic, namely the home office, will not be available. Assuming that public transport and even mobility along the infrastructure of roads will be harmed: How will be people work then? How can schools and universities fulfil their functions then? Looking at one country, Norway, assuming that Germany will follow many of the Norwegian examples, e.g. in the use of electric cars, we can immediately see a breakdown of many essential services. In Bergen all home-care services are run with the help of electric cars. All public transport is to a very large extent already electrified. Services such as garbage removal or mail delivery is already based on vehicles driven by electric power. Traffic regulation will stop and cause chaos on the roads for those cars which are still run on combustion engines (the others may not be on the road anymore). *Home-life and food security*: We might assume that smart home technologies will be widely distributed. They may cover areas such as heating, kitchen, entertainment and information, and energy supply. If energy supply fails, refrigerating and cooking may become a big challenge. Food supply in the neighbourhood shops and at home will be affected. What helped us during the pandemic, namely internet-based home-deliveries, will not be possible. Home-schooling is not an option to the same degree when there is no internet. *Communication*: The mobile phone networks will be affected and can be out of function for a while. This can affect all emergencies like medical first help, and communication between health-care units, but also other emergencies like warnings to the police, fire-brigade etc. *Social vulnerability*: Those sections of the populations which are less privileged than most, and often dependent on institutional support will face particularly large challenges. What happens when crisis centres (e.g. about family violence) are not functioning because of personnel that cannot come to these centres, or when people in need do not know how to contact them?

I do not think we will be able to spell out in detail all possible impacts of such a hazard. A lot will depend on regional specifics and also on the time and the development of a thoroughly digitized society (for more scenarios of disruptions in a digitized world, see Schrijvers et al 2021). Yet, what we already can foresee is that how one will get through such a crisis will also to some significant extent depend on how people will react in such a crisis and how they can seek out alternative functions and solutions. This is the question of social resilience.

6.7 Steps to Increase Social Resilience in Crisis.

We have argued that personal and social values, and by extension social identities, will be a crucial factor for a strong social resilience. Taking the lead from Lam (2021) we can assume that social identities will respond in a varying degree. What assumedly is most relevant is how they respond in terms of compliance, defection or revolt. These response patterns will in turn relate to trust in societal institutions and politics. Similarly, the willingness to react with solidarity, respect and empathy is highly value-based. Response patterns will also be varying in relation to the time the disruption will last, and whether certain thresholds for particular responses can be observed in the peer groups.

For scenarios like the above one may surmise that neighbourhood networks might be of crucial importance to come up with intermediary solutions for fulfilling the most basic needs of people. Understanding what challenges and needs the people in the immediate vicinity face because of such a crisis is crucial to organize and prioritize help. Innovative thinking to re-create simple solutions to the crisis of non-availability of digitized functions or tools based on electric energy supply, might be important as well. Flexibility to re-define work-tasks in such a way that they can be performed individually, rather than with colleagues who are distanced, may be called for. Service- and care-institutions would have to make hard choices how to manage a triage of what is possible versus what is needed.

Given the basic unpredictability of the nature of all vulnerabilities and the resulting social challenges, one may conclude that it would be hard to prepare for all these eventualities. While a risk-register (where such exists) serves the purpose to provide possible institutional responses to, and mitigation of classes of risks[6] and help risk communication, social responses to crisis are typically not included in risk-registers. The question is therefore if we have any way to prepare the ground for adaptivity, innovation, solidarity and compliance among people.

Here it is suggested that the above-mentioned features in a society will be based on the existing value-landscapes in society. We assume, furthermore, that those value-landscapes will not necessarily comprise or highly prioritize those particular values which become crucial in a crisis situation. One may therefore safely assume that preparing the value sets of people for crisis might imply some learning processes. Aiming for social resilience would then mean to trigger reflection on the value-landscapes and to look for mechanisms how to strengthen pro-social values of particular importance in crisis management. As indicated earlier, legal regulations may have little impact on social attitudes in the long run, while soft-law, based on voluntary identifications with values and resulting heuristics for behaviour may underpin response patterns which are called for in crisis management. In other words, we assume that social resilience needs to be prepared by pre-crisis processes of social learning.

---

[6] As a potentially negative by-product, risk-registers may also provide the illusion of control, which in turn can hamper flexibility and adequate realization of complexity.

Let us try to be more concrete. One condition for such learning is obviously a better grasp and better representation of values as they occur in society. This is not an easy task, given the state of value research we mentioned earlier. However, some pragmatic moves can be indicated. One could imagine an organized learning process in several steps:

- Authorities encourage the building of neighbourhood networks. This can be stimulated by information campaigns, small start-up stipends, and online-guidance and networks.
- Selected neighbourhood exercises, arranged by local authorities or NGOs, to exercise best-practice communities.
- Research efforts to map the variety of value-landscapes in society (see next paragraph).
- Institutional reflection about how to learn, reward or encourage value-based behaviour patterns, in particular in regard local neighbourhoods and for selected crisis scenarios.

6.8 Learning About Our Values

Can we get more insights into actually held values through some form of structured research, e.g. the grounded theory approach? In the project "Sustaining ethical aquaculture trade" (SEAT-project; cf. https://www.universiteitleiden.nl/en/research/research-projects/science/cml-sustaining-ethical-aquatic-trade-seat), the Bergen group, led by Kaiser, was given exactly this task. They were to chart the ethical terrain between four aquaculture-producing countries in Asia and the European consumers. To this end, the group designed a three-tiered empirical research method, which consisted of 1. a survey, 2. a number of qualitative interviews based on narratives and scenarios, and 3. a deliberative workshop to arrive at considered—and possibly consensual—judgments among the stakeholders. The results of the research were illuminating and noteworthy (cf. Bremer et al. 2012, 2013, 2014, 2015). However, one important finding was also that the ethical landscapes in the four countries under study differed significantly in their ranking of values, while this was not to the same extent reflected in the (ethical) principles people seemed to accept. They also differed significantly in their preferred futures which they tried to bring in line with their values.

Given the difficulties with studying and charting existing values, the group took great care to design the survey protocols in intensive processes of pilots and preparatory workshops with local researchers.

Similarly, the second step, the qualitative interviews were also prepared with local researchers. Narratives were designed which exemplified different norms and different rules of behaviour, based on certain values, like equity, animal welfare, justice, care and precaution etc. The respondents were asked to give their preference for these types of behaviour (or not to prefer them).

The final step was the deliberation in the workshops with people (ca 50–60 participants in each country), while "experts" from research were excluded from participation (they did the moderation). An introductory video was followed by discussions of future scenarios (developed by the experts) where they had to place a preference, which then had to be adjusted according to how the preference fitted the values they had embraced.

In spite of our rather superficial and rough approach to values, interesting and relatively robust results emerged. Thus, one could clearly see cultural differences, often due to religious traditions.

The difference in value landscapes is coherent with the differences we found in the choice of preferred future scenarios, while the precise outcome of these preferences could admittedly not be predicted on the basis of the values alone. Similar results have later been obtained by Mimi Lam who adopted a variant of our method of value mapping for the case of a herring fisheries conflict between the Haida and the federal government in Canada (cf.: Lam et al.2019; Pitcher et al. 2017).

We present this as an example of an empirical approach to values and ethics, in this sense being an example of practical ethics. Other tools are certainly available as well. One now widely discussed tool is the participatory use of an ethical matrix (Mepham et al. 2006; Kaiser et al 2007; Kaiser and Forsberg 2001) which offers condensed views and opinions on how research results contribute positively or negatively to normative principles as they are interpreted and viewed by different stakeholder groups. Another method is the multicriteria mapping (MCM) approach, mainly developed by Andy Stirling (Stirling and Mayer 2001; Coburn and Stirling 2016). Valuable data may be supplied by some of the Eurobarometers as well. Perhaps the approach here advocated could be cast as a version of constructivist grounded theory (Charmaz 2000), while a major principle should be the combination of quantitative with qualitative data-sets in a multi-model and multi-data research design, in a transdisciplinary setting, co-constructed with stakeholder participation.

The upshot of this section is the following:

- Social resilience can be strengthened through pre-crisis policy measures.
- These measures presuppose an adequate grasp on existing value-landscapes in society.
- In spite of difficulties with a general conception of values, one may gain insights into existing value-landscapes by utilizing pragmatic research approaches (of practical ethics) and co-constructed in a transdisciplinary manner.
- From a research point of view, the most promising approaches may be combinations of bottom-up quantitative and qualitative approaches, utilizing various frameworks and lenses.
- Measures to inspire social learning for crisis management in communities and selected neighbourhoods may serve to benchmark best practices and communicate moral standards.

6.9 Conclusion

This paper aimed at a conceptual discussion of how values inform behaviour, and how this pens out in times of an imminent crisis. As a side, we noted that science-for-policy may still be ill-prepared for crisis management of complex societal disruptions. Our first observation was based on recent experiences, but also supported theoretically, namely that social identities matter how people respond to a given crisis affecting their daily lives. A crisis can trigger social identities which are normally more on the backburner or dormant. We observed furthermore that social identities are developed on a bedrock of individual and social values. The existence of such values is rarely disputed, but when it comes to a systematic understanding of values, our sciences are still struggling to provide a comprehensive and generally accepted theory of values. Values are apparently very elusive for empirical research, and the danger of values-as-truisms is always present. Despite this state of the art, we expressed an optimistic view that some lower-level insights can be gained if contextualized appropriately. We then added some well-known features of systems theory and systems thinking, as we believe that valuable insights can be won if we take note of the multiple modelling approaches to human behaviour. Since no model is true as such, and certainly not in regard complex systems, we propagated a multi-model approach to capture the various modes of behaviour and behavioural change. This has implications for how we approach social learning. The traditional contrast between rational actor theories versus cultural heuristics and automated responses was here interpreted as a constant tension between response patterns, and not necessarily as a permanent either-or, but rather as capabilities which are triggered differently and can both be realized within individuals. In line with the multi-model approach mentioned earlier, we also noted that from a policy point of view learning mechanisms aiming at compliance, trust, solidarity and empathy seem to be stepping-stones towards social resilience. We then concluded by examining some basic features of a crisis in our digitized energy society and put forward the hypothesis that a major contribution to strengthen social resilience might be to raise awareness of the importance of neighbourhood networks and the pro-social values that might prepare them for adverse stressors. In the final section we hinted at some research methods and approaches which might help us link our behavioural patterns, social identities, and learning processes more explicitly to our value-landscapes.

We want again to underline that the reflections above are tentative conceptual reflections, with many uncertainties and working hypotheses, obviously open for new input and learning. But we justify this by the conviction that many current approaches which disregard the underlying value dimensions of people fail because of their narrow tunnel-view and one-dimensional thinking. We do not believe we can find pure techno-fixes for the complex challenges our societies face, and certainly not prepare us for managing the potentially big disruptions our digitized society may have to face in the future, rare as they may be. We also justify putting forward tentative thoughts about social resilience by a moral conviction: the belief that humans have the capacity to learn from failures—if only in the long run and with ups and downs—, and to collectively work towards designing more just and better futures. With Aristotle we hold that ethics is the vision of the good life, and that we all are united by our striving towards that goal.

References

Bernard, M.M., Maio, G.R., Olson, J.M.: Effects of introspection about reasons for values: extending research on values-as-truisms. Soc. Cogn. **21**(1), 1–25 (2003)

Bremer, S., Haugen, A.S., Kaiser, M.: Mapping core values and ethical principles for livelihoods in Asia. In: Potthast, T., Meisch, S. (eds.) Climate Change and Sustainable Development - Ethical Perspectives on Land Use and Food Production. Wageningen Academic Publishers, Wageningen, pp. 419–424 (2012)

Bremer, S., Haugen, A.S., Kaiser, M.: Whose sustainability counts? Engaging with debates on the sustainability of Bangladeshi shrimp? In: Röcklinsberg, H., Sandin, P. (eds.) The Ethics of Consumption: The Citizen, The Market and The Law. Wageningen Academic Publishers: Wageningen (2013)

Bremer, S., Johansen, J., Øyen, S., Kaiser, M., Haugen, A.S.: Mapping the ethical terrain of Chinese. Aquaculture. In: Brautaset, C., Dent, C.M. (eds.) The Great Diversity: Trajectories of Asian Development. Wageningen Academic Publishers, Wageningen (2014)

Bremer, S., Millar, K., Wright, N., Kaiser, M.: Responsible techno-innovation in aquaculture: employing ethical engagement to explore attitudes to GM salmon in Northern Europe. Aquaculture **437**, 370–381 (2015). https://doi.org/10.1016/j.aquaculture.2014.12.031

Brosch, T., Sander, D. (eds.): Handbook of Value: Perspectives from Economics, Neuroscience, Philosophy, Psychology and Sociology. Oxford University Press (2015)

Campbell, D.T.: Assessing the impact of planned social change. Eval. Program Plan. **2**(1), 67–90 (1979). https://doi.org/10.1016/0149-7189(79)90048-X

Charmaz, K.: Grounded theory: objectivist and constructivist methods. In: Denzin, N.K., Lincoln, Y. (eds.) Handbook of Qualitative Research, 2nd edn. SAGE Publications Inc., Thousand Oaks, London, New Delhi (2000)

Coburn, J., Stirling, A.: Multicriteria mapping manual-version 2.0. SPRU Working Paper Series SWPS 2016–21, November 2016 (2016). https://www.researchgate.net/profile/Andy-Stirling/publication/310746262_Multicriteria_Mapping_Manual_-_Version_20/links/58359fe508ae102f073d5ad9/Multicriteria-Mapping-Manual-Version-20.pdf

Fischhoff, B.: Value elicitation – is there anything in there? Am. Psychol. **46**, 835–847 (1991)

Funtowicz, S.O., Ravetz, J.R.: Science for the post-normal age. Futures **25**(7), 739–755 (1993)

Funtowicz, S., Ravetz, J.: Post-normal science: how does it resonate with the world of today? In: Science for Policy Handbook, pp. 14–18. Elsevier (2020)

Gluckman, P.: Policy: the art of science advice to government. Nature **507**, 163–165 (2014)

Gluckman, P-D., Bardsley, A., Kaiser, M.: Brokerage at the science-policy interface: from conceptual framework to practical guidance. Humanit. Soc. Sci. Commun. **8**(84), 1–10 (2021). https://doi.org/10.1057/s41599-021-00756-3

Gunderson, L.H., Holling, C.S. (eds.): Panarchy: Understanding Transformations in Human and Natural Systems. Island Press (2002)

Hills, M.D.: Kluckhohn and Strodtbeck's values orientation theory. Online Read. Psychol. Cult. **4**(4), 2307–2919 (2002)

Holling, C.S., Gunderson, L.H., Peterson, G.D.: Sustainability and panarchies. In: Gunderson, L.H., Holling, C.S. (eds.) Panarchy: Understanding Transformations in Human and Natural Systems, pp. 63–102. Island Press (2002)

Inglehart, R.: The Silent Revolution. Princeton University Press, Princeton (1977)

Inglehart, R.: Culture Shift in Advanced Industrial Society. Princeton University Press, Princeton (1990)
Inglehart, R.: Modernization and Post-Modernization. Princeton University Press, Princeton (1997)
Kaiser, M.: The idea of a theory of values and the metaphor of value-landscapes. Humanit. Soc. Sci. Commun. (2024). Springer/Nature
Kaiser, M., Forsberg, E.-M.: Assessing fisheries – using an ethical matrix in a participatory process. J. Agric. Environ. Ethics **14**, 191–200 (2001)
Kaiser, M., Gluckman, P.: Looking at the future of transdisciplinary research. International Science Council, April 2023 (2023). https://doi.org/10.24948/2023.05, https://council.science/publications/looking-at-the-future-of-transdisciplinary-research/
Kaiser, M., Buklijas, T., Allen, K., Bardsley, A., Kiro, C., Peter Gluckman, P.: Blog: "Good information is required for good decision making: why transdisciplinarity matters" (2020). https://informedfutures.org/why-transdisciplinarity-matters/
Kaiser, M., Buklijas, T., Gluckman, P.: «Models and numbers: Representing the world or imposing order?». Perspect. Sci. **30**(4), 525–548 (2022). https://doi.org/10.1162/posc_a_00373
Kaiser, M., Chen, A.T.Y., Gluckman, P.: «Should policy makers trust composite indices? A commentary on the pitfalls of inappropriate indices for policy formation». Health Res. Policy Syst. **19**, 40 (2021). https://doi.org/10.1186/s12961-021-00702-4
Kaiser, M., Millar, K., Thorstensen, E., Tomkins, S.: Developing the ethical matrix as a decision support framework: GM fish as a case study. J. Agric. Environ. Ethics **20**, 65–80 (2007)
Kluckhohn, F.R., Strodtbeck, F.L.: Variations in Value Orientations. Row, Peterson, Evanston, IL (1961)
Lam, M.E.: United by the global COVID-19 pandemic: divided by our values and viral identities. Humanit. Soc. Sci. Commun. **8**, 1–6 (2021). https://www.nature.com/articles/s41599-020-00679-5
Lam, M.E., Pitcher, T.J., Surma, S., Scott, J., Kaiser, M., White, A.S., Pakhomov, E.A., Ward, L.M.: Value-and ecosystem-based management approach: the Pacific herring fishery conflict. Mar. Ecol. Prog. Ser. **617**, 341–364 (2019)
Lee, J.A., Sneddon, J.N., Daly, T.M., Schwartz, S.H., Soutar, G.N., Louviere, J.J.: Testing and extending Schwartz refined value theory using a best–worst scaling approach. Assessment **26**(2), 166–180 (2019)
Levins, R.: The strategy of model building in population biology. Am. Sci. **54**(4), 421–431 (1966)
Maio, G.R., Olson, J.M.: Values as truisms: evidence and implications. J. Pers. Soc. Psychol. **74**(2), 294–311 (1998)
Marteau, T.: Towards environmentally sustainable human behaviour: targeting non-conscious and conscious processes for effective and acceptable policies. Philos. Trans. R. Soc. A 375: 20160371 (2017). https://doi.org/10.1098/rsta.2016.0371
Maslow, A.H.: A theory of human motivation. Psychol. Rev. **50**(4), 379–396 (1943)
Maslow, A.H.: New Knowledge in Human Values. Harper and Row, New York (1959)
Maslow, A.H.: Motivation and Personality, 2nd edn. Harper & Row, New York (1970)
Mepham, B., Kaiser, M., Thorstensen, E., Tomkins, S., Millar, K.: Ethical matrix manual. LEI, onderdeel van Wageningen UR (2006)
OECD Science, Technology and Policy Papers: Addressing societal challenges using transdisciplinary research. June 2020, no. 88 (2020). https://www.oecd-ilibrary.org/science-and-technology/addressing-societal-challenges-using-transdisciplinary-research_0ca0ca45-en
Oreskes, N.,, Shrader-Frechette, K., Belitz, K.: Verification, validation, and confirmation of numerical models in the earth sciences. Science **263**(5147), 641–646 (1994)
Ostrom, E.: Collective action and the evolution of social norms. J. Econ. Perspect. **14**(3), 137–158 (2000)
Page, S.E.: The Model Thinker. Basic Books, New York (2018, 2021)
Pitcher, T.J., Lam, M.E., Kaiser, M, White, A. (S.-J.), Pakhomov, E. (2017) Hard of herring. In: Tortell, P., Young, M., Nemetz, P. (eds.) Reflections of Canada: Illuminating Our Opportunities

and Challenges at 150+ years, pp. 112–119, 309pp. Peter Wall Institute of Advanced Studies, Vancouver, Canada
Powell, R.: Absoliute and relative gains in international relations theory. Am. Polit. Sci. Rev. **85**(4), 1303–1320 (1991)
Renn, O.: Transdisciplinarity: synthesis towards a modular approach. Futures **130**, 102744 (2021). https://doi.org/10.1016/j.futures.2021.102744
Rokeach, M.: The Nature of Human Values. Free Press, New York (1973)
SAPEA (Renn et al.): Making sense of science: for policy under conditions of complexity and uncertainty (2019). https://doi.org/10.26356/MASOS, www.sapea.info/making-sense-of-science
Schrijvers, E., Prins, E., Passchier, R.: Preparing for Digital Disruption. Series: Research for Policy: Studies by the Netherlands Council for Government Policy (2021). https://doi.org/10.1007/978-3-030-77838-5
Spates, J.L.: The sociology of values. Ann. Rev. Sociol. **9**, 27–49 (1983)
Schwartz S.H. (1992) Universals in the content and structure of values: theoretical advances and empirical tests in 20 countries. In: Zanna, M. (ed.) Advances in Experimental Social Psychology, vol. 25. Academic, New York
Schwartz, S.H.: An overview of the schwartz theory of basic values. Online Read. Psychol. Cult. **2**(1), 11 (2012). https://doi.org/10.9707/2307-0919.1116
Schwartz, S.: Basic individual values: sources and consequences (Brosch & Sander eds.) (2016)
Stirling, A., Mayer, S.: A novel approach to the appraisal of technological risk: a multicriteria mapping study of a genetically modified crop. Eviron. Plan. C. Gov. Policy **19**(4), 529–555 (2001)
Strack, F., Deutsch, R.: Reflective and impulsive determinants of social behavior. Pers. Soc. Psychol. Rev. **8**(3), 220–247 (2004)
Ullmann-Margalit, E.: The Emergence of Norms. Oxford (1977)

Chapter 7
Governance of Energy Infrastructures: Comparing Planning Cultures in the Netherlands, the United Kingdom, and Germany

Jörg Knieling

Abstract The transition to 100% renewable energy requires the development of new technical infrastructures. In the light of this challenge, the governance of energy infrastructures becomes a pivotal issue, and the paper discusses how conflicts in the planning and implementation process can be resolved more effectively. It describes different approaches to communicative governance and planning in three European countries, Germany, the Netherlands and the United Kingdom, and outlines the specific characteristics and outlines the advantages and disadvantages of each. A comparative analysis of these three approaches reveals a strong correlation between the governance of energy infrastructure and national planning cultures. This encompasses both visible artefacts of energy planning and the planning and social environment, which includes traditions, path dependencies and current challenges. The paper concludes with a discussion of the potential contributions of the convergence of planning systems in Europe to enhanced governance of energy infrastructure, along with a consideration of the limitations that must be taken into account. It is recommended that future research address the following open questions regarding the governance of energy infrastructure: firstly, the zero option in decision-making; secondly, the binding effect of early conflict resolution in the subsequent planning process; and thirdly, the neutrality of the organisation responsible for public participation.

Keywords Energy infrastructure · Governance · Communicative planning · Planning culture · Germany · Netherlands · United Kingdom

---

J. Knieling (✉)
HafenCity University Hamburg, Hamburg, Germany
e-mail: joerg.knieling@hcu-hamburg.de

D. van Doren et al. (eds.), *Towards Energy System Resilience*, Ethics of Science and Technology Assessment 55, https://doi.org/10.1007/978-3-032-03187-7_7

7.1 Introduction

The energy transition to 100% renewable energy requires the development of new technical infrastructures, such as energy pipelines, wind and solar farms, or reservoirs for energy storage. For example, the EU and many European countries are seeking to provide innovative power plants and trans-European energy networks to distribute renewable energy. Although many of these ambitious projects such as energy pipelines stretching from offshore wind farms in the North Sea to areas in the South are under time pressure, they must respect the democratic rights to information, participation, and legal objection.

In light of this situation, there is heterogenous discussion as to which solution is most adequate to integrating public participation into the implementation process of renewable energy projects has recently been initiated. Informed by two contrasting positions, this discussion is characterised by the following arguments: whereas on the one hand it is proposed that the planning process be shortened resp. accelerated and participation opportunities reduced, on the other hand it is advocated that participation be taken more seriously in order to achieve both earlier and more efficient mechanisms of conflict resolution.

This paper considers scientific discourses on infrastructure planning related to complexity, insecurity, and social acceptance (Marshall 2013; Priemus 2007; Salet et al. 2012; Wolsink 2007). The empirical basis constitutes three case studies on different participation policies introduced in Germany, the Netherlands, and the United Kingdom. Whereas in the UK early informal participation has become obligatory, the Netherlands has set up a public service unit to train to various administrative bodies for better infrastructure participation processes. In Germany, legislation has reduced the scope of infrastructure planning whilst at the same time including early, yet non-binding forms of participation. By comparing these three approaches, the paper draws conclusions for further research.

7.2 Theoretical Approaches to Planning Large Infrastructures

What distinguishes conflicts surrounding large infrastructure projects from familiar land use conflicts? (cf. Glasl 2024; Schiller 2009). When it comes to aspects of change or interventions in existing structures, conflicts over large infrastructures are similar to other land use issues. In most cases, they relate to conflicts of interest over a perceived negative impact on the part of affected communities, institutions, landowners, or citizens. While the local community is often negatively affected by a variety of disruptions, higher political bodies at federal or European level see such projects as contributing to common good. Although some local arguments may be critically characterised as NIMBY (Not in My Backyard) positions, reflecting individual or institutional refusal to accept change, such resistance essentially articulates democratically legitimised interests. However, large infrastructure projects often provoke a more fundamental debate about competing development models. In transport, the construction of new motorways can be seen as a necessary solution to traffic congestion and poor connectivity, but it can also lead to increased traffic and hinder innovative mobility strategies. In the energy sector, politicians and environmental NGOs advocate for a new pan-European macro grid to facilitate the transition to renewable energy. However, when it comes to nature conservation issues, the same actors can become opponents of infrastructure projects at the local level. Moreover, the debate on renewable energy development is overlaid with a debate on the need for energy transition and scepticism or uncertainty about the reality of climate change. If local politicians or citizens are not convinced that projected climate changes will affect them, it is unlikely that they will accept potential disadvantages for their property or quality of life. Similar effects can be highlighted in the discussion of the temporal dimension of climate change. If the risks of climate change do not become visible in the next 50, 70, or 100 years, affected stakeholders rarely see the need to accept a compromise. While politicians fear losing future elections, local people fear the loss of their quality of life. This short overview of practical dilemmas related to the planning of large infrastructure projects may serve as an explanation of some specific challenges in this field. To gain a better understanding of these challenges, theoretical aspects addressing uncertainty, complexity, and social acceptability may be helpful (Giezen 2012; Priemus 2007; Salet et al. 2012; Wolsink 2007). Uncertainty relates primarily to the long duration of large infrastructure projects. Although it is certain that projects will affect various areas of public and private development, any prediction is constrained by limited knowledge of the future. The challenge for planning is therefore to adapt to this fundamental framework of uncertainty.

In contrast to "normal" planning conflicts, large infrastructure projects are characterised by a high degree of complexity. This complexity is primarily due to the large scale of infrastructure projects, which have spatial, social, and temporal impacts. As a result, planning processes need to take into account a wider range of actors and interests across multiple entities (e.g. of resorts or public institutions) and policy levels (the multi-level dimension).

In terms of social acceptance, the way in which opinions are formed and decisions are taken plays a key role. Spatial planning has a long tradition of communicative planning at the local level. Communication and the "communicative turn in planning" (Healey 1992, 1996) provide the opportunity to integrate divergent opinions into decision-making. This seems all the more important given the characteristics of uncertainty and complexity described above. Apart from the fact that public planning can hardly take into account all aspects relevant to a complex infrastructure project, any large project is based on assumptions about future developments that may be unrealistic or to some extent hypothetical. Communication not only provides a methodological means of engaging conflict actors and creating a broader basis of common understanding, but also allows opponents to gain publicity and transparency. Although communication may appear to fuel the conflict, the resulting discussion and the inclusion of differ opinions in decision-making can lead to better solutions in the long run.

In the field of energy infrastructure, another aspect related to social acceptance may become relevant. Energy planning is traditionally a technically oriented field that is characterised by engineering rationality. The experts involved are usually professionally socialised as part of an engineering community accustomed to finding the best solution from a technical point of view. From this perspective, citizens' protest appears as a disturbance and an obstacle to the selection of the preferred solution. This specific self-image of energy planning can limit the achievement of a compromise between different and conflicting actors.

7.3 Netherlands: Service Unit for Public Participation at the National Level

Due to the scarcity of land, infrastructure planning has always been a bottleneck for development in the Netherlands. As an industrialised country, the Netherlands has to organise public and private land use efficiently. While the country is renowned for gaining new land by diking and draining former sea areas, new infrastructure projects are often subject to a high level of potential conflict with both communities and environmental or other specific interests.

In 1997 the Dutch government introduced a centralised procedure for the planning process of large infrastructure projects. As the ministerial department responsible for the field of infrastructure, "Inspraak" was responsible for organising the approval procedures, hearings, etc., which are related to public participation. In its key role as a central 'mailbox' for the infrastructure planning processes, the idea was that this experienced and specialised central unit could be able to handle these processes more efficiently. As Inspraak was the common point of reference for all objections to infrastructure projects, the new organisation offered greater transparency to the citizens involved.

In a subsequent phase, the responsibilities of this "clearing unit" were extended to encompass early public participation and provide support to the transport administration in the organisation of these participation processes. This resulted in a significant change for the newly named Centre for Public Participation (CPP), as it now acted as an advisor and was tasked with qualifying the responsible administrators. As a result, the CPP published a range of information on how to manage public participation effectively. The emphasis on conflict resolution laid the foundation for a new sector of planning services. Recognising the potential, the private sector began to develop the skills to undertake these as contractors.

Following the acquisition of considerable experience in the field of consultation with neighbouring administrative units, a new phase of development has become subject of discussion. Following a shift in emphasis of the CPP from formal to informal processes, the current idea is that the CPP could be requested by responsible administrations to take over the organisation of participation processes. The anticipated advantage of this approach resides in the possibility for the CPP to focus its experience and develop case-specific solutions depending on the context at hand.

The Dutch approach to large infrastructure projects, as evidenced by their internal administrative strategy, is one of several methodologies employed in addressing this issue. The establishment of a dedicated unit tasked with orchestrating participation processes and qualifying associated entities represents an organisational innovation aimed at enhancing the efficiency of essential procedures. Given the emergence of conflict resolution as a new market segment for consultancies, this innovative administrative strategy has also catalysed the development of novel planning services within the private sector. The Dutch approach, as evidenced by its handling of intricate infrastructure planning procedures, appears to have contributed to enhanced efficiency. Nevertheless, there may also be a number of unsolved aspects that can perhaps be viewed critically. Given that the CPP is a public agency and is thus perceived by citizens and other involved parties as being directly connected to the interests of public bodies, the first issue relates to the existence of sufficient neutrality. Despite the CPP's efforts to maintain independence, citizens may harbour scepticism due to the hierarchical nature of public administration, which could potentially result in the minister responsible making decisions that align with the prevailing governmental agenda. A further critical issue pertains to the centralised nature of the organisational structure for participation. While this centralised structure may offer certain advantages, such as the pooling of knowledge and legal expertise, it may also hinder the adoption of more suitable solutions that might emerge from a more decentralised and multi-task arrangement.

7.4 United Kingdom: Obligatory Early Public Participation

The United Kingdom is renowned for extreme changes in public planning policy, depending on the political majority (cf. Marshall 2013; Swain et al. 2013). Historically, the Labour party has been more or less in favour of strong public planning and a differentiated organisational structure of spatial planning, whereas the Conservatives have primarily promoted market forces and limited planning competencies. The latter perspective perceives public planning as an impediment to market-driven development, and it was during this period that the government initiated the establishment of a coordinated planning procedure for public infrastructure. A new bill was introduced, stipulating that project initiators must organise a reliable public participation process, the details of which were to be documented and attached to the formal application for project permission. This obligation to integrate early participation was motivated by the desire to achieve a better and more efficient planning process, and was viewed as an opportunity to negotiate compromises while still possible.

The introduction of this obligatory regulation has resulted in the establishment of several consultancies specialising in conflict resolution, as well as the emergence of a new field of professional planning services driven by the private sector. While citizens were previously reliant on the willingness of the project initiator to implement a participation process, the legal obligation provides a reliable basis for their involvement in the planning process. Nevertheless, the sole responsibility of the project initiator for the participation process is also potentially critical This raises questions regarding the adequacy of existing quality standards and the feasibility of achieving a fair and neutral process. The utilisation of a private consultancy to organise the participation and conflict resolution process, funded by the project initiator, can give rise to concerns regarding the consideration of all interests. However, the UK approach demonstrates a viable alternative by imposing sufficient responsibility on the project initiator to pay for the obligatory participation process. This approach is particularly noteworthy in light of the substantial criticism voiced by investors within the German context, underscoring the importance of a balanced and comprehensive review of the pertinent issues.

7.5 Germany: Voluntary Early Participation and Accelerating Planning Processes

Germany's history with public infrastructure has been marked by significant struggles, with the 1970s witnessing a contentious discourse surrounding the utilisation of nuclear power. This discourse, characterised by profound disagreements, centred on the establishment of nuclear power stations and waste disposal sites. Concurrently, a robust environmental protection movement emerged, serving as a counter-expertise within numerous infrastructure-related conflicts. Examples in this context include to the construction of motorways or railways, the construction or extension of airports (Berlin, Frankfurt, Munich), as well as the enlargement and straightening of river beds (Danube, Elbe).

As of the year 2000, the complexity of these conflicts increased further, and in light of accelerating climate change and further nuclear catastrophes (1986: Chernobyl, 2011: Fukushima), Germany initiated an energy transition towards renewable energies. This transition encompassed a wide range of infrastructure projects focused on wind and solar energy, as well as biogas, along with the development of storage facilities and energy pipelines. Consequently, local environmental concerns began to impede the broader objective of achieving the complete energy transition, which environmentalists sought to achieve. Despite a shared fundamental understanding of environmental protection's objectives, both sides began to pursue divergent interests. Moreover, the pursuit of the overarching goal of achieving an energy transition imposed significant time constraints on planning processes.

In addition to the conflicts surrounding energy infrastructure, prominent conflicts related to transportation infrastructure have functioned as an impulse to the German discussion on public participation with the "Stuttgart 21" project being perhaps the most notable example in this context. This project advocates for the reconstruction of the central railway station in the city of Stuttgart. Following an over ten-year-long formal planning process, the conflict emerged at a later stage in the approval process when citizens recognised the impact the project would have on the central city and regional transportation system. The protest, which was quite intense and severe, induced a fundamental political shift in the federal state of Baden-Württemberg, with the conservatives soon having lost their absolute majority and the first Minister-President from the Green Party coming into power. The newly constituted government appointed a Secretary of State with explicit responsibility for the development of new public participation models for planning processes in Baden-Württemberg.

Consequently, a broad discussion on public participation has emerged in both scientific and political discourse. Whilst some argued that Germany was being blocked by the so-called "Wutbürger" (angry citizens), others advocated a new form of democracy whereby citizens assumed responsibility for and objected the established political decision-making processes. In response, the Federal Ministry of Transport, Building and Urban Development published a "Handbook on Good Public Participation" which includes many ideas and recommendations on how to address public participation when it comes to the planning of large projects in the transportation sector (BMVBS 2012). Nevertheless, critics insisted that informal procedures were still insufficient. Further encouraged by a new law on public participation that was simultaneously launched by the Federal Ministry of the Interior and primarily strived to accelerate planning processes that do not require early public participation (BMI 2013), these critics have induced a controversial debate on the following three key issues (e.g. BfR 2012; Groß 2011; Wulfhorst 2011; Ziekow 2012): In addition to deliberations concerning the optimal timing for engagement and the imperative to take into account the outcomes of early participation in subsequent phases of the approval process, the third facet pertains to the incorporation of diverse interests and perspectives in participation. This latter facet explores the question of whether the so-called zero-option (i.e. maintenance of the status quo) should be incorporated into public participation.

The primary issue that has been the subject of considerable controversy is that of the optimal point in time for the initiation of public participation. While the majority of experts and stakeholders agree that public participation should be incorporated into the earlier phases of the planning process, there is a divergence of opinion with regard to the most appropriate implementation of participation at such an early stage. On the one hand, bureaucratic positions advocate for the provision of information and dialogue on the existing options for a new project. On the other hand,, citizen-oriented arguments advocate for the greater mobilisation of citizens to form an opinion on proposed projects early on in the planning process.

A closely related issue concerns the question of whether early participation should be obligatory or optional. The Ministry contends that an obligatory solution would result in unnecessary complications and an extension of the planning process for small consensual projects. Conversely, the opposing perspective asserts that ensuring the full confidence of citizens in planning processes is paramount. If early participation remains voluntary, non-participatory procedures may well go unnoticed by citizens, potentially fostering a distrustful atmosphere between the public and project developers.

Another point of contention pertains to the question of whether the maintenance of the status quo should be incorporated within the scope of early public participation. In light of the prevailing constraints on the range of options, the citizens' perspective is markedly in favour of the zero-option. Conversely, the counter-argument underscores the potential implications for Germany's credibility as a modern industrial nation, emphasising the encouragement of novel and innovative advancements.

The German debate on shaping public participation in large infrastructure projects is characterised by a high degree of complexity, yet a major shift in addressing this question has become observable in recent years. Previous claims for greater public participation have been characterised as counter-productive to economic development and global competition; however, emerging discussions are somewhat more open to participation. Indeed, the search for innovative ways to include citizens in formal public procedures has become a broad issue of concern. Alongside the ongoing Stuttgart 21 project, numerous other initiatives such as airport expansions, railways, motorways, and nuclear waste sites have underscored the necessity for novel and inclusive decision-making processes that preserve the advanced and complex societies of the present age. In the context of the impending energy transition towards renewable energies, new conflicts that necessitate anticipatory processes are already becoming apparent. In many regions, the construction of energy conduits, wind and solar parks, and storage infrastructure is anticipated to provoke substantial protest, given the presence of both NIMBY sentiments and divergent opinions.

It is evident that the majority of stakeholders have come to acknowledge the necessity for increased effort in addressing these challenges, despite their associated costs and time demands that exceed those of the "good old days" (as some may nostalgically recall). Germany, from this perspective, appears to be making progress in developing a more inclusive planning system. However, it is essentially still in the process of experimenting with novel regulations, forms, and methods of participation. Examples for this development include the "Handbook on good participation" in major transportation projects by the Federal Ministry of Transport (BMVBS 2012) and a guideline for early participation in industrial or infrastructure projects by the Association of German Engineers (VDI 2015). Despite ongoing resistance from administration and politics still remains high, citizens have become a well-organised and highly qualified force in public planning, gradually transforming the societal framework for the planning of large infrastructure projects.

7.6 Comparison of Governance in Energy Infrastructure in Germany, the Netherlands and United Kingdom

The differing strategies pursued in the Netherlands, the UK, and Germany to address the complexity, uncertainty, and divergence of interests in large infrastructure projects are outlined in Table 7.1, which provides an overview of selected aspects.

All three cases demonstrate that early participation has become an accepted method of engaging competing interests in infrastructure planning. The involvement of the public at such an early stage not only facilitates the consideration and discussion of diverse and still relatively open interests, but also provides sufficient scope for the identification of alternatives or compromises. However, the comparison reveals that there is no common strategy for addressing public participation. Indeed, a number of fundamental differences and/or contradictory approaches exist, the identification of which allows for a discussion of the pros and cons related to the ideal conceptualisation of a comprehensive and democratically sound public participation strategy in infrastructure planning.

The comparison indicates that the German government is sceptical regarding obligatory participation in infrastructure planning, whereas the UK has introduced such a regulation to ensure early conflict resolution and to encourage planning applicants to confirm the serious effort made to reach a compromise. The comparison also demonstrates that planning legislation exerts a direct impact on planning in practice, and that the implementation of an obligatory participation process by the government will create a new market segment for private consultancies. If this market segment remains stable, it is characterised by positive development and creates specific qualification requirements that impact the planning community. Although also the US example highlights the necessity of this development to achieving high process quality of conflict resolution, the differing legislative context renders a comparison difficult (cf. Raab and Susskind 2009). As demonstrated in the Dutch context, an alternative strategy for developing competencies involves the establishment of an

Table 7.1 Public participation in infrastructure planning in the Netherlands, the UK, and Germany

	The Netherlands	United Kingdom	Germany
Time of participation	Early participation	Early participation	Early participation
Obligation of public participation	Non-binding	Obligatory and binding	Non-binding
Responsibility for participation	Administration	Project promotor	Administration
Funding of participation	Public	Project promotor	Public
Role of private sector (planning agencies)	New market segment	New market segment	Still unclear if it will become a relevant new market segment
Type of innovation	intra-administrative	External	Various approaches

administrative qualifying system, which contributes to a gradual shift in perspective that is opposed to public participation, particularly by technical and engineering sectors of the administration, both of which are characterised by a primarily rational self-understanding of planning processes. The CPP's portfolio serves to confront this specific mindset by providing new perspectives. In comparison to the Netherlands and the UK, Germany is still lagging when it comes to the establishment of comparable structures for public conflict resolution, although not primarily driven by planning but by juridical questions related to the process efficiency of legal cases. Nevertheless, a new mediation law was recently passed, and to stimulate the creation of "mediation" as a new market segment for planning services, infrastructure planning could nevertheless make use of this legislative framework.

However, this preliminary analysis of varying public participation methods in infrastructure planning still leaves several questions unanswered, all of which are relevant to understanding the specific approaches. Indeed, a key question relates to whether the regulations allow for the discussion of the zero-option, i.e. the possibility that the proposed infrastructure project will not be implemented at all. Theoretically, the zero-option is regarded as a highly relevant consideration by opposing interest groups, as the decision-making process would otherwise have already been restricted by a hierarchical, top-down approach. Secondly, the question arises as to how binding the results of early conflict resolution are for subsequent stages of decision-making. This is a matter of crucial importance to the reliability of the participation process, as it engenders a sense of commitment and investment from participants in the process, which is essential for the success.

7.7 Conclusion

The comparison of the various approaches of public participation in infrastructure planning across the Netherlands, the United Kingdom, and Germany reveals several salient questions pertinent to a comprehensive understanding of these specific approaches. Primarily, it is crucial to ascertain whether the regulatory framework permits deliberations on the zero-option, which pertains to the possibility that the proposed infrastructure project will not be realised. In principle, the zero-option is regarded as a significant mechanism for addressing concerns of opposing interest groups. If the zero-option is absent, the decision-making process would already have been significantly constrained by a hierarchical, top-down approach. Secondly, the question of the impact of early conflict resolution on subsequent decision-making processes is pivotal to the reliability of the participation process. The question must be posed: why should individuals engage and invest effort in a process if the results are not binding in subsequent phases of infrastructure project approval?

From a European standpoint, the heterogeneity of planning regulation and organisation gives rise to the question of whether greater convergence is appropriate or whether the prevailing divergent status quo is sufficient. In the context of transnational infrastructure projects within Europe, a more coherent planning process may well contribute to strengthening a joint European development. However, variable and place-specific planning traditions and cultures (Knieling and Othengrafen 2009) must be considered when contemplating more coherent approaches to infrastructure planning. Additionally, the question of how positive experiences can be integrated into the concept of "European infrastructure planning" remains unanswered. Examples in this regard include the obligation of early participation, the fair and transparent nature of the participation process, and the duty of the project initiator to bear the costs of participation. Furthermore, other relevant aspects such as the neutrality of the responsible organisation and the binding nature of the early participation results for the subsequent planning process are not yet sufficiently organised. This underscores the necessity for further discourse within both scientific and practical contexts.

It is evident that a number of significant aspects have yet to be organised in a satisfactory manner for implementing high quality planning of energy infrastructures. These include the binding character of the results of early participation in the subsequent planning process, and the neutrality of the responsible organisation. The latter aspect pertains to the question of how to establish an intermediate institutional solution that is responsible for the participation process but is not dependent on one of the pro-project stakeholders, be it the investor, developer or the public sector. In urban planning, analogous scenarios have been addressed by the concept of advocacy planning, wherein the municipality engages an urban planner or a comparable expert, independent of hierarchical content and organisational structures. The planner's role is to assist citizens in articulating their interests and advocating for them in the face of the municipal administration and political processes. In the context of infrastructure planning, the project developer may be obligated to allocate the requisite financial resources, and the establishment of a public foundation as an intermediary entity could be contemplated. Such an institution could be constituted for individual major infrastructure projects or as a comprehensive framework overseeing all contentious cases. The introduction of such an intermediary entity in the domain of participation and conflict resolution represents a departure from existing methodologies, such as the Dutch CPP (as a service unit of the ministry) or the UK approach, where the organisation of the participation process is delegated to the project developer. However, a comprehensive analysis of the merits and drawbacks of these approaches, considering the unique characteristics of each, could yield valuable insights leading to the identification of suitable solutions. In summary, the comparison of practices in the three selected countries offers further aspects for the conceptualisation of participation governance in the planning of large energy infrastructures. The international framework of the comparison recommends further research in this direction (e.g. in the form of an EU-COST action). However, the longer experience of environmental conflict resolution in the US calls for a broadening of the scope to include environmental mediation approaches. The analysis highlights several unresolved issues that should be addressed in future debates on practice and research.

References

BfR (Advisory Board for Territorial Development at the Federal Ministry of Transportation, Building and Urban Development), Knieling, J., Battis, U., Beckmann, K., Kilper, H., Magel, H., Schmitz, H., Troeger-Weiß, G., Wieth-Knudsen, L., Wiese-von Ofen, I., Wulfhorst, R.: Öffentlichkeitsbeteiligung bei Planungs- und Zulassungsverfahren großer Infrastrukturvorhaben, Recommendations of the Advisory Board for Territorial Development, Berlin (2012)

BMI (Federal Ministry of the Interior): Gesetz zur Verbesserung der Öffentlichkeitsbeteiligung und Vereinheitlichung von Planfeststellungsverfahren, Bundestagsbeschluss 28 February 2013, Berlin (2013)

BMVBS (Federal Ministry of Transportation, Building and Urban Development): „Handbuch für eine gute Bürgerbeteiligung" zur Planung von Großvorhaben im Verkehrssektor, Berlin (2012)

Giezen, M.: Navigating mega projects through uncertainty and complexity: strategic and adaptive capacity in planning and decision-making. Dissertation at University of Amsterdam, Amsterdam (2012)

Glasl, F.: Konfliktmanagement, 13th updated edition, Haupt, Stuttgart (2024)

Groß, T.: Stuttgart 21: Folgerungen für Demokratie und Verwaltungsverfahren. Die Öffentliche Verwaltung **64**(13), 510–516 (2011)

Healey, P.: Planning through debate: the communicative turn in planning theory. Town Plan. Rev. **63**(2), 143–162 (1992)

Healey, P.: The communicative turn in planning theory and its implications for spatial strategy formation. Environ. Plan. B. Plan. des. **23**(2), 217–234 (1996)

Knieling, J., Othengrafen, F. (eds.): Planning Cultures in Europe: Decoding Cultural Phenomena in Urban and Regional Planning. Routledge, Farnham/Burlington (2009)

Marshall, T.: Planning Major Infrastructure: A Critical Analysis. Routledge, Abingdon (2013)

Priemus, H.: Development and design of large infrastructure projects. Environ. Plan. B. Plan. des. **34**(4), 626–644 (2007)

Raab, J., Susskind, L.: New approaches to consensus building and speeding up large-scale energy infrastructure projects. In: Conference paper, The Expansion of the German Transmission Grid, 23 June 2009, Göttingen University, Göttingen (2009)

Salet, W., Bertolini, L., Giezen, M.: Complexity and uncertainty: problem or asset in decision making of mega infrastructure projects? Int. J. Urban Reg. Res. **37**(6), 1–17 (2012)

Schiller, G.: Praxisprobleme bei der Planfeststellung von Energiefreileitungen. Umwelt und Planungsrecht **7**, 245–253 (2009)

Swain, C., Marshall, T., Baden, T. (eds.): English Regional Planning 2000–2010: Lessons for the Future. RTPI Library Series. London, RTPI (2013)

VDI (Association of German Engineers): Frühe Öffentlichkeitsbeteiligung bei Industrie- und Infrastrukturprojekten (Early public participation in industrial and infrastructure projects), VDI Richtlinie 7000, Düsseldorf (2015)

Wolsink, M.: Planning of renewable schemes. Energy Policy **35**, 2692–2704 (2007)

Wulfhorst, R.: Konsequenzen aus „Stuttgart 21": Vorschläge zur Verbesserung der Bürgerbeteiligung. Die Öffentliche Verwaltung **64**(15), 581–590 (2011)

Ziekow, J.: Neue Formen der Bürgerbeteiligung? München, C.H. Beck (2012)

Chapter 8
Energy Resilience as a Participative Process

Gabriele Fohr

Abstract In energy system analysis, we usually employ models—most often techno-economic, optimising models—to create possible future scenarios. Linked to the models are often qualitative storylines: narratives of plausible socio-economic future pathways that serve as input assumptions for the models. However, despite this approach of integrating social, behavioural and political elements into the models, the ability of present energy system models to measure resilience is limited. We know that in case of complex, uncertain, or even ambiguous problems, we cannot adequately assess, and legitimately decide upon, a system without involving a wider public to answer critical, sometimes ethical questions, such as: can we imagine and agree upon a minimum system functionality or an allocation of resources in case of a blackout? In this paper, a process view on resilience is highlighted that brings together both perspectives—modelling and participative—in an analytic-deliberative approach (National Research Council 1996; Renn 2008). Within a process that complements analytic findings with deliberation, the quality and legitimacy of decisions, as well as the decision-making capacity of a social system would improve. A rough idea of where to implement deliberative elements into an energy resilience monitoring process is addressed.

Keywords Energy resilience · Resilience monitoring · Analytic-deliberative approach · Multi-stakeholder process · Participative scenario-building · Social resilience · Self-organization

---

The original version of the chapter has been revised. A correction to this chapter can be found at https://doi.org/10.1007/978-3-032-03187-7_18

G. Fohr (✉)
IQIB—Institut für qualifizierende Innovationsforschung und–beratung, Bad Neuenahr-Ahrweiler, Germany

D. van Doren et al. (eds.), *Towards Energy System Resilience*, Ethics of Science and Technology Assessment 55, https://doi.org/10.1007/978-3-032-03187-7_8

8.1 Introduction

As much as it is commonly acknowledged that our energy system is a socio-technical system, we do look for a way to monitor (and by this, improve) the socio-technical resilience of these systems.

In energy system analysis, we usually employ models—i.e., most notably, technical, techno-economic, optimizing models—to simulate and react to present, and possible future pathways. Although these models consider the system´s context (the setting, the “world”) in terms of framework conditions, storylines or scenarios, they define it as “external” to the system or “different” from the—largely quantitative—model parameters.

However, in the case of stress or disaster to the system, this understanding might not be enough. As shown in the previous chapters, the resilience of a complex system draws on elements, and has implications that might not be quantifiable or even modelable. Its monitoring should therefore comprise more than a model-based scenario analysis. In addition to the narrative element of scenario-building and the analytic element of modelling, a deliberative element should be applied, based on the understanding of resilience monitoring and evaluation as a process.

We argue that some elements of social resilience—in addition to existing values, attitudes, norms, or even civic engagement that we can collect and measure ex ante (Datola et al. 2021)—can only be adequately considered and possibly included into the analysis if the analytic component is complemented by a deliberative setting. In the following sections, we will take a closer look at this combination of analytic and deliberative elements, why it is important, and at which point in the process deliberative formats can be implemented favourably.

8.2 The Analytic-Deliberative (AD) Approach

8.2.1 Reasons for Choosing the AD Approach

To seek public participation when dealing with problems of high ambiguity might seem counter-intuitive at first. Why should we draw on multiple opinions and take up the investment of valuable resources, especially time, for dealing with a complex, highly ambiguous task? Would that not be an issue for experts? The opposite is true: the problem is that complex and ambiguous that science and expert knowledge alone cannot grasp it; instead, other types of knowledge and creativity have to be taken into account to be able to cover and deal with it; society as a whole needs to be involved to legitimise process and outcome, and to build capacity for current and future challenges.

Even if we apply appropriate models that integrate necessary and diverse disciplinary aspects, the question of how, and who is to, define the problem, raise important issues, appraise models and evaluate model outcomes, remains. At this point, the need for participation is clear. But of whom, when, and to what extent?

A well-practiced approach to complement analytic findings with deliberation is the analytic-deliberative (AD) approach for risk assessment and environmental decision making (National Research Council 1996; Dietz and Stern 2008; Renn 2008). It is promising for energy systems in that it explicitly considers elements of deliberation within a process of monitoring and evaluation of system states and interventions. It would also match the claim of the International Risk Governance Council (IRGC) to tackle problems of high ambiguity with a discourse-based strategy (see (Renn 2005).[1] By involving all interested and affected parties, it would add essentially to build social resilience, especially people´s participative capacity—a key factor for "being able to respond adequately".[2]

In addition to the often mentioned and unquestionably valid reasons for participation –improving the quality and legitimacy of assessments and decisions –, deliberation may also enhance the decision-making capacity of both the agency and the incorporated representatives of stakeholders or the public at large (National Research Council 1996, p. 43) © 1996 by the National Academy of Sciences. All Rights Reserved. Used with permission of The National Academies Press, from Understanding Risk: Informing Decisions in a Democratic Society, Committee on Risk Characterization, National Research Council, 1996; permission conveyed through Copyright Clearance Center, Inc..

In the field of social resilience, (Lorenz 2013) and (Saja et al. 2021) distinguish between adaptive, coping, and participative capacity as dimensions of social resilience, with participative capacity as a unique property of social systems that is "decisive to whether a social system can use its own adaptive and coping capacities" (Lorenz 2013, p. 17 © 2010 Springer Nature. Reproduced with permission from SNCSC). If reduced, e.g. through the unequal distribution of power, it hinders the ability of social systems to self-organise (Lorenz 2013, p. 17). It can on the other hand be enhanced by deliberation, i.e. by the active involvement in an analytic-deliberative process, as suggested by the (National Research Council 1996).

8.2.2 Fields of Application

Originated in environmental public participation research (Dietz and Stern 2008), applications of the AD approach can be found in different research areas, most notably in environment-related issues, from natural resources management, especially forest management (St-Laurent et al. 2020), agricultural water management (Allain et al. 2020) and agroecology (Bellocchi et al. 2015), climate adaptation planning (Webler et al. 2016), sustainable energy futures (Bellamy et al. 2022), sustainable cities (Gollagher and Hartz-Karp 2013), and disaster recovery (Miles 2018).

[1] See also Chap. 5.

[2] See Chap. 2.

Not every paper or application explicitly refers to the AD approach—some only use the terms “participation”, “public participation”, or “civic engagement”, and do not mention “deliberation”—but employ deliberative elements in a way that matches our description.

The list shows a diffusion of the approach from its use in natural sciences and the management of natural resources to its application in the fields of sustainability, climate change and built environments. This is seemingly quite similar to the diffusion path of the resilience notion (see Norris et al. 2008) for the varying levels of analysis in resilience definitions, from physical to ecological to individual or (Alexander 2013) for the evolution of the term resilience or, more recently, for resilience in risk and disaster management (Graveline and Germain 2022).

In the following sections, an overview of important elements of the AD approach is given.

8.2.3 Elements of the AD Approach, Participants, and Process

“Risk characterization is the outcome of an analytic-deliberative process.” This normative statement was postulated by the U.S. National Research Council in 1996 (National Research Council 1996, p. 3 © 1996 by the National Academy of Sciences. All Rights Reserved. Used with permission of The National Academies Press, from Understanding Risk: Informing Decisions in a Democratic Society, Committee on Risk Characterization, National Research Council, 1996; permission conveyed through Copyright Clearance Center, Inc.). It is further described as structured by two components: analysis and deliberation, which complement and inform each other, with a “systematic analysis that is appropriate to the problem, responds to the needs of the interested and affected parties, and treats uncertainties of importance to the decision problem in a comprehensible way”, and “deliberations that formulate the decision problem, guide analysis to improve decision participants’ understanding, seek the meaning of analytic findings and uncertainties, and improve the ability of interested and affected parties to participate effectively in the risk decision process.” (National Research Council 1996, p. 3 © 1996 by the National Academy of Sciences. All Rights Reserved. Used with permission of The National Academies Press, from Understanding Risk: Informing Decisions in a Democratic Society, Committee on Risk Characterization, National Research Council, 1996; permission conveyed through Copyright Clearance Center, Inc.).

In this definition, we get an idea of the mutual character of the process, the way in which both components are connected with, and refer to, each other. Whereas “analysis provides answers to factual questions, deliberation is “any formal or informal process for communication and for raising and collectively considering issues.” (National Research Council 1996, p. 73 © 1996 by the National Academy of Sciences. All Rights Reserved. Used with permission of The National Academies

Press, from Understanding Risk: Informing Decisions in a Democratic Society, Committee on Risk Characterization, National Research Council, 1996; permission conveyed through Copyright Clearance Center, Inc.).

It is also recursive: analysis "involves estimating the likelihood of occurrence and possible severity of particular kinds of harm. Analysis can also be used to examine risk problems to characterize their history and analyse possible outcomes of different decisions, strategies or policies" (National Research Council 1996, p. 97 © 1996 by the National Academy of Sciences. All Rights Reserved. Used with permission of The National Academies Press, from Understanding Risk: Informing Decisions in a Democratic Society, Committee on Risk Characterization, National Research Council, 1996; permission conveyed through Copyright Clearance Center, Inc.).; deliberation considers "each aspect of an issue and it may revisit earlier discussion on the basis of new knowledge or insights." (National Research Council 1996, p. 73 © 1996 by the National Academy of Sciences. All Rights Reserved. Used with permission of The National Academies Press, from Understanding Risk: Informing Decisions in a Democratic Society, Committee on Risk Characterization, National Research Council, 1996; permission conveyed through Copyright Clearance Center, Inc.).

It is effective when it arrives "at understandings that most participants consider adequate or acceptable within the existing limits of time or effort and that all recognize to be subject to reconsideration in the future." (National Research Council 1996, p. 73f © 1996 by the National Academy of Sciences. All Rights Reserved. Used with permission of The National Academies Press, from Understanding Risk: Informing Decisions in a Democratic Society, Committee on Risk Characterization, National Research Council, 1996; permission conveyed through Copyright Clearance Center, Inc.).

This is not to say that analysis and deliberation proceed one after the other—that deliberation is used to make decisions after, and on the basis of, a risk analysis—or that analysis creates knowledge, but deliberation does not. In fact, both elements contribute to each step of the process leading to risk decisions (National Research Council 1996, p. 118 © 1996 by the National Academy of Sciences. All Rights Reserved. Used with permission of The National Academies Press, from Understanding Risk: Informing Decisions in a Democratic Society, Committee on Risk Characterization, National Research Council, 1996; permission conveyed through Copyright Clearance Center, Inc.).

Within the AD process, the participation of interested and affected parties is considered imperative. It "includes any of a variety of mechanisms and processes used to involve and draw on members of the public or their representatives in the activities of public- or private-sector organizations that are engaged in informing or making environmental assessments or decisions." (Dietz and Stern 2008, p. 12 © 2008 by the National Academy of Sciences. All Rights Reserved. Used with permission of The National Academies Press, from Public Participation in Environmental Assessment and Decision Making, Stern, Paul C., Dietz, Thomas, Committee on the Human Dimensions of Global Change, National

Research Council 2008; permission conveyed through Copyright Clearance Center, Inc.).

"“The public” may consist of organized interests, sometimes referred to as stakeholders; people selected by a systematic process to create a representative sample, as is done in survey research; people selected purposively to represent particular perspectives, knowledge bases, or interests; or individuals who themselves choose to engage in processes that are open to all. Which of these versions of “the public” are the actual participants can make a difference to participatory processes, because different selections are likely to represent different sets of interests or concerns in the process." (Dietz and Stern 2008, p. 12 © 2008 by the National Academy of Sciences. All Rights Reserved. Used with permission of The National Academies Press, from Public Participation in Environmental Assessment and Decision Making, Stern, Paul C., Dietz, Thomas, Committee on the Human Dimensions of Global Change, National Research Council 2008; permission conveyed through Copyright Clearance Center, Inc.).

We can therefore distinguish between different “publics”—different kinds of participants:

- Stakeholders—organised groups that are or will be affected by or that have a strong interest in the outcome of a decision;
- A directly affected public—individuals and nonorganised groups that will experience positive or negative effects from the outcome;
- An observing public—the media, cultural elites, and opinion leaders, the media, cultural elites, and opinion leaders who may comment on the issue or influence public opinion; and
- A general public—all individuals who are not directly affected by the issue but may be part of public opinion on it. (Dietz and Stern 2008, p. 15 © 2008 by the National Academy of Sciences. All Rights Reserved. Used with permission of The National Academies Press, from Public Participation in Environmental Assessment and Decision Making, Stern, Paul C., Dietz, Thomas, Committee on the Human Dimensions of Global Change, National Research Council 2008; permission conveyed through Copyright Clearance Center, Inc.).

You may note that within each of the above categories, different fractions and sub-groups may be identified which should be involved. Who should be involved, the total breadth of the involvement, should match the issue (Dietz and Stern 2008, p. 15 © 2008 by the National Academy of Sciences. All Rights Reserved. Used with permission of The National Academies Press, from Public Participation in Environmental Assessment and Decision Making, Stern, Paul C., Dietz, Thomas, Committee on the Human Dimensions of Global Change, National Research Council 2008; permission conveyed through Copyright Clearance Center, Inc.).

The public can be involved at different points in the process, and to different degrees. Their roles, as well as their degrees of power or influence onto the actual decision making, can change during the process. The National Research Council provides a structure for the process that comprises nine steps: five steps that prepare decisions (process design, problem formulation, generation of options), the decision making itself, and two activities that

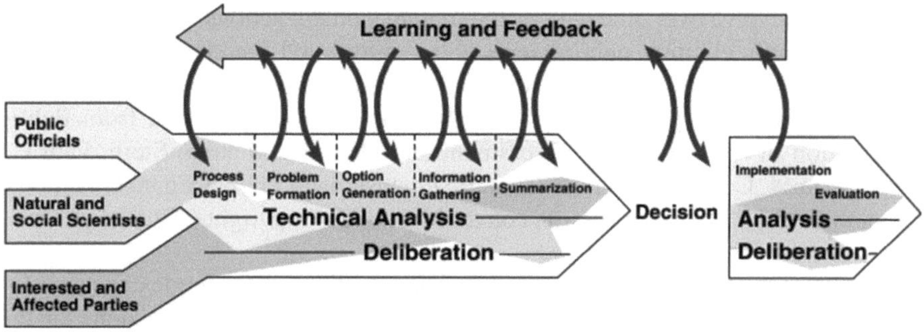

Fig. 8.1 A schematic representation of environmental decision processes (National Research Council 1996, p. 28 © 1996 by the National Academy of Sciences. All Rights Reserved. Used with permission of The National Academies Press, from Understanding Risk: Informing Decisions in a Democratic Society, Committee on Risk Characterization, National Research Council, 1996; permission conveyed through Copyright Clearance Center, Inc.)

follow (Dietz and Stern 2008, p. 16 © 2008 by the National Academy of Sciences. All Rights Reserved. Used with permission of The National Academies Press, from Public Participation in Environmental Assessment and Decision Making, Stern, Paul C., Dietz, Thomas, Committee on the Human Dimensions of Global Change, National Research Council 2008; permission conveyed through Copyright Clearance Center, Inc.). Figure 8.1 shows a flow chart of this process.

Although the figure suggests a linear, unidirectional process, learning and feedback loops are integrated at every process stage. After the last process stage, evaluation, the whole process may start again.

Of course, the scheme will not apply for every participation process. There is no single best format or method for participation to be used, or procedure to be applied; instead, as with the participants, the process needs to be adapted to the specific problem and purpose (Dietz and Stern 2008, p. 111, 149f © 2008 by the National Academy of Sciences. All Rights Reserved. Used with permission of The National Academies Press, from Public Participation in Environmental Assessment and Decision Making, Stern, Paul C., Dietz, Thomas, Committee on the Human Dimensions of Global Change, National Research Council 2008; permission conveyed through Copyright Clearance Center, Inc.). In fact, it is quite context- and place-specific (cf. Dubois et al. 2019).

The AD approach draws essentially on public participation. It emphasises that "both the public (..) and the scientific community have substantial expertise, but expertise of different kinds and on different matters (Dietz 1987): On one hand, (..) the public often has detailed knowledge of the local context and everyday practices that is not readily available to the scientific analyst. And of course the public, rather than the scientific community, is the legitimate source of information about public values and preferences. On the other hand, scientific analysis is essential for understanding the dynamics of complex systems and assessing the uncertainty in how such systems evolve over time with different management actions. It is also valuable in systematically eliciting public views about values and preferences (..).” (Dietz and Stern 2008, p. 148 © 2008 by the National Academy of Sciences. All Rights Reserved. Used with permission of The National Academies Press, from Public Participation in Environmental Assessment and Decision Making, Stern, Paul C., Dietz, Thomas, Committee on the Human Dimensions of Global Change, National Research Council 2008; permission conveyed through Copyright Clearance Center, Inc.).

Uncertainty assessment is unavoidable when dealing with complex systems for decision making as the uncertainties may be more important for the decision than the presumed certainties. The issue of elicitation of public or social values is beset with many methodological and conceptual difficulties as values are notoriously elusive. There is no widely accepted or cross-disciplinary theory of values. Methods to elicit them are faced with the problem of values-as-truisms (as in: “Friede, Freude, Eierkuchen”), opening up a space for empirical methodologies which shift between (a) articulated values, (b) partially recognised values, and (c) basic and deep values (cf Kaiser 2024). In view of these challenges to systematic analytic approaches, participatory processes in tackling complexity for decision making become particularly important.

It is clear that this mutual exchange and the activities related to it need to take their time. It is a process that assesses and decides upon (interventions into) a system that in itself is not static but progressing, and thus will require continuous decision-making and evaluation. But if we understand social resilience as an integral part of energy resilience, the process in itself might contribute an enhanced system state, respectively system performance, in terms of resilience.

8.3 Monitoring Energy Resilience

8.3.1 Resilience as a Process

A widely used definition of resilience by the United Nations Office for Disaster Risk Reduction describes it as “the ability of a system, community or society exposed to hazards to resist, absorb, accommodate, adapt to, transform and recover from the effects of a hazard in a timely and efficient manner, including through the preservation and restoration of its essential basic structures and functions through risk management.” (UNDRR, n.d.). Whether this ability should arise from a set of properties and characteristics of a community, or rather, as one linked to processes unfolding within communities, is subject of scholarly discussion (Comes et al. 2019, p. 3). Figure 8.2 from Comes et al. (2019) provides an overview of the different views of the term.

Authors in the field of disaster preparedness, e.g. (Norris et al. 2008), define resilience as “a process linking a set of adaptive capacities to a positive trajectory of functioning and adaptation after a disturbance.” (Norris et al. 2008, p. 130 © 2008

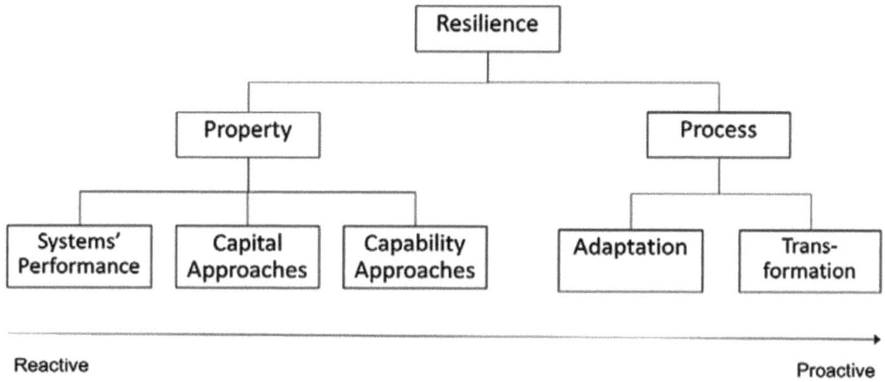

Fig. 8.2 Lenses on resilience. From Comes et al. (2019, p. 2) © 2017 Informa UK Limited, trading as Taylor & Francis Group. Reprinted by permission of the publisher, Taylor & Francis Ltd.

John Wiley & Sons - Books. Used with permission of John Wiley & Sons - Books, from *Community resilience as a metaphor, theory, set of capacities, and strategy for disaster readiness*, Fran H Norris, Susan P Stevens, Betty Pfefferbaum, Karen F Wyche, Rose L Pfefferbaum, American Psychological Association (Ed.), American Journal of Community Psychology, Volume 41, Issue 1-2, 1973; permission conveyed through Copyright Clearance Center, Inc.). With this, the authors highlight the understanding of resilience as a process, in which someone or something—any actor or set of actors—actively links capacities (properties, resources,..) in a purposeful way.

Lorenz (2013) and Saja et al. (2021), in social resilience literature, both stress the importance of participative capacities for resilience building, but mix the terms. Lorenz (2013) defines participative capacity as a measure of the ability of a social system to self-organize and use its adaptive and coping capacities (Lorenz 2013, p. 17). According to Lorenz, however, participative capacity is "a unique property of social systems: it manifests within social systems and between them (social–social systems)." (Lorenz 2013, p. 17 © 2010 Springer Nature. Reproduced with permission from SNCSC). Lorenz (2013) and Saja et al. (2021) define participative, or transformative capacities respectively, as capacities of self-organisation and participation in pre-disaster decision-making processes (Saja et al. 2021, p. 795 f).

In the field of community resilience, Pfefferbaum et al. (2013) outline that resilience-supporting resources, just like other types of knowledge, can be acquired and developed, and resilience-enhancing skills can be learned and practiced—individually and collectively. If we want to evaluate and monitor resilience, these resources can be measured for one point in time—we can draw e.g. on indicators for citizen participation. But there is a difference between the characteristics of a community as is exists today, and the change it is subject to until tomorrow, if it is exposed to threats and faced with the need for making decisions. This means that participative or transformative capacities have to be built up and practiced to give them a positive direction towards resilience—built up and practiced before, to be applied in, times of disaster. When an adverse event occurs, they to be already in existence.

In the face of time-sensitive decisions, the process of actively linking resources and capacities towards resilience can be supported (and accelerated) by ICT applications, including AI systems. "We believe that the focus on capabilities and process is especially useful when debating ICT and resilience. To be resilient, a community depends on infrastructures, the capabilities to use it (..) and a continuous process to turn the learning enabled through ICT into adaptation and transformation." (Comes et al. 2019, p. 3 © 2017 Informa UK Limited, trading as Taylor & Francis Group. Reprinted by permission of the publisher, Taylor & Francis Ltd.). A good example of how an ICT-supported process can enhance resilience is shown by Čaušević et al. (2021). The authors define power systems as consisting of three layers: the physical, the ICT and the social layer. They analyse interventions based on self-organization in and between these layers, and explain how this can contribute to lessening the consequences of power outages. In particular, a self-determined distribution of local energy resources where producers, prosumers and consumers can decide for themselves in what way to prioritise and allocate resources (in contrast to existing techniques, where loads are prioritised by a central authority), can be used beneficially beforehand and after an outage. In cases where data are unavailable, a machine learning technique, federated learning, can be used to forecast demand, supply, and flexibilities for balancing the grid (Čaušević et al. 2021).

The example shows opportunities as well as ethical and democratic challenges. Though fast and useful in face of disaster, an ICT-enabled process would require some preconditions. Which resources to prioritise and allocate, and where to, are political decisions that have to be made in accordance with democratic rules. Most of them—e.g., in which case, to what extent and in what time frame a self-determined distribution can take place—will have to be specified on a national or EU-level. Local and regional authorities might then further design the practice.

We can assume that capacities of self-organization and self-determination do not arise by themselves—that they have to be learned and practiced continuously. If meant to enhance resilience, they should be well-informed, inclusive, and characterised by learning—aspects we best provide for by applying the AD process for monitoring and evaluation.

8.3.2 Energy System Modelling

In energy system analysis, we usually employ models to evaluate the change of certain parameters, or the system as a whole, over time—e.g. of prices, critical materials or other resources, under different framework conditions. In order to be calculable, the model parameters need to be largely quantitative—whereas the framework conditions are at first qualitative descriptions of possible future developments, based on expert opinions.

As for the evaluation of the resilience of a system, even this complementation of narrative elements to models does not seem sufficient due to the models´ focus on technical and financial aspects, while placing less emphasis on social, cultural, or psychological drivers. This is also true for ReMoDigital, where time- and cost-optimizing models are employed—although we know that resilience is enhanced by abundance rather than efficiency.[3] In ReMoDigital, the problem is resolved to a certain point with the use of stress tests: the models analyse system behaviour under stress, like extreme weather or hacker attacks, in different context scenarios. Their output can then be compared in terms of degree of system functionality and therefore provides us with some information about the parameters of a more or less resilient energy system.

The context scenarios that are linked to the models are narratives—qualitative storylines of plausible socio-political-economic future pathways, that serve as input assumptions for the models. According to Hirt et al. (2020), three strategies exist for linking models and transitions theories (Hirt et al. 2020; Trutnevyte et al. 2019):

- Iterating: Storylines serve as inputs for models (define input assumptions)
- Merging: Storylines are integrated into the model
- Bridging: Storylines and models run in parallel and interact only at certain “bridges”, which requires finding “common ground” (Hirt et al. 2020, pp. 172–174 © 2020 Elsevier B.V. All rights reserved. Reprinted from Environmental Innovation and Societal Transitions, Volume 35, Léon F. Hirt, Guillaume Schell, Marlyne Sahakian, Evelina Trutnevyte, *A review of linking models and socio-technical transitions theories for energy and climate solutions*, 162–179, 2020, with permission from Elsevier). (Fig. 8.3).

In ReMoDigital, the cross-impact balances method was used.[4] It precedes the analysis and feeds the models with input assumptions. It was not repeated due to the short project duration and because of the project´s focus on technological energy system analysis—but in theory it could be. We can subsume it under the first, iterating, strategy, following the categories of Hirt et al. (2020).

One way to address the shortcomings of techno-economic, optimising models would be to apply other types of models too, such as agent-based models, that more explicitly focus on social, relational, or behavioural factors (cf. Krumm et al. 2022; Li et al. 2015; Trutnevyte et al. 2014, 2019). Roege et al. (2014) for example, distinguished physical, information, cognitive, and social aspects of energy systems and suggested metrics (Roege et al. 2014, p. 5 f). This would align with the second, merging, strategy of (Hirt et al. 2020).

Two problems arise with this wider view on the subject on the part of the modellers. By applying diverse models and parameters, where do we find the information, knowledge and data and who is to evaluate it? If we broaden the boundaries or integrate more diverse, qualitative data, our system gets even more complex. New individual and collective actors respectively agents come up. (On their side, in the

[3] See Chap. 2 in this book.

[4] See Chap. 4 in this book.

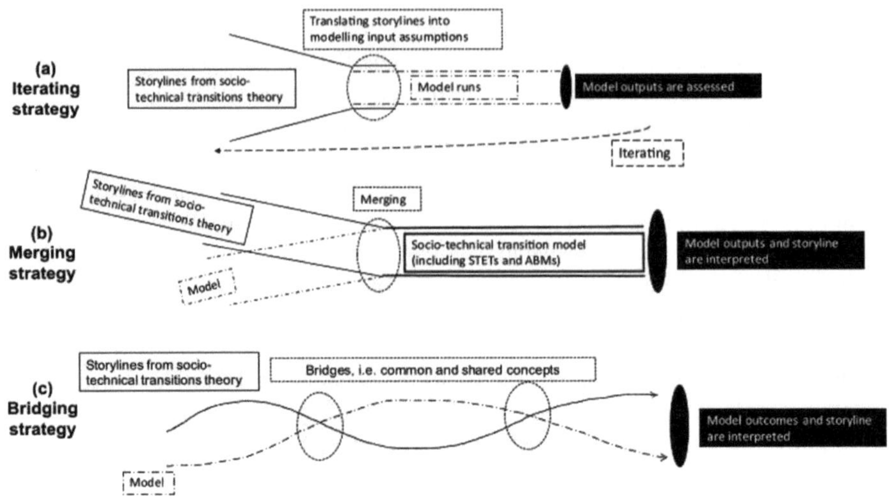

Fig. 8.3 Methodological strategies for applying models and analytical frameworks, and linking them. Figure taken from: (Hirt et al. 2020 © 2020 Elsevier B.V. All rights reserved. Reprinted from Environmental Innovation and Societal Transitions, Volume 35, Léon F. Hirt, Guillaume Schell, Marlyne Sahakian, Evelina Trutnevyte, A review of linking models and socio-technical transitions theories for energy and climate solutions, 162–179, 2020, with permission from Elsevier)

real world, questions might come up too on matters of resilience: where do we find information and data and who is to analyse it?). Both modellers and the public have to find answers to difficult, sometimes ethical questions regarding energy resilience, such as—in case of shortages—what is a minimum system functionality, a fair allocation of resources, or can we agree upon a prioritisation of certain consumers, and which? These questions can only be raised, and answers found, in collaboration with each other. It requires finding common ground, a common language—which needs to be learned and practiced. This would fit well with the characteristics of the analytic-deliberative process, where analysis and deliberation support and inform each other in raising publicly relevant questions, framing and structuring dialogue, and making methodological value judgements. It would therefore add to build social resilience, especially people´s participative capacity. This would match with the bridging strategy, where models and storylines run side by side and meet at certain "bridges"—points in time, at which they need to communicate to inform and learn from each other (Hirt et al. 2020).

Within such a process, models and storylines would be in a dialogue with stakeholders, respectively the public, and adjust to them. Webler and Tuler (2021) also emphasise the learning aspect of an analytic-deliberative process: "AD does not advocate that scientists only do analysis or that political actors be limited to deliberation (Webler and Tuler 1998). This has implications for public participation. To the degree that it is reasonable, both need to participate reciprocally in all activities. Consequently, learning is inherent as scientists learn to deliberate and non-scientists learn forms of analysis. The model also recognizes that policymaking is a step-wise process, and that learning is inherently iterative." (Webler and Tuler 2021, p. 508 © 2018, Society for Risk Analysis).

Our view is supported by recent research on co-producing energy strategies that highlights "the need for co-production/ action research efforts that can support real-world transition processes and provide a better understanding of local contexts as an alternative to efforts that would seek to simply improve societal representations within energy system models." (McGookin et al. 2022, p. 1). We therefore suggest to place analytic findings into a participative process where they not only can be informed, but repeatedly re-evaluated, reflected and reviewed by stakeholders, respectively interested and affected parties (see 2.3).

8.3.3 Local Situatedness and Local Learning

The understanding of local contexts is essential for all involved parties. The involvement of stakeholder groups, however, varies on different spatial levels according to a review of scientific papers on participative methods in energy system modelling and planning by McGookin et al. (2021): At the national level, academic and expert actors dominate; while on the subnational level, much more civil society organisations and other non-expert actors were engaged, with larger focus on understanding local perceptions and priorities (McGookin et al. 2021) (Fig. 8.4).

In their case study on forest carbon management, St-Laurent et al. (2020) developed recommendations that "(..) highlight the importance of involving multiple actors, views and worldviews. We also emphasize the importance of inclusive deliberation that is based on the best available science, but also on other forms of expertise,

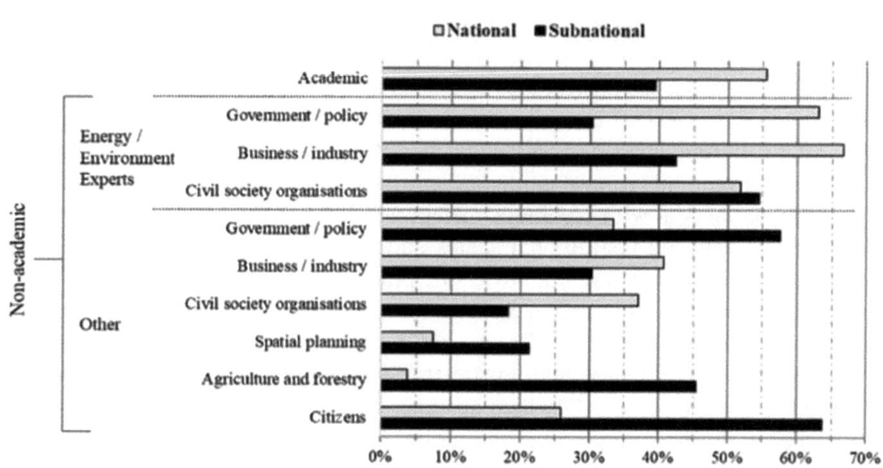

Fig. 8.4 Range of different stakeholders by share of papers that involved each group. Diagram from McGookin et al. (2021)

including lay and traditional knowledge. Perhaps most importantly, our recommendations are consistent with others who call for opening-up analysis, deliberation and appraisal in participatory engagement. This means acknowledging that a one-size-fits-all solution does not always exist, but rather that plural and conditional policy options are often more advised in the context of complex environmental issues." (St-Laurent et al. 2020, p. 1). This argument against the distribution of best practices, and for individual, specific appraisals that involve diverse local stakeholders, also holds true for the evaluation and monitoring of energy resilience, where social factors play crucial parts.

In which way assessments and pathways elaborated with local stakeholders can be upscaled to higher spatial levels is unclear, however. In their study on catchment planning in Australia, Folke et al. (2010) found that "Deliberate transformational change can be initiated at multiple scales, and perhaps gradually, as suggested by recent experience with applying resilience thinking to catchment planning (..). Deliberate transformational change at the scale of the whole catchment, of all the component parts at the same time, is likely to be too costly, undesirable or socially unacceptable. Transformational changes at lower scales, in a sequential way, can lead to feedback effects at the catchment scale, which is a learning process, and facilitate eventual catchment-scale transformational change. Actors and organizations that bridge the local to higher social–ecological scales are often involved in such processes (..)." (Folke et al. 2010, p. 5).

If one applies the adaptive cycle and panarchy framework by Gunderson and Holling (2002) and links the levels of the framework to spatial scales, local initiatives for decision making in matters of energy resiliency would relate to the small-scale levels of a nested, hierarchical system. According to Gunderson and Holling, each level of the panarchic system, from small to large, is subject to a continuous adaptive cycle of mutations and rearrangements (Gunderson and Holling 2002)—a process that unfolds within each level, but also between them. This exact dynamic might also work for the upscaling of local or grassroots approaches to higher political levels. Local approaches and initiatives can then be understood as creating diverse, evolutionary approaches.[5] Even if not broadly applicable or transferable to another place, very different, instead of only a few, best practices should then be encouraged.

But what would this re-evaluation and up-dated assessment mean in actual practice? How do we deal with continuous adaptive cycles of mutations and rearrangements in and between institutional scales and including grassroot units? What is theoretically wished for may not always be practically achievable within restricted resources and ongoing democratic processes about other development goals (see e.g. the complex issues related to the SDGs). Funding of scientific projects and assessment is always limited in time and seldom stretches beyond their lifetime. One may have to settle for a pragmatic proxy which absorbs the main characteristics of the above insights, assesses the current status and proposes further actions to strengthen social resilience. This could for instance be a select group composed of a small group of stakeholders—scientific experts, industry, and NGOs—which updates information

[5] See Chap. 3 in this book.

on relevant community activities and assesses contributions to social resilience in the energy sector on a bi-annual basis. Experiences from shocks and other disturbances (as e.g. discussed in Kaiser et al. 2021) would be integrated in updated assessments and recommendations to policy.

8.3.4 Synthesis: AD Approach for Energy System Resilience

A lot of studies cover the appropriateness of specific models, or mix of models, for specific purposes (e.g. Lopion et al. 2018; Ringkjøb et al. 2018), including ABMs (Klein et al. 2019), or call for the integration of social and environmental factors into scenario and model building (Krumm et al. 2022; Süsser et al. 2022)—see also (Trutnevyte et al. 2014) who link a UK-wide, highly detailed storyline for the power system transition until 2050 with multiple models.

However, recent literature indicates that analysis, respectively scenario building through models alone does not provide the best answers in terms of quality and value representation (Baard 2021; Venturini et al. 2019; Xexakis et al. 2020). (Xexakis and Trutnevyte 2022) illustrate this with model-based electricity supply scenarios, which are not aligned with citizens′ perspectives. The authors highlight that stakeholder and public engagement would enrich model-based scenario ensembles—an approach that gains increasing momentum in energy system modelling.

State-of-the-art scenario-building often combines qualitative narratives of storylines with quantitative models. These storylines are developed through the engagement of experts and stakeholders, but the approaches (in terms of the extent and purpose of the participation) vary considerably. Hirt et al. (2020) describe the different methodological strategies to link models and storylines, resp. socio-technical transition theories, for energy and climate solutions: In the iterating strategy, narrative storylines serve as input assumptions for the models; in the merging strategy, storylines are integrated into the models, which then also cover social and behavioural aspects. For both strategies, models are the central analytical elements; narratives serve and frame them. In contrast to that, the third, bridging, strategy, where models and frameworks run in parallel, perceived as independent from each other, only interact at certain bridges. At these bridges, multiple actors, views and worldviews can be involved; a participative process that involves a wider public can unfold. The equal value of models, narrative frameworks and bridges in a participative process where—in addition to expert opinions—a wider public can be involved, is essential when we think of their informing and legitimating power (cf. Dietz and Stern 2008). Within this process, and by applying a deliberative approach (National Research Council 1996; Renn 2005), social capacities for energy resiliency, especially participative or transformative capacity, would be increased.

8.4 Conclusion

There is a growing interest in, and need for, participative approaches for energy system modelling to increase the quality of model-based outputs. In contrast to this, only a few examples of well-structured, participative processes exist (McGookin et al. 2024). One reason might be that models are perceived as too complicated and impractical and not fit for public, non-expert participation (Ben Amer et al. 2020). At times, very different perceptions and expectations of stakeholders have to be taken into account. Göke et al. (2023, p. 1) for example, show that "to impact decision-making, scenarios do not have to be accurate—but credible and evoke coherent expectations in diverse stakeholders". (Reprinted from Energy Research & Social Science, Volume 97, Leonard Göke, Jens Weibezahn, Christian von Hirschhausen, A collective blueprint, not a crystal ball: How expectations and participation shape long-term energy scenarios, 102957. © 2023, with permission from Elsevier.)

For this purpose, digital support tools might be of use, that have been proved effective for local-scale energy transitions (Hewitt et al. 2020). However, there is evidence that the longer-term effectiveness of these tools remains limited: "In terms of methodologies in decision science, our factsheets, Riskmeter, and group discussions all proved effective in forming the preferences and improving knowledge, but we also intriguingly found that in a longer run the participants tended to revert back to their initial opinions. The latter finding opens up multiple new research questions on the longer-term effectiveness of informational tools and stability of informed preferences." (Volken et al. 2018, p. 11,478) Reprinted with permission from Sandra P. Volken, Georgios Xexakis, Evelina Trutnevyte, Perspectives of Informed Citizen Panel on Low-Carbon Electricity Portfolios in Switzerland and Longer-Term Evaluation of Informational Material, Environmental Science & Technology 2018, 52, 20, 11478–11489. © 2018 American Chemical Society.

There are indications too that energy models and energy policy influence each other, instead of models informing decision-making (Süsser et al. 2021). Süsser et al. (2022) propose model transparency and co-creative, inclusive approaches with different stakeholders to counteract this.

The analytic-deliberative approach from environmental management with its character of participation and deliberation and a longer-term process therefore seems the most promising way to ensure value representation in energy system modelling and building up social capacities that are needed for enhancing resilience.

References

Alexander, D.E.: Resilience and disaster risk reduction: an etymological journey. Nat. Hazard. **13**(11), 2707–2716 (2013). https://doi.org/10.5194/NHESS-13-2707-2013

Allain, S., Plumecocq, G., Leenhardt, D.: Linking deliberative evaluation with integrated assessment and modelling: a methodological framework and its application to agricultural water management. Futures **120**, 102566 (2020). https://doi.org/10.1016/J.FUTURES.2020.102566

Baard, P.: Knowledge, participation, and the future: epistemic quality in energy scenario construction. Energy Res. Soc. Sci. **75**, 102019 (2021). https://doi.org/10.1016/j.erss.2021.102019

Bellamy, R., Chilvers, J., Pallett, H., Hargreaves, T.: Appraising sociotechnical visions of sustainable energy futures: a distributed deliberative mapping approach. Energy Res. Soc. Sci. **85**, 102414 (2022). https://doi.org/10.1016/j.erss.2021.102414

Bellocchi, G., Rivington, M., Matthews, K., Acutis, M.: Deliberative processes for comprehensive evaluation of agroecological models. A review. Agron. Sustain. Dev. **35**(2), 589–605 (2015). https://doi.org/10.1007/S13593-014-0271-0/FIGURES/7

Ben Amer, S., Gregg, J.S., Sperling, K., Drysdale, D.: Too complicated and impractical? An exploratory study on the role of energy system models in municipal decision-making processes in Denmark. Energy Res. Soc. Sci. **70**, 101673 (2020). https://doi.org/10.1016/J.ERSS.2020.101673

Čaušević, S., Saxena, K., Warnier, M., Abhyankar, A.R., Brazier, F.M.T.: Energy resilience through self-organization during widespread power outages. Sustain. Resilient Infrastruct. **6**(5), 300–314 (2021). https://doi.org/10.1080/23789689.2019.1666341

Comes, T., Meesters, K., Torjesen, S.: Making sense of crises: the implications of information asymmetries for resilience and social justice in disaster-ridden communities. Sustain. Resilient Infrastruct. **4**(3), 124–136 (2019). https://doi.org/10.1080/23789689.2017.1405653

Datola, G., Bottero, M., De Angelis, E.: Enhancing urban resilience capacities: an analytic network process-based application. Environ. Clim. Technol. **25**(1), 1270–1283 (2021). https://doi.org/10.2478/rtuect-2021-0096

Dietz, T.: Theory and method in social impact assessment. Sociol. Inq. **57**(1), 54–69 (1987). https://doi.org/10.1111/J.1475-682X.1987.TB01180.X

Dietz, T., Stern, P.C.: Public participation in environmental assessment and decision making. In: Public Participation in Environmental Assessment and Decision Making, pp. 1–305 (2008). https://doi.org/10.17226/12434

Dubois, A., Holzer, S., Xexakis, G., Cousse, J., Trutnevyte, E.: Informed citizen panels on the Swiss electricity mix 2035: Longer-term evolution of citizen preferences and affect in two cities. Energies **12**(22) (2019). https://doi.org/10.3390/EN12224231

Folke, C., Carpenter, S.R., Walker, B., Scheffer, M., Chapin, T., Rockström, J.: Resilience thinking: integrating resilience, adaptability and transformability. Ecol. Soc. **15**(4) (2010). https://doi.org/10.5751/ES-03610-150420

Göke, L., Weibezahn, J., von Hirschhausen, C.: A collective blueprint, not a crystal ball: how expectations and participation shape long-term energy scenarios. Energy Res. Soc. Sci. **97**, 102957 (2023). https://doi.org/10.1016/J.ERSS.2023.102957

Gollagher, M., Hartz-Karp, J.: The role of deliberative collaborative governance in achieving sustainable cities. Sustainability (Switzerland) **5**(6), 2343–2366 (2013). https://doi.org/10.3390/SU5062343

Graveline, M.H., Germain, D.: Disaster risk resilience: conceptual evolution, key issues, and opportunities. Int. J. Disaster Risk Sci. **13**(3), 330–341 (2022). https://doi.org/10.1007/S13753-022-00419-0

Gunderson, L.H., Holling, C.: Resilience and adaptive cycles. In: Gunderson, L.H., Holling, C.S. (eds.) Panarchy: Understanding Transformations in Human and Natural Systems, pp. 25–62 (2002)

Hewitt, R.J., de Boer, C., Flacke, J.: Participatory development of digital support tools for local-scale energy transitions: lessons from two European case studies. Glob. Transit. **2**, 138–149 (2020). https://doi.org/10.1016/J.GLT.2020.07.003

Hirt, L.F., Schell, G., Sahakian, M., Trutnevyte, E.: A review of linking models and socio-technical transitions theories for energy and climate solutions. Environ. Innov. Soc. Trans. **35**, 162–179 (2020). https://doi.org/10.1016/J.EIST.2020.03.002

Kaiser, M.: The idea of a theory of values and the metaphor of value-landscapes. Humanit. Soc. Sci. Commun. **11**(1), 1–10 (2024). https://doi.org/10.1057/s41599-024-02749-4

Kaiser, M., Buklijas, T., Gluckman, P., Matthias Kaiser, P.: Models and numbers: representing the world or imposing order? (2021). https://arxiv.org/abs/2104.00029v1
Klein, M., Frey, U.J., Reeg, M.: Models within models—agent-based modelling and simulation in energy systems analysis. JASSS **22**(4) (2019). https://doi.org/10.18564/JASSS.4129
Krumm, A., Süsser, D., Blechinger, P.: Modelling social aspects of the energy transition: what is the current representation of social factors in energy models? Energy **239**, 121706 (2022). https://doi.org/10.1016/J.ENERGY.2021.121706
Li, F.G.N., Trutnevyte, E., Strachan, N.: A review of socio-technical energy transition (STET) models. Technol. Forecast. Soc. Chang. **100**, 290–305 (2015). https://doi.org/10.1016/J.TECHFORE.2015.07.017
Lopion, P., Markewitz, P., Robinius, M., Stolten, D.: A review of current challenges and trends in energy systems modeling. Renew. Sustain. Energy Rev. **96**, 156–166 (2018). https://doi.org/10.1016/J.RSER.2018.07.045
Lorenz, D.F.: The diversity of resilience: Contributions from a social science perspective. Nat. Hazards **67**(1), 7–24 (2013). https://doi.org/10.1007/s11069-010-9654-y
McGookin, C., Mac Uidhir, T., Ó Gallachóir, B., Byrne, E.: Doing things differently: bridging community concerns and energy system modelling with a transdisciplinary approach in rural Ireland. Energy Res. Soc. Sci. **89**, 102658 (2022). https://doi.org/10.1016/J.ERSS.2022.102658
McGookin, C., Ó Gallachóir, B., Byrne, E.: Participatory methods in energy system modelling and planning – a review. Renew. Sustain. Energy Rev. **151**, 111504 (2021). https://doi.org/10.1016/J.RSER.2021.111504
McGookin, C., Süsser, D., Xexakis, G., Trutnevyte, E., McDowall, W., Nikas, A., Koasidis, K., Few, S., Andersen, P. D., Demski, C., Fortes, P., Simoes, S. G., Bishop, C., Rogan, F., Ó Gallachóir, B.: Advancing participatory energy systems modelling. Energy Strat. Rev. **52**, 101319 (2024). https://doi.org/10.1016/J.ESR.2024.101319
Miles, S.B.: Participatory disaster recovery simulation modeling for community resilience planning. Int. J. Disaster Risk Sci. **9**, 519–529 (2018). https://doi.org/10.1007/s13753-018-0202-9
National Research Council: Understanding risk: informing decisions in a democratic society. In: Understanding Risk. National Academies Press (1996). https://doi.org/10.17226/5138
Norris, F.H., Stevens, S.P., Pfefferbaum, B., Wyche, K.F., Pfefferbaum, R.L.: Community resilience as a metaphor, theory, set of capacities, and strategy for disaster readiness. Am. J. Community Psychol. **41**(1–2), 127–150 (2008). https://doi.org/10.1007/S10464-007-9156-6
Pfefferbaum, R.L., Pfefferbaum, B., Van Horn, R.L., Klomp, R.W., Norris, F.H., Reissman, D.B.: The communities advancing resilience toolkit (CART): an intervention to build community resilience to disasters. J. Public Health Manag. Pract. **19**(3), 250–258 (2013). https://doi.org/10.1097/PHH.0B013E318268AED8
Renn, O.: Risk governance - towards an integrative approach. IRCG Risk Governance Framework; IRGC Risk Governance Framework - IRGC (2005). https://irgc.org/risk-governance/irgc-risk-governance-framework/
Renn, O.: White Paper on Risk Governance: Toward an Integrative Framework, pp. 3–73 (2008). https://doi.org/10.1007/978-1-4020-6799-0_1
Ringkjøb, H.K., Haugan, P.M., Solbrekke, I.M.: A review of modelling tools for energy and electricity systems with large shares of variable renewables. Renew. Sustain. Energy Rev. **96**, 440–459 (2018). https://doi.org/10.1016/J.RSER.2018.08.002
Roege, P.E., Collier, Z.A., Mancillas, J., McDonagh, J.A., Linkov, I.: Metrics for energy resilience. Energy Policy **72**, 249–256 (2014). https://doi.org/10.1016/J.ENPOL.2014.04.012
Saja, A.M.A., Teo, M., Goonetilleke, A., Ziyath, A.M.: A critical review of social resilience properties and pathways in disaster management. Int. J. Disaster Risk Sci. **12**(6), 790–804 (2021). https://doi.org/10.1007/S13753-021-00378-Y/FIGURES/5
St-Laurent, G.P., Hoberg, G., Sheppard, S.R.J., Hagerman, S.M.: Designing and evaluating analytic-deliberative engagement processes for natural resources management. Elementa **8**, 8 (2020). https://doi.org/10.1525/elementa.402

Süsser, D., Ceglarz, A., Gaschnig, H., Stavrakas, V., Flamos, A., Giannakidis, G., Lilliestam, J.: Model-based policymaking or policy-based modelling? How energy models and energy policy interact. Energy Res. Soc. Sci. **75**, 101984 (2021). https://doi.org/10.1016/J.ERSS.2021.101984

Süsser, D., Martin, N., Stavrakas, V., Gaschnig, H., Talens-Peiró, L., Flamos, A., Madrid-López, C., Lilliestam, J.: Why energy models should integrate social and environmental factors: assessing user needs, omission impacts, and real-word accuracy in the European Union. Energy Res. Soc. Sci. **92**, 102775 (2022). https://doi.org/10.1016/J.ERSS.2022.102775

Trutnevyte, E., Barton, J., O'Grady, Á., Ogunkunle, D., Pudjianto, D., Robertson, E.: Linking a storyline with multiple models: a cross-scale study of the UK power system transition. Technol. Forecast. Soc. Chang. **89**, 26–42 (2014). https://doi.org/10.1016/J.TECHFORE.2014.08.018

Trutnevyte, E., Hirt, L.F., Bauer, N., Cherp, A., Hawkes, A., Edelenbosch, O.Y., Pedde, S., van Vuuren, D.P.: Societal transformations in models for energy and climate policy: the ambitious next step. One Earth **1**(4), 423–433 (2019). https://doi.org/10.1016/J.ONEEAR.2019.12.002

UNDRR. Resilience (n.d.). https://www.undrr.org/terminology/resilience. Accessed 31 Aug 2024

Venturini, G., Hansen, M., Andersen, P.D.: Linking narratives and energy system modelling in transport scenarios: a participatory perspective from Denmark. Energy Res. Soc. Sci. **52**, 204–220 (2019). https://doi.org/10.1016/J.ERSS.2019.01.019

Volken, S.P., Xexakis, G., Trutnevyte, E.: Perspectives of informed citizen panel on low-carbon electricity portfolios in Switzerland and longer-term evaluation of informational materials. Environ. Sci. Technol. **52**(20), 11478–11489 (2018). https://doi.org/10.1021/ACS.EST.8B01265

Webler, T., Tuler, S.: How to do environmental decision making: varying perspectives on the U.S. National Research Council's understanding risk report. Hum. Ecol. Rev. **5**(1) (1998). https://cedar.wwu.edu/envs_facpubs/9

Webler, T., Tuler, S.: Four decades of public participation in risk decision making. Risk Anal. **41**(3), 503–518 (2021). https://doi.org/10.1111/RISA.13250

Webler, T., Tuler, S., Dow, K., Whitehead, J., Kettle, N.: Design and evaluation of a local analytic-deliberative process for climate adaptation planning. Local Environ. **21**(2), 166–188 (2016). https://doi.org/10.1080/13549839.2014.930425)

Xexakis, G., Hansmann, R., Volken, S.P., Trutnevyte, E.: Models on the wrong track: model-based electricity supply scenarios in Switzerland are not aligned with the perspectives of energy experts and the public. Renew. Sustain. Energy Rev. **134**, 110297 (2020). https://doi.org/10.1016/J.RSER.2020.110297

Xexakis, G., Trutnevyte, E.: Model-based scenarios of EU27 electricity supply are not aligned with the perspectives of French, German, and Polish citizens. Renew. Sustain. Energy Transit. **2**, 100031 (2022). https://doi.org/10.1016/J.RSET.2022.100031

Chapter 9
Resilience in the System of Infrastructure Law

Martin Wickel

Abstract Resilience is of increasing importance in the development and operation of technical infrastructure. In this chapter, we show how resilience can be embedded in the structure of infrastructure law. We analyse substantive law as the legal basis for responding to resilience requirements. And we look at the legal instruments for implementing these substantive requirements. The example we have chosen to explore this issue is electricity infrastructure. So, we look at electricity generation plants, transmission lines and electricity networks. Due to the heterogeneous structure of infrastructure law, different areas of law such as energy law, emission control law and the law of planning approval will be covered. Different substantive requirements, such as dynamic basic obligations (emission control law) or the NOVA principle (energy law) will be addressed. In the same way, different instruments of implementation, such as permit requirements, ex-post interventions, monitoring, planning and procedures will be analysed. The analysis is guided by the question of whether these legal concepts are able to accommodate the elements of resilience, in particular resistance, robustness and adaptability.

Keywords Infrastructure law · Resilience · Emission control law · Energy law · Planning approval

9.1 Introduction

In this chapter we investigate how the concept of resilience can be embedded in infrastructure law, using specific parts (see Sect. 9.3.) of the electricity supply infrastructure as an example. For the purpose of this chapter infrastructure encompasses only technological (hard) infrastructure thereby excluding social (soft) infrastructure. The regulation of different kinds of technological infrastructure share many characteristics, whereas the regulation of social infrastructure (e.g. health, child care, housing) follow other rationalities and requirements. Since different legal systems each have

M. Wickel (✉)
HafenCity Universität Hamburg, Hamburg, Germany
e-mail: martin.wickel@hcu-hamburg.de

D. van Doren et al. (eds.), *Towards Energy System Resilience*, Ethics of Science and Technology Assessment 55, https://doi.org/10.1007/978-3-032-03187-7_9

their own characteristics, the chapter will focus on German law, specifically federal law, since the power to legislate in the relevant areas—energy law, air pollution control—lies at the federal level (Art. 70 et seq. basic law (*Grundgesetz—GG*)).

The analysis does not address the question of which requirements an infrastructure must meet to be resilient and whether these are fulfilled. Often these requirements are the subject of detailed technical regulations.[1] Rather this chapter will examine which structural features of infrastructure law the legislator and the regulator should concentrate on in regulating for resilience.

Infrastructure law is not a systematically coherent field of law (Schulze-Fielitz 2015, 9 et seq.). There is no "infrastructure code" laying out general principles and requirements or creating instruments applicable to all infrastructure. Infrastructure is often subject to legal requirements originating in different fields of law. This is true even if the analysis is limited to one infrastructure system, such as electricity supply infrastructure in this case (Franzius 2022, 302). Accordingly, consideration of the question how the concept of resilience can be embedded in infrastructure law requires consideration of various fields of law. In the case of electricity supply this includes energy law, emission control law, and the law of sectoral planning (*Planfeststellung*). The laws governing electricity supply and infrastructure provision are generally characterized by a high degree of heterogeneity. The different applicable laws pursue different objectives, have different legislative histories and originate from different times. This means that they have different structures, use different instruments and follow different rationales.

9.2 Anchoring Resilience in the Structure of Infrastructure Law

Renn, Kaiser & von Gleich have shown that there is no single definition of resilience (See Chap. 2 in this book (Sect. 2.1)). Referring to the International Risk Governance Council they describe resilience "as a normative goal for risk management systems to deal with highly uncertain events or processes (surprises)" (See Chap. 2 in this book (Sect. 2.1)). It can be categorized as a facet of the precautionary principle (See Chap. 5 in this book (Sect. 5.1)). Recently the EU has issued the Directive (EU) 2022/2557 on the resilience of critical entities (CER-Directive). According to Art. 2 "'resilience' means a critical entity's ability to prevent, protect against, respond to, resist, mitigate, absorb, accommodate and recover from an incident". It is likely that this definition will exert major influence in the legal field, at least in Europe. For example, the draft of an umbrella law for strengthening the physical resilience of operators of critical facilities in Germany (Dachgesetz zur Stärkung der physischen Resilienz

[1] For a comprehensive overview of the necessary requirements and relevant design principles, see Chap. 2 in this book (Sect. 2.3.3).

von Betreibern kritischer Anlagen—KRITIS-DachG[2]) has adopted this definition in Section 2. However, this chapter is based on an understanding of resilience that goes further than the CER-Directive in including aspects of adaptivity and transformation as it is developed by Renn, Kaiser and von Gleich earlier in this book (See Chap. 2 in this book (Sect. 2.1)).

The starting point of this analysis shall be three characteristics that have been described as core elements of resilience: resistance, robustness, and adaptivity.[3] Referring to the diagram of Kröger (See Chap. 2 in this book (Sect. 2.3.3)), resistance and robustness represent the capability to absorb a shock and to recover. After the disrupting event the infrastructure system returns to its original state. Adaptivity goes further. Here it may be necessary to change the system. Adaptivity might have an evolutionary facet. The system bounces forward.

As already mentioned, this chapter does not try to identify the requirements of resilience for electricity infrastructure, but focuses on how to accommodate those requirements in the structure of the law, namely the legal provisions governing infrastructure. The guiding question is how the named elements of resilience (resistance, robustness and adaptivity) can be embedded in infrastructure law. For example, the overall goal of the project is to explore how digital solutions can be used to facilitate greater resilience. In this context, the question becomes: how can law be used to ensure that the respective digital solutions are applied, and which parts of the current infrastructure law need to be addressed? This chapter considers different areas of the regulative landscape. The analysis follows the distinction between substantive law on the one hand and regulatory instruments for implementing substantive law on the other. We will first describe how these parts of the regulatory landscape can be used to facilitate resilience in general. We will then turn to specific parts of the infrastructure and their regulation. The distinction between substantive law and instruments for its implementation will again structure the analysis.

9.2.1 Resilience in Substantive Law

At the outset, the three core elements of resilience present themselves as normative requirements for substantive law. For example, substantive law describes the technical requirements that must be met by the parts of the infrastructure or the obligations that must be fulfilled by the operators. Drawing on the case of electricity supply, resilience requires that electricity networks and their infrastructure components must be designed such that they are resistant, robust, and adaptable. Technical specifications, such as reliability (e.g. n-1-principle), are best formulated as substantive rules.

---

[2] https://www.bmi.bund.de/SharedDocs/gesetzgebungsverfahren/DE/Downloads/referentenentwuerfe/KM4/KRITIS-DachG-2.pdf;jsessionid=3704A1CB124D2B0EA856A967CD9E20E7.live892?__blob=publicationFile&v=4.

[3] See Chap. 2 in this book (Sect. 2.1).

One question that needs to be answered in the field of substantive law is the relationship between resilience and its elements and the system of danger prevention and precaution. The dichotomy is described by Renn (See Chap. 5 in this book). Averting danger is a classic administrative task. As a rule, it requires sufficiently reliable knowledge about the potentially damaging event and the extent of the possible damage (cf. Ramsauer 2018, para 33). Even if uncertainties make it impossible to determine probabilities, the precautionary principle may be used to justify measures to be taken. These measures may, for example, serve to minimize risks. However, any such preventative action by the state is constrained by a requirement of proportionality to the threat of harm. Thus, precautionary measures are more difficult to justify than measures taken to prevent imminent danger in terms of proportionality (cf. Ramsauer 2018, para 35).

If the damaging event is known and its probability can be determined, the requirement to absorb a shock and recover from it can generally be assigned to the area of danger prevention. If the probability is not known, the precautionary principle may have to be invoked (Kloepfer 2016, Section 4 para 43). In this context, therefore, danger prevention and precaution may be seen as a continuum in which the precautionary principle serves to enable measures to be taken despite low probabilities that would not be justified from the point of view of danger prevention. In this context, the question of adaptation would generally not arise, since in the case of foreseeable events and damage, it should be possible to design the facilities from the outset in such a way that the resilience requirements are met.

The picture is different if the damaging event and the damage are not foreseeable or the probabilities cannot be determined. Requirements are then essentially assigned to the area of precautionary measures. Renn describes the connection between precaution and the avoidance of irreversibility.[4] In this situation, adaptivity becomes important because this element of resilience can facilitate reversibility.

Resilience is not static but should encompass a dynamic element. In particular, adaptivity requires that electricity networks or their components must be modified with the times as the relevant circumstances change. Just as the networks and their components must not remain in their original state, so too should substantive law be dynamic and adaptive. This highlights the importance of coordinating substantive requirements with other instruments of the law, with particular emphasis on procedural requirements, monitoring and evaluation.

9.2.2 Instruments for Implementing the Elements of Resilience

Substantive law is not usually self-executing. Although the operators of networks or their components are bound by the applicable law, in most cases regulatory authorities

[4] See Chap. 5 in this book (Sect. 5.5).

are responsible for implementation and enforcement. Therefore, infrastructure law must provide for instruments that can facilitate implementation.

The range of possible instruments is wide. They can be distinguished according to the underlying mechanism used to reach the relevant policy goals. Among others, market-based economic instruments are typically distinguished from regulatory instruments described as 'command-and- control'. The former use economic incentives to facilitate the desired behaviour. The latter concentrate on setting legally enforceable standards. This chapter concentrates on regulatory instruments. This does not mean that there are no economic instruments in the field of energy law. On the contrary, the decarbonization of electricity generation is essentially controlled by a 'cap-and-trade' system, the EU ETS, based on Directive 2003/87/EC establishing a system for greenhouse gas emission allowance trading within the Union. However, with respect to electricity supply, i.e. networks and their components, resilience requires at least in part the definition of detailed requirements and standards that allow little leeway in their implementation. Here, we assume that regulatory instruments are best suited to provide for this high degree of exactness in achieving performance standards in the electricity system.

In the following sections, we first focus on the tools of permit requirements (1.), ex-post intervention (2.), and monitoring (3.), as part of the command-and-control mechanisms. We then turn to planning, since substantive requirements are often formulated in plans (4.). And finally, we look at procedural requirements that often accompany in particular permit requirements and plans. (5.).

9.2.2.1 Preventive Control: Permit Requirements

Taking a closer look at the portfolio of regulatory instruments, two kinds of instruments seem particularly suitable to enhance resilience. First, preventive control comes to mind. One example is the requirement to apply for a permit for the construction and operation of an electricity network or its components, like generation facilities or transmission lines. A permit requirement seems particularly well suited to support the resistance and robustness elements of resilience, at least to the extent that these can be part of the initial design of the network or its components. Of course, that does not preclude questions of resistance and robustness from also arising later during the operation of networks or their components. This may be the case when either technical standards advance or as a reaction to disruptions. In the latter case, the element of adaptivity once again comes to the fore.

9.2.2.2 Ex-Post Interventions

Disruptions that cannot be reasonably foreseen, are not taken into account in the initial design process. The same applies to the normal evolution of technical standards. The precautionary principle in particular requires that the facilities are at least to some extent open to change. To facilitate adaptivity, instruments are needed that can initiate alterations in networks and their components already in operation. It might be possible to define some contingent reactions to disruptions and the corresponding adaptation in the initial design process and control them in the permission process. Overall, however, the situation is likely to be characterized by considerable uncertainty. Therefore, from a legal point of view, it seems more important here to design and establish mechanisms to ensure that the permit system for the network and its components, once licensed and operational, is flexible enough to allow for adjustments and that these adjustments can be implemented through fair and effective procedures. The focus here is therefore on instruments of monitoring and ex-post intervention (see above).

Preventive control, as embodied in the requirement to apply for a permit, has advantages over ex-post intervention. First, if the electricity generation facility or transmission line is still in the design and permit stage, i.e. not yet built, it is easier to adapt the proposed facility or transmission line to enhance resilience. Further, any asserted legal claim by the developer is weaker ex ante, than for a network or components with a valid permit, fully developed, and in operation. The granting of a permit creates legal rights on the part of the licensee. The licensee has a legitimate—and constitutionally protected (Bundesverfassungsgericht 2010, Rn. 28)—expectation in favour of continuing lawful use and enjoyment of the permitted activity, here the operation of the network and its components. If interventions are subsequently ordered by the licensor that affect rights granted under the permit, this would require justification on the basis of proportionality (Jarass 2022, Section 6 para 56; Bundesverfassungsgericht (2010) Rn. 48). For similar reasons, new developments in technical standards are easier to implement during the planning and construction phase than attempts to mandate retrofitting or upgrades at a later stage. Second, the developer's motivations change. Prior to obtaining the permit, the developer has an interest in cooperating with the authorities that require resilience, as this is a condition for obtaining the permit that allows the project to be built. After the project is built, the developer's main interest is to operate without interference from ex-post interventions (Wickel 1996, 281).

Nonetheless, ex-post-interventions are an essential element of a resilience-enhancing legal system. In the case of a disruptive event, resistance and recovery are facilitated by the original permit. Since the operator has to run the installation according to the conditions of the permit, they are bound to return to the original state (recover). Adaptation, however, might require the installation to develop to a new level and this may require changes to the installation that are in conflict with the permit. Ex-post interventions are needed to modify the permit so that it allows for the new state of the installation.

9.2.2.3 Monitoring

Monitoring is part of the suite of enforcement tools available to the relevant authorities. Firstly, monitoring of electricity supply and demand is required to assess compliance with substantive law and a prerequisite for gathering sufficient evidence to justify any subsequent lawful intervention in networks and or their components. At the same time, it is part of general environmental monitoring (Kloepfer and Durner 2020, p. 106).

In terms of resilience, monitoring of current developments can play an essential role. Energy systems are currently transformed with the goal of decarbonization, primarily to mitigate climate change. Other goals include reducing the use of resources and protecting the environment. However, enhancing resilience is not a primary goal. In energy system transformation, the standards of resilience must be reconsidered. Therefore, it is necessary to see resilience as a separate subject of monitoring.

9.2.2.4 Planning

With respect to a network-based infrastructure resilience can only be analysed by looking at the components that are part of the network. At the same time, system-wide resilience is best facilitated and assessed through plans that coordinate the system as a whole. Planning is also an important part of infrastructure law (cf. Hermes 2022, p. 99). Plans can define resilience requirements for the whole network and describe how these requirements must be met by parts of the network. Therefore, plans describe preconditions for the granting of permits for generation facilities or transmission lines as well as situations that may lead to ex-post interventions.

Higher-level plans serve a central legal function because they integrate the required components of the networks and the requirements for the developer into an overall concept. Reference to overarching interests embedded in the plan can help to justify the need to build the components of the networks, e.g. generation facilities and transmission lines, on the one hand and the requirements for the specific components on the other. The construction of generation facilities and transmission lines may affect third parties' rights which must be justified according to the principles of the rule of law. Often the proportionality of the permit to construct and operate cannot be fully assessed by looking at the generation facilities or transmission line in isolation, if these also serve an essential function within the wider system. The same proportional approach applies to any attempts to impose mandatory preconditions on the issuance of a permit. Here the authorities often need a strong justification for otherwise refusing or restricting the permit, since the developer very often has a presumptive right to build.

The burden of proof is greater for any requirements sought to be implemented through ex-post interventions by regulatory authorities. Here the legal position of the developer is stronger—under statutory law as well as constitutional law (see above)—than at the permit application stage. Requirements that may be imposed on new generation facilities or transmission lines may not be imposed on existing ones. In order to demonstrate the proportionality of requirements that restrict the construction and operation of the components of the networks, overall concepts that make wider long-term interests visible and relate these interests to existing networks and their components are useful and sometimes even necessary. Therefore, it seems important to define and describe resilience requirements in higher-level plans as well.

9.2.2.5 Procedure

If we assume that public participation has resilience-enhancing effects,[5] then procedural law requirements come into view. In particular, the approval of generation facilities or transmission lines is often linked to a formal approval or permit procedure (e.g. Section 10 Federal Emission Control Act (*Bundes-Immissionsschutzgesetz—BImSchG*; Section 73 Administrative Procedure Act (*Verwaltungsverfahrensgesetz—VwVfG*). These procedures are not uniform across different fields of law, but they have similar structures. A key component of these procedures is public participation. The composition of the public may change during the process. While in the early stages the general public has access to the procedure, in some cases (e.g. Section 73 (4) VwVfG; Section 21 (1) Environmental Impact Assessment Act (*Umweltverträglichkeitsprüfungsgesetz—UVPG*)), but not all (e.g. Section 10(3) sentence 4 BImSchG), this is later narrowed down to the public actually affected by the proposed development. NGOs are often considered to be part of the affected public (e.g. Section 3(9) UVPG). A second key component is the participation of other authorities as representatives of public interests.

The procedure is characterized by different phases. Particularly important steps include that the proposal documents be publicly accessible through open inspection, that the members of the public have the opportunity to comment, and that a public hearing is held prior to an openly publicized decision. However, not all relevant public participation procedures follow the same rules.

Such procedures primarily serve to inform both the authority that has to make the decision and the public that will be ultimately affected by the decision (e.g. Section 4a(1) Federal Building Code (*Baugesetzbuch—BauGB*)). In addition, participation is considered to increase the public acceptance and legitimacy of the infrastructure projects as well as enhance the protection of affected rights (Erbguth 2018, 121).

In many cases, infrastructure projects and plans are subject to environmental impact assessment under the Environmental Impact Assessment Act (*Umweltverträglichkeitsprüfungsgesetz—UVPG*). The UVPG implements Directive 2011/92/EU on the assessment of the effects of certain public and private projects on the environment and Directive 2001/42/EC on the assessment of the effects of certain plans and programmes on the environment. Those laws do not directly mention resilience, but it can be argued that the effect of the assessments may also have resilience-enhancing effects (cf. Baldurson et al. 2024, p. 29).

[5] See Chap. 8 in this book; for different participatory approaches see Chap. 7 in this book.

9.2.3 Summary: Structural Elements of Infrastructure Law

So far, we have considered various structural elements of infrastructure law. We have shown that resilience elements should be formulated as substantive law requirements. However, this must be accompanied by the right set of instruments to facilitate the implementation of these requirements. In the next section we examine how these elements are represented in specific regulations governing different parts of electricity infrastructure.

9.3 Electricity Supply: The Relevant Parts of Infrastructure Law

When analysing electricity supply infrastructure, two levels must be considered: the network as a whole and the individual components that make up the network, particularly generation facilities and transmission lines. Depending on the level to be analysed, different areas of law come into focus. Generation facilities are usually subject to the BImSchG. High-voltage lines, even while regulated by the Energy Industry Act (*Energiewirtschaftsgesetz—EnWG*), are subject to sectoral planning law requiring planning approval (*Planfeststellung*). Finally, the networks as a whole are regulated under the EnWG.

This section will examine how elements of resilience can be embedded in the named regulatory regimes for generation facilities (Sect. 9.3.1), transmission lines (Sect. 9.3.2) and finally networks (Sec. 9.3.3). It should be noted that the infrastructure of electricity supply contains many more components; the ones selected here serve as examples. Therefore, this section does not aim to provide a complete picture of resilience in the electricity supply sector, but rather to highlight key features of regulation as relevant examples.

9.3.1 Generation Facilities

The regulatory regime that covers facilities that generate electricity is set forth by the BImSchG. Most facilities in question require a permit under the BImSchG according to the Fourth Ordinance under the BImSchG (*4. BImSchV*). Smaller facilities may be exempt from the permit requirement but are still regulated by the BImSchG. Additional permit requirements may arise under other laws, such as building permits according to each state's Building Acts (*Landes-Bauordnungen*).

9.3.1.1 Substantive Legal Requirements for Generation Facilities

For facilities that require a permit, the central substantive legal requirements of the BImSchG are the dynamic basic obligations of Section 5 BImSchG. These provide that certain standards must be met not only at the time a permit is granted, but also during ongoing operation (Roßnagel/Hentschel 2019, para 2). This includes protection against environmental harm and other hazards, substantial impairment and substantial nuisance to the public and neighbours (Section 5(1) no. 1 and 2 BImSchG); the avoidance, recycling, and disposal of waste (Section 5(1) no. 3 BImSchG); and the economical and efficient use of energy (Section 5(1) no. 4 BImSchG). These basic obligations are dynamic in the sense that the relevant standards are liable to change over time. Despite the permit, the facility must meet higher standards over time (Jarass 2022, Section 5 para 2; Roßnagel/Hentschel 2019, para 25). Further, the task of adapting a facility as a result of a disruption may form a part of the dynamic obligations of the operator. In particular, the obligation to protect against environmental harm may justify dynamic adaptations to an existing facility.

The standards of Section 5(1) BImSchG are in part very abstract. For example, Section 5(1) no. 2 BImSchG is an embodiment of the precautionary principle, requiring precautions against environmental harm, other hazards, significant impairment and significant nuisance in accordance with the state of technology. What constitutes the "state of technology" is not obvious. Although Section 3(6) BImSchG provides a definition, determining what constitutes the state of technology is a complex task. In order to avoid legal uncertainty, in many cases the standard must be specified by regulatory authorities (Jarass 2022, Section 5 para 1). Accordingly, Section 7 BImSchG allows the government to issue ordinances in order to specify the basic obligations in more detail. The same applies to administrative regulations based on Section 48 BImSchG.

The requirements of the BImSchG focus primarily on the environmental effects caused by electricity generation facilities. The safety of the energy supply plays a subordinate role. This must be seen in the context of the main purpose of the BImSchG, which is to protect against environmental harm and to prevent the occurrence of environmental hazards.

By contrast, an electricity operator's obligation under Section 49 EnWG applies specifically to energy systems. It is formulated in the same way as the basic obligations of Section 5(1) BImSchG, i.e. it refers to the construction and operation of facilities. Section 49(1) sentence 1 EnWG stipulates that energy systems are to be operated in such a way that technical safety is ensured. In this context, Section 49(1) sentence 2 EnWG refers in particular to the "generally accepted rules of technology". According to Section 49(2) EnWG, compliance with the generally accepted rules of technology is presumed if it complies with the applicable technical rules of the VDE (Association of German Electrical Engineers) for the generation, transmission and supply of electricity. The legislature therefore defers to technical rules and regulations for the stipulation of more detailed requirements. At the same time, Section 49(1) 4 EnWG empowers the federal government to impose certain requirements with regard to the technical safety of electricity generating facilities (Bourwieg 2015a, b, para 12).

Here, too, it is argued that a need for adaptation could be triggered ex post for existing systems. However, it is unclear whether this could lawfully occur unless accompanied by technical standards (in contrast, Bourwieg 2015a, b, para 5c).

Clearly the basic obligations according to Section 5 BImSchG and the additional obligations under Section 49 EnWG are dynamic in the sense that the requirements are likely to change over the lifetime of a generation facility. Not only advances in the state of technology or the generally accepted technical rules, but also disruptive events may lead to a reassessment of the safety of facilities, which may, in turn, require adaptations to be made to existing generation facilities. Since the obligations apply not only to construction but also throughout the operational life of the facility in accordance with the relevant requirements, all three elements of resilience can be embedded in law. Resistance and robustness may assume greater prominence during the original design and construction of the facility. However, adaptations rendered necessary by unanticipated disruptions cannot be taken into account in the original design. The facility may need to be adapted subsequently as requirements evolve. The legislature has responded to this requirement by formulating dynamic basic obligations. Thus, the adaptivity element of resilience is also embedded in the substantive rules. The substantive law governing electricity generating facilities can be seen as a good example of a legal regime that addresses all three elements of resilience considered here.

9.3.1.2 Instruments for Implementing Substantive Requirements for Generation Facilities

The dynamic basic obligations as described above are not self-executing (Jarass 2022, Section 5 para 1). This means that the operator of a generation facility does not have to comply with the requirements on its own initiative, unless ordered to do so by the authorities (Roßnagel/Hentschel 2019, para 26). The available instruments differ depending on whether the facility is yet to be built or whether it is already in operation. If it is a matter of enforcing the basic obligations during the construction of the facility, a permit requirement (a.) may suffice in many cases. Thus, the resistance and robustness elements of resilience could be most effectively enforced through permit requirements imposed on the initial design of the generation facility.

Permit requirements are not the first choice of instrument for responding to unforeseeable disruptions. The ability to react to disruptions and adapt accordingly can only be assessed in advance to a limited extent, if at all. Enforcement against the operator here depends on effective monitoring and any ex-post intervention (b.). This poses the question of openness of the facility to change, including the reversibility of its construction and operation decisions. Finally, the granting of a permit is in many cases subject to a permit procedure (c.).

- **Permit requirements**

As already seen, the BImSchG will in most cases require that a permit be obtained for electricity-generating facilities. Any successful grant of such a permit is subject to the fulfilment of various preconditions. The permit requirements are stipulated in Section 6 BImSchG. According to Section 6(1) no. 1 BImSchG, the first precondition is that the requirements of Section 5 BImSchG and the government ordinances based on Section 7 BImSchG are fulfilled. This essentially refers to substantive legal requirements described above and compliance with the basic obligations. Furthermore, Section 6(1) no. 2 BImSchG refers to other regulations under public law. This also opens up the permit process for further consideration under Section 49 EnWG (see above Sect. 9.3.1.1). Section 6(1) BImSchG provides that the permit is to be granted if the preconditions are met. The provision therefore creates a right to a permit.

This means that aspects of resilience will be considered in the decision to the extent that they are anchored in the statutory permit requirements. However, there is no room for additional aspects to be included in the decision through the exercise of discretion; if the permit requirements are fulfilled, a permit must be issued as of right (Jarass 2022, Section 6 para 45).

- **Ex-post interventions and monitoring**

As described above, the dynamic basic obligations are to be observed not only during the construction of the facilities, but also during their operation. From a purely substantive point of view, this requires continuous adaptation of the facilities (Roßnagel/Hentschel 2019, para 23). However, the dynamic basic obligations are not self-executing. The facility operator is not obliged to continuously adapt the facility to the changing requirements on its own initiative. Usually, administrative orders are necessary to specify what action is required under the dynamic basic obligations, which are otherwise very abstract and not specific enough to be complied with by the operator. Therefore, the dynamic basic obligations are usually given effect by order of the authorities. The primary instrument for doing so is through the issue of ex-post orders in accordance with Section 17 BImSchG. Other, more intrusive measures include the prohibition of operations according to Section 20 BImSchG, revocation of the permit according to Section 21 BImSchG, or withdrawal of the permit according to Section 48 of the Administrative Procedure Act (*Verwaltungsverfahrensgesetz—VwVfG*). However, an ex-post order under Section 17 BImSchG appears particularly suitable for ensuring reasonable adaptations to an existing facility, because it allows for modifications without fundamentally calling the continued operation of the facility into question.

Pursuant to Section 49(5) EnWG, compliance with the obligation under Section 49(1) EnWG is monitored by the competent authorities under state law, which are empowered to take enforcement measures. Meanwhile, the facilities' operators have continuous obligations to provide these authorities with information pursuant to Section 49(6) EnWG. This is important, as the competent authorities would otherwise not have the necessary information to take action (Bourwieg 2015a, b, para 15).

- **Procedure**

As previously discussed, legal procedure also plays a role in anchoring resilience. The BImSchG provides for two 'tracks' of permit procedures: the formal procedure according to Section 10 BImSchG and the simplified procedure according to Section 19 BImSchG. The formal procedure has the essential components described above (see 2.2.5), in particular public participation and referral of the proposal to relevant authorities. The simplified procedure does not require public participation. In certain cases the European directives 2010/75/EU (Industrial Emissions) and 2011/92/EU (Environmental Impact Assessment) also modify the procedural requirements.

Changes to facilities for the purpose of adaptation as a reaction to unforeseen disruptions are sometimes also subject to notification or even approval requirements. The same applies to evolving technical standards which trigger the dynamic basic obligations (Führ 2019, para 8). Depending on the severity of the required changes the procedure may take different forms (Führ 2019, para 41). If the changes do not constitute modifications in the sense of Section 15 et seq. BImSchG, they are not subject to notification. Non-substantial changes only have to be notified in accordance with Section 15 BImSchG. Finally, substantial changes are subject to approval requirements in accordance with Sections 16 and 16a BImSchG. According to Section 16(2) BImSchG, public participation may be waived under certain conditions. Overall, it is apparent that the procedural requirements for ex-post interventions tend to be less onerous than those applicable to the initial application for a facility permit.

9.3.1.3 Summary: Generation Facilities

Looking at the legal regime governing electricity generation facilities, the dynamic operator obligations are the first to stand out. They seem to be suitable to take into account the elements of resilience. When it comes to the implementation of these substantive requirements, the instruments at the preventive stage seem particularly well suited to enforce the requirements of resilience. The permit requirement seems to be particularly suitable as an enforcement instrument. However, it should be noted that there is no general planning approach and that there is no discretion to deny the permit if the regulatory requirements are met. The instruments for ex-post intervention are also well established. Due to the protection of existing installations, resilience requirements are more difficult to enforce at this stage in the life of the installation.

9.3.2 Transmission Lines

The issue of resilience also arises in the regulation of the transmission lines that make up electricity networks. The focus here is on transmission lines requiring planning approval. The categories of transmission lines subject to the requirement of planning approval under the EnWG are contained in Section 43(1) sentence 1 EnWG. According to that provision, high-voltage lines with a nominal voltage of 110 kV or more require planning approval. In addition, extra-high-voltage lines within the meaning of Section 2(1) Grid Expansion Acceleration Act (*Netzausbaubeschleunigungsgesetz Übertragungsnetz—NABEG*) also require planning approval.

9.3.2.1 Requirements of Substantive Law for Transmission Lines

Projects that require planning approval are also subject to substantive legal requirements. In general, there must be a clear justification for the development, compliance with strictly binding legal requirements, and application of the principle of balancing, all of which are cited as preconditions for the granting of any planning approval (Steinberg et al. 2012, Section 3 para 5–7).

The most relevant provisions with regard to resilience are the strictly binding legal requirements, which are also referred to—somewhat misleadingly—as planning guidelines (*Planungsleitsätze*, see Steinberg et al. 2012, Section 3 para 6). These provisions offer promising potential for anchoring the elements of resilience in substantive law. They can exactly describe the standards the transmission lines have to fulfil. In considering transmission lines that require planning approval, one must again refer in particular to Section 49 EnWG. As noted earlier, according to Section 49(1) EnWG, the technical safety of energy systems must be ensured in accordance with the generally recognized technical rules. This requirement also applies to high-voltage lines requiring planning approval, during both their construction and ongoing operations.

Unlike permits under the BImSchG, the law of planning approval does not impose any dynamic basic obligations (Wickel 2020a, para 36). This does not mean that the requirements for the operation of transmission lines cannot change over time. However, unlike the dynamic basic obligations under the BImSchG, planning approval law does not automatically oblige electricity distributors to comply with new standards.

The planning approval process is characterized by a discretion in decision-making that allows a balance to be struck between the various interests affected by a proposed transmission line. In contrast to permits under the BImSchG, there is no equivalent 'right' to planning approval, simply because certain preconditions have been met (Wickel 2020a, para 35). Rather, the planning approval process opens up a wider range of possible decisions. The choice between the possible outcomes is governed by the principle of balancing. Here, four requirements are distinguished: the balancing must be carried out (1.); all relevant interests must be identified and considered in the balancing task (2.); all important interests concerned must be correctly identified and weighted accordingly (3.); and finally, a balance must be struck between the interests, which must not be out of proportion to the objective weight of the interests (4.) (Steinberg et al. 2012, Section 3 para 111).

In the planning system, the three elements of resilience can be directly taken into account as relevant factors to be balanced, to the extent that they are not already part of the strictly binding requirements for granting any planning approval as described above. In this context, it is also possible to give more or less weight to each element, i.e. to give them more or less priority as a planning objective. However, the balancing process cannot establish resilience "at any price", since the respective requirements are always in competition with other interests, which must be appropriately balanced.

9.3.2.2 Instruments for Implementing Substantive Requirements for Transmission Lines

- **Preventive Control: Planning approval**

Since the transmission lines are subject to planning approval, a system of preventive control applies. The requirement of planning approval serves the same purpose as the requirement for a permit under the BImSchG. In contrast to a permit under the BImSchG, there is no 'right' to planning approval and the planning decision-maker enjoys a certain degree of discretion. Therefore, the elements of resilience could be considered even if all strictly binding requirements are met. However, to ensure that elements of resilience are effectively taken into account, it makes sense to define them as part of binding preconditions. If they are otherwise left as the subject of planning discretion and the principle of balancing, then they may be set aside in favour of other interests (Wickel 2020b, para 128).

- **Ex-post interventions and monitoring**

Section 75(2) VwVfG regulates the circumstances in which ex-post interventions may be made to existing transmission lines with a valid planning approval. This provision grants transmission lines an especially high degree of legal protection (Wickel 2020a, para 36). Section 75(2) sentence 1 VwVfG stipulates that once the planning decision can no longer be appealed, any claim to refrain from constructing the project, to remove or modify the facilities, or to refrain from using them are excluded. Here, the so-called obligation to 'exclude and tolerate' the plan-approved project is established (Wickel 2020c, para 29). With regard to resilience, this is of minor importance insofar as the elements of resistance and robustness should already have been reflected in the decision on the planning approval.

Whether the electricity distributor operating the transmission line has to react to disruptions is important for the project's adaptability. Here, too, the legal provisions are very restrictive. Section 75(2) sentence 2 VwVfG stipulates that the developer may be subject to interventions in response to unforeseen impacts of the project. However, this only applies if the rights of third parties are affected, not for the benefit of public interests. A change in the assessment of the resistance and robustness of a transmission line can only lead to its adaptation if the rights of third parties are affected. Moreover, after 30 years, the transmission line is insulated against any kind of ex-post intervention (Section 75(3) sentence 2 VwVfG).

Section 43i EnWG contains a monitoring provision that is specifically tailored to transmission lines that require planning approval. These projects must be monitored to ensure they are carried out in accordance with the environmental provisions of the approval. The authorities may also take enforcement measures to ensure compliance. However, this provision applies only to the environmental provisions of the approval and so it is of secondary importance for ensuring the resilience of electricity supply. In addition, transmission lines operated under a planning approval are subject to network-wide monitoring.

- **Planning**

The planning approval of extra-high-voltage lines and high-voltage lines is preceded by a multi-level planning process, which is outlined here as an example of a planning regime. With this system, the legislator aims to speed up the urgently needed expansion of electricity grids. At the same time, the participation of companies, authorities and the public must be guaranteed. The following analysis only outlines the process and does not claim to cover all details. Some of the steps in the planning process may not apply to all transmission lines. The focus here is on specific elements of the planning system that can have a resilience-enhancing effect, rather than accurately describing the planning system as a whole.

The first step in the planning process for new high-voltage lines is the development of a scenario framework (Section 12a EnWG). The scenario framework is drawn up every two years by the transmission network operators and approved by the regulatory authority. It comprises at least three scenarios which cover the probable development of the next 10–15 years within the framework of the climate and energy policy goals of the Federal Government. Three additional scenarios must look to the year 2045 and describe the probable development within the framework of the statutory targets and goals set by the Federal Government. The draft scenario framework is submitted to the regulatory authority. The regulatory authority publishes the draft and organizes the participation of the public and public authorities. The regulatory authority approves the scenario framework, taking into account the results of the public participation.

Based on the scenario framework, the transmission system operators prepare the network development plan for the transmission networks (Section 12b EnWG). This follows the NOVA principle (cf. Sötebier 2015, para. 2). This principle is defined in Section 11(1) sentence 1 EnWG. It stipulates that operators of energy supply networks are obliged to operate, maintain and optimize, reinforce and expand an energy supply network on a secure, reliable, efficient and non-discriminatory basis in line with demand, insofar as this is economically reasonable. The network development plan must therefore contain all effective measures for the demand-oriented optimization, reinforcement and expansion of the network that are necessary for secure and reliable network operation. The public and public authorities are also participating in the preparation of the network development plan. The regulatory authority reviews the network development plan. It then consults other authorities and the public. Finally, it confirms the network development plan (Section 12c EnWG).

Section 12d EnWG stipulates that the planning and implementation status of measures to optimize, strengthen and expand the transmission network is subject to a continuous monitoring and controlling. The regulatory authority is responsible for monitoring, while the Federal Ministry of Economics and Technology is responsible for controlling.

The network development plan also serves as the basis for the Federal Requirements Plan (Section 12e EnWG). At least every four years, the regulatory authority submits the network development plan to the federal government, which then submits the draft of the Bundesbedarfsplan (plan designating necessary projects) to the federal legislature. The federal legislature then enacts the Bundesbedarfsplan in the form of a law (Bundesbedarfsplangesetz—BBPlG). The BBPlG comprises projects that serve to adapt, develop and expand the transmission grids for the integration of renewable energy sources, the interoperability of the electricity grids within the European Union, the connection of new power plants, and the avoidance of structural bottlenecks in the transmission grid. Having identified and listed these important projects, the Act also determines the necessity and urgency of each project to ensure the safe and reliable operation of the grid. This is binding on transmission network operators and licensing authorities when planning approval is granted for transmission lines. The BBPlG also states that the implementation of the listed projects is necessary for reasons of overriding public interest and in the interest of public safety. This means that these projects are given special weight when balanced against other interests and concerns.

The BBPlG serves as the basis for federal sectoral planning (*Bundesfachplanung*), a procedure which determines the route corridors for the extra-high voltage lines (Section 4 NABEG). The Federal Network Agency (*Bundesnetzagentur*) decides on this at the request of the project developer (Section 6 NABEG). The Agency must examine whether there are overriding public or private interests that conflict with the preferred route corridor for a given project. Among other things, the requirements of regional planning, urban development concerns and possible alternative route corridors must be considered. The federal sectoral planning procedure first requires an application conference to discuss the subject and scope of planning (Art. 7 NABEG). The project developer, the relevant authorities and public stakeholders as well as companies must be invited to the application conference. The application conference is open to the public. Once the scope of the federal sectoral planning has been determined (Art. 7(4) NABEG) and the necessary documents have been prepared (Art. 8 NABEG), the public and other relevant authorities become involved (Art. 9 NABEG), culminating in a public discussion (Art. 10 NABEG). If successful, the process then concludes with the planning approval of the extra-high-voltage lines in accordance with Section 18 NABEG.

This planning regime can promote the resilience of transmission networks. The networks themselves must be designed in accordance with the NOVA principle, which is presented here as a substantive legal requirement. In particular, the regular preparation of scenario frameworks and network development plans allows for adjustments, thereby meeting the need for adaptivity. At various planning levels, there is scope for decision-making that allows resilience aspects to be incorporated into planning. Finally, the assumption that the resilience-enhancing effect of extensive participation processes finds some support at each planning level, in particular through the repeated involvement of the public and other authorities.

9.3.2.3 Summary: Transmission Lines

The substantive requirements to which transmission lines are subject prior to planning approval allow all elements of resilience to be taken into account. In contrast to generation plants, there is a planning discretion that allows resilience to be considered even in the absence of explicit regulation. With regard to the implementation of substantive requirements, a distinction needs to be made between the stages in the life cycle of transmission lines. Planning approval is a preventive instrument that allows resilience to be taken into account. However, the instruments of ex-post intervention are particularly weak in the context of transmission lines. It should be emphasised that transmission lines are subject to a multi-level planning system which allows resilience requirements to be considered at many stages.

9.3.3 Supply Networks

The provision and operation of energy distribution networks falls within the scope of public services (Sötebier 2015, para. 6 ff). However, following privatizations in recent decades, the state no longer carries out tasks in the area of energy networks itself, with these tasks instead transferred to private companies. However, the obligation to provide services in the public interest remains. The state's responsibility for providing electricity infrastructure has transformed into a responsibility to ensure that this task is carried out by the private sector. The state fulfils this responsibility by regulating private network operators (Bundesverwaltungsgericht 2002, para. 18).

The holistic assessment of networks is particularly important with respect to ensuring resilience. In many cases, resilience will not be achieved by a single generation facility or transmission line within a network and so attention must turn to the network's parts and the extent of their cooperation. It was shown above how resilience is embedded in the federal emission control law (see Sect. 9.3.1) and planning law (see Sect. 9.3.2), both of which regulate only parts of energy networks. The next section analyses the regulation of the network as a whole.

9.3.3.1 Requirements of Substantive Law for Networks

As mentioned before (see 2.5.3.2.2.3) Section 11(1) sentence 1 EnWG stipulates that operators of energy supply networks are obliged to operate, maintain and optimize, reinforce and expand an energy supply network on a secure, reliable, efficient and non-discriminatory basis in line with demand, insofar as this is economically reasonable (NOVA-principle). The NOVA-principle covers the operation, maintenance and development of the networks (Sötebier 2015, para. 17). This general obligation is further specified in other provisions, which will not be considered in detail here. For example, Section 11(1) sentence 2 EnWG already stipulates that the duties set out in Sections 12–16a EnWG must be fulfilled.

The key terms in this context are the safety, reliability and performance of the electricity network as well as its optimization, reinforcement and expansion in line with demand. The restrictive addition of economic reasonableness mostly serves to satisfy considerations of proportionality.

Several distinctions can be made between these requirements. The first requirements—safety, reliability, and performance—refer to the networks in their current state. The safety requirement can be further split into two elements. Firstly, technical safety and secondly, security of supply (Sötebier 2015, para. 23–24). Theobald also sees the maintenance of network operations, for example through decentralized supply structures, as measures to ensure security (Theobald 2024, para 16).

Sötebier describes reliability in the technical sense as the ability of a unit to meet the requirements of its intended use within specified tolerances. Parameters of reliability are lifetime, failure rate and availability. With respect to power supply, possible metrics are the duration and frequency of power supply interruptions, the amount of power not transmitted, and the load not covered. Essentially, network reliability aims to deliver energy in the promised quantity and quality (Sötebier 2015, para 29). Performance concerns the ability to deliver power at the required quality. There is therefore some overlap between reliability and performance (Sötebier 2015, para 30).

Resistance and robustness are primarily reflected in the regulatory requirements for safety, reliability, and performance. As seen above resilience is among other aspects the ability of infrastructure to resist, absorb, and recover from an adverse occurrence. From this perspective, reliability and performance ensure that the supply network can deliver energy in the required quantity and quality even in the event of disruptions.

The second part of Section 11(1) sentence 1 EnWG requires the optimization, reinforcement and expansion of the networks. The obligation to maintain, which is also listed, is not considered here and instead the focus lies on the development of the network (Sötebier 2015, para 56). Optimization includes measures that increase the effectiveness and efficiency of the existing network, while leaving the network essentially unchanged. Network reinforcements are measures that relate to the existing network without expanding it. Only expansion then leads to the creation of new network capacities (Sötebier 2015, para 57 ff.). In any event, the network cannot remain in a static state, but must be further developed as necessary in line with demand.

The obligation to optimize, reinforce, and expand networks points to the adaptation element of resilience. The requirement of adaptivity poses the question of how a system responds to disruptions. Can it respond to the disruption with adjustments and improvements to the system? Can it be optimized in response to the disruption ("bounce forward")? If the operator of a network were only required to operate it in its initial state, this would not adequately ensure the adaptivity element of resilience. However, this can be covered through the extension of the operator's obligations to include optimization, reinforcement and expansion.

It can be shown that the different elements of resilience can be represented as different substantive requirements. Resilience and robustness appear as technical properties that networks must meet at all times and these requirements can be specified in advance. With respect to energy networks, these are the safety, reliability and performance of the network. Adaptivity, on the other hand, contains a dynamic element. The aim is to react to unforeseen disruptions and to modify the network as necessary. In the context of Section 11(1) EnWG, this is expressed above all in the requirements for optimization, reinforcement and expansion. These requirements go beyond the original state of the network and therefore represent dynamic obligations.

9.3.3.2 Instruments for Implementing Substantive Requirements for Networks

Section 11(1) sentence 1 EnWG sets requirements for networks. These requirements can in part be assigned to the three different elements of resilience, as discussed above. However, the question then arises as to how these requirements are enforced by the authorities.

- **Permit requirement**

The EnWG does not provide for any permit requirement for the construction of a network. However, as discussed earlier, such permit requirements exist for the generation facilities and transmission lines that are connected to form a network.

Section 4 EnWG itself only contains a permit requirement for the commencement of network operations. It is a license referring to the person of the operator, usually a company, that can only be refused on grounds lying in that person (Assmann 2024, para 37). However, in effect this can be seen as supplementary to the facility-related permit conditions. Hermes points out that the permit related to facilities is essentially subject to preconditions also related to facilities. The personal and entrepreneurial qualities of the applicant, on the other hand, are reviewed in the procedure pursuant to Section 4 EnWG (Hermes 2015a, para 9).

Section 4(2) EnWG requires that the operator has the necessary personnel, technical and economic capacity and reliability to ensure network operation in the long term. In this context, network operation must meet the requirements of the EnWG, including Section 11(1) EnWG (Hermes 2015b, para 20). Indirectly, the substantive requirements of Section 11(1) EnWG also play a role here. Nevertheless, the requirements of Section 11(1) EnWG are not reviewed separately in this permit procedure.

In summary, the requirements of Section 11(1) EnWG are not subject to a preventive control procedure.

- **Ex-post interventions**

The absence of preventive control places particular focus on supervision and intervention if substantive legal requirements are violated. Violations of the substantive obligations, in particular of Section 11(1) EnWG, can be remedied by ex-post interventions.

Section 65 EnWG contains a general clause for the enforcement of the EnWG (Kühling et al. 2022, 337). First, under Section 65 (1) sentence 1 EnWG, the authority can oblige companies to cease conduct that is contrary to the provisions of the EnWG. This decision aims to ensure that any infringements are not continued (Kühling et al. 2022, 338). Behavioural or even structural remedial orders may be considered (Section 65(1) sentence 2 EnWG).

Section 65(2) EnWG provides for stronger remedies. If a company does not comply with its obligations under the EnWG, the regulatory authority can issue compliance orders. The prerequisites are ultimately the same as for measures under Section 65(1) EnWG (Hanebeck 2015, para. 4). Thus, Section 65(2) EnWG supplements Section 65(1) EnWG and permits more extensive compliance orders (Hanebeck 2015, para 11; Kühling et al. 2022, 338).

- **Monitoring**

Pursuant to Section 51(1) EnWG, the Federal Network Agency conducts ongoing monitoring of the security of supply. Section 51(3) EnWG specifies the objectives of this monitoring for the field of electricity supply. Accordingly, the monitoring comprises:

- the current and future relationship between supply and demand in European electricity markets (No. 1);
- existing generation capacities as well as those in planning or under construction (No. 2);
- existing interconnection lines as well as projects in planning or under construction (No. 3);
- the expected development of demand (No. 4);
- network maintenance (No. 5);
- an analysis of network disturbances and of measures taken by electricity system operators to ensure the security and reliability of the electricity system in the short and longer term (No. 6); and
- measures to serve peak demand and to manage outages by one or more suppliers (No. 7).

The monitoring results can also be used to justify measures pursuant to Section 65(1) and (2) EnWG. This applies, for example, if monitoring reveals that the operator is not operating the network in accordance with the requirements of Section 11(1) EnWG (Hanebeck 2015, para 4).

In addition, there are other instruments for obtaining information. Pursuant to Section 52(1) EnWG, operators of energy supply networks must submit an annual report on supply disruptions that have occurred. The operators must also describe the measures taken to prevent future supply disruptions. On the one hand, this reporting obligation supports the monitoring of Section 51 EnWG. On the other hand, it provides information on possible violations of the requirements of Section 11(1) EnWG (Bourwieg 2015c, para 2).

- **Planning**

As previously discussed, plans can serve as particularly effective instruments for ensuring resilience, as they are able to provide an integrated view of electricity supply. Planning instruments exist in the field of electricity supply at various levels.

Section 14d EnWG regulates the instrument of network expansion plans, which was added to the EnWG in 2021 and serves to implement Art. 32(3) of Directive (EU) 2019/944 on common rules for the internal market for electricity. Pursuant to Section 14d(1) EnWG, the operators of electricity distribution networks must submit plans for their respective electricity distribution networks, so-called network expansion plans, to the regulatory authority. This must be done for the first time by April 30, 2024, and every two years thereafter. The instrument is intended to enable integrated and forward-looking network planning.

The plans are based on the aforementioned NOVA principle (Bundesregierung 2021, 101), i.e. the obligations of the network operators described in detail above. Pursuant to Section 14d(4) no. 3 EnWG, they contain a description of the expected development of the distribution task up to the year 2045, including the measures likely to be required for the optimization, reinforcement, renewal and expansion of the network, as well as the anticipated energy efficiency and demand management measures. In addition, pursuant to Section 14d(4) no. 4 EnWG, the plans must contain intended measures for the optimization, reinforcement, renewal and expansion of the network, as well as the necessary energy efficiency and demand management measures for the next five and ten years.

9.3.3.3 Summary: Networks

Networks are subject to the NOVA principle. Operators of energy supply networks are obliged to operate, maintain and optimise, reinforce and expand an energy supply network in a secure, reliable, efficient and non-discriminatory basis in accordance with demand, insofar as this is economically reasonable. We have been able to show that the various elements of resilience are reflected in this substantive requirement. However, these substantive requirements are not enforced through preventive control. There is no permit requirement for operating a network. The focus therefore shifts to instruments of ex-post intervention.

9.4 Conclusion

Resilience is of growing importance for the legislator in the field of infrastructure law. Increasing digitalization, the decarbonization and transformation of energy systems, and new threats from digital or physical attacks on electricity infrastructure pose new challenges in the regulation of infrastructure. Derendorf et al. give detailed consideration to the wide variety of potentially disruptive events.[6] These challenges now

[6] See Chap. 14 in this book.

confront a patchwork of infrastructure laws that have evolved over decades, encompassing different legal regimes governing the various elements of infrastructure. In this chapter, we identified electricity generating facilities, high voltage transmission lines, and electricity networks as distinct subjects of regulation. But similar examples could certainly be found in other legal regimes.

Despite its heterogeneity, the various parts of infrastructure law are firmly established. New requirements of resilience may challenge this structure and, therefore, may be difficult to accommodate without substantial reforms. However, the legislator must not only change existing provisions in various laws, but also deal with vested legal rights and interests, such as valid permits held by operators, some of which enjoy constitutional protection.

In this chapter we showed which elements of infrastructure law must be addressed in rethinking regulatory regimes to enhance system resilience. The normative elements of resilience—resistance, robustness, and adaptivity—must be first anchored in the substantive law defining the legal basis of infrastructure systems. We showed that in particular the dynamic character of the basic obligations under the BImSchG renders them suitable not only to ensure that facilities absorb and recover from shocks, but to be able to adapt to future shocks in an evolutionary way. Another resilience-enhancing form of regulation in substantive law is the NOVA-principle laid out in the Energy Industry Act (EnWG).

In any case, substantive legal requirements are insufficient to enhance resilience alone. Any legal requirement must be implemented and enforced, which shifts the focus to the instruments addressed at the operators of electricity generation facilities, transmission lines and networks. There is a wide choice among available instruments. In this chapter we looked at permit requirements, monitoring and ex-post interventions as part of a regulatory command-and-control system. We also considered planning, as plans are instruments that help to relate different parts of a system to each other. Finally, procedural laws appear to play a role as well, in particular since they are crucial for giving effect to public participation, which is thought to increase a system's resilience-enhancing capabilities.

The law of resilience has developed only recently and is still emerging. In 2022, the European Union issued Directive (EU) 2022/2557 on the resilience of critical entities (CER Directive). The directive must be transposed by EU Member States by 17 October 2024. The German Federal Ministry of the Interior and Community is preparing a law to implement the CER Directive and strengthen the resilience of critical installations. At the time of writing, the Ministry plans to enact an umbrella law that addresses a broad range of critical infrastructure, including energy, transportation and traffic, finance and insurance, health system, drinking water, wastewater, food, information technology and telecommunications, outer space, municipal waste disposal, (Bundesministerium des Innern und für Heimat 2024). How all of this will affect fragmented existing infrastructure laws remains to be seen.

References

Assmann: § 4 in: Assmann; Peiffer (eds.) Beck'scher Online-Kommentar EnWG (Beck'scher Online Commentary Energy Industry Act), 12th edn. (2024)

Baldurson; Banet; Chyong: Building resilience in Europe's energy systems (2024). https://cerre.eu/wp-content/uploads/2023/06/01062023_CERRE_REPORT_RESILIENCE.pdf. Accessed 27 Feb 2024

Bourwieg: § 49 in Britz; Hellermann; Hermes (eds.), Energiewirtschaftsgesetz (Energy Industry Act), 3rd edn. (2015a)

Bourwieg: § 51 in Britz; Hellermann; Hermes (eds.), Energiewirtschaftsgesetz (Energy Industry Act), 3rd edn. (2015b)

Bourwieg: § 52 in Britz; Hellermann; Hermes (eds.), Energiewirtschaftsgesetz (Energy Industry Act), 3rd edn. (2015c)

Bundesministerium des Innern (Federal Ministry of the Interior): https://www.bmi.bund.de/SharedDocs/gesetzgebungsverfahren/DE/KRITIS-DachG.html. Accessed 20 Feb 2024

Bundesregierung (Federal Government): Gesetzentwurf der Bundesregierung, Entwurf eines Gesetzes zur Umsetzung unionsrechtlicher Vorgaben und zur Regelung reiner Wasserstoffnetze im Energiewirtschaftsrecht (Draft bill of the Federal Government, Draft law on the implementation of EU requirements and the regulation of pure hydrogen networks in energy industry law), Deutscher Bundestag (German Bundestag), Drucksache (printed paper) 19/27453

Bundesverfassungsgericht (Federal Constitutional Court): 1. Senat (Senate) 3. Kammer (Chamber), Beschluss vom (decision of) 14.01.2010 – 1 BvR 1627/09 (2010)

Bundesverwaltungsgericht (Federal Administrative Law): Urteil vom (judgement of) 11. Juli 2002 – 4 C 9/00 –, juris (2002)

Franzius: Infrastrukturen zwischen Regulierung und Planung, Herausforderungen des Energieinfrastrukturrechts (Infrastructures Between Regulation and Planning, Challenges of Infrastructure Law), Zeitschrift für das gesamte Recht der Energiewirtschaft (EnWZ) 2022, p. 302 (2022)

Erbguth: Zur Fortentwicklung der Öffentlichkeitsbeteiligung im räumlichen Planungs- und Zulassungsrecht (On the Further Development of Public Participation in Spatial Planning and Licensing Law), Umwelt- und Planungsrecht (UPR) 2018, pp. 121–127 (2018)

Führ: § 15 in Führ (ed.) GK-BImSchG, 2nd edn. (2019)

Hanebeck: § 65 in Britz; Hellermann; Hermes (eds.), Energiewirtschaftsgesetz (Energy Industry Act), 3rd edn. (2015)

Hermes: Die Systementwicklungsplanung – Instrument zur klimagerechten Transformation des Energiesystems (System development planning - a tool for climate-just transformation of the energy system), Zeitschrift für das gesamte Recht der Energiewirtschaft (EnWZ) 2022, p. 99 (2022)

Hermes: § 4 in Britz; Hellermann; Hermes (eds.), Energiewirtschaftsgesetz (Energy Industry Act), 3rd edn. (2015a)

Hermes: § 11 in Britz; Hellermann; Hermes (eds.), Energiewirtschaftsgesetz (Energy Industry Act), 3rd edn. (2015b)

Jarass: BImSchG (Federal Immission Control Act), 14. Auflage 2022 (2022)

Kloepfer: Umweltrecht (Environmental Law), 4th edn. (2016)

Kloepfer; Durner: Umweltschutzrecht (Environmental Protection Law) (2020)

Kühling; Rasbach; Busch: Energierecht (Energy Law), 5th edn. (2022)

Ramsauer: § 3 Allgemeines Umweltverwaltungsrecht (General Environmental Law) in Koch; Hofmann; Reese (eds.) Handbuch Umweltrecht (Handbook on Environmental Law), 5th edn. (2018)

Roßnagel; Hentschel: § 5 in Führ (ed.) GK-BImSchG (Federal Immission Control Act), 2nd edn. (2019)

Schulze-Fielitz: Energie-Infrastrukturrecht im Prozess der Wissenschaftsentwicklung (Energy Infrastructure Law in the Process of Scientific Development). In: Sabine Schlacke/Mathias Schubert, Energie-Infrastrukturrecht (2015)

Sötebier: § 11 in Britz; Hellermann; Hermes (eds.), Energiewirtschaftsgesetz (Energy Industry Act), 3rd edn. (2015)
Steinberg; Wickel; Müller: Fachplanung (Sectoral Planning), 4th edn. (2012)
Theobald: § 11 EnWG in Theobald; Kühling (eds.) Energierecht (Energy Law) (2024)
Wickel: Bestandsschutz im Umweltrecht (Protection of Existing Rights in Environmental Law) (1996)
Wickel: § 72 VwVfG in Fehling/Kastner/Störmer (eds) Verwaltungsrecht (Administrative Law) 5th edn., (2020a)
Wickel: § 74 VwVfG in Fehling/Kastner/Störmer (eds) Verwaltungsrecht (Administrative Law), 5th edn. (2020b)
Wickel: § 75 VwVfG in Fehling/Kastner/Störmer (eds) Verwaltungsrecht (Administrative Law), 5th edn. (2020c)

Chapter 10
Resilience, Shock Prevention, and Interacting Systems: Economic Aspects of ReMoDigital

Thomas H. W. Ziesemer

Abstract This chapter starts with the development of a resilience index covering prevention, vulnerability, and prevention. A graphical presentation shows that it covers high-impact rare events (HIRE) as well as low impact high probability cases. The index is then compared to those in engineering, geography, and economics. Reducing shock probabilities and vulnerability and speeding up recovery is costly. Maximization of the index under a resilience-budget constraint shows how expenditures should be well balanced over the three aspects. Numerical and historical examples are constructed and discussed to illustrate the principles. Interactions of systems like digitalization and electricity are considered more broadly for eleven interacting systems. Incentives and policies regarding investment in resilience are described also regarding the international dimension. We provide a simple empirical vector-error correction model for German data in electricity, consumption, GDP per capita and oil prices, which is stable in growth rates. Shock analysis through impulse-response functions show cases of no return to baseline after negative shocks, and long-run overshooting for all variables after a negative shock to GDP per capita, supporting the possibility of the idea of coming out stronger out of a crisis, called super resilience. Policy suggestions from the literature are collected in a separate section.

Keywords Resilience costs · Interacting systems · Vector-error-correction model · Shock analysis

T. H. W. Ziesemer (✉)
Maastricht Economic and Social Research Institute on Innovation and Technology, United Nations University (UNU-MERIT), Maastricht, The Netherlands
e-mail: Ziesemer@merit.unu.edu; T.Ziesemer@maastrichtuniversity.nl

D. van Doren et al. (eds.), *Towards Energy System Resilience*, Ethics of Science and Technology Assessment 55, https://doi.org/10.1007/978-3-032-03187-7_10

10.1 Introduction

The resilience literature presented previously in this book[1] focusses mostly on the post-shock phase (Mahzarnia et al. 2020). Thereby it takes the probabilities of the shocks as given. This is justified to the extent that the probabilities for events like earthquakes and hurricanes cannot be changed. However, the probabilities for heavy rainfalls and software mistakes, which have become important most recently, can be changed through international climate policies and firm policies regarding quality control. Changing the probabilities is costly though.[2] Similarly, Renn et al. (2025) have indicated that damage reduction—reducing vulnerability or improving recovery—is also costly, either when done preventively or as adaptation after the shocks. Given limited budgets for these types of costs we are confronted with the problem of allocating the scarce means over these aspects of resilience. For this purpose, we provide a small first-principles model formalizing the main line of the argument of Renn et al. (2025). It allows to maximize resilience through balancing a given budget over shock prevention, damage reduction and, in a second modelling step, vulnerability reduction.[3]

In Sect. 10.2 of this chapter we (i) present the generic model, (ii) provide the conditions for a well-balanced allocation of means over the aspects of resilience, (iii) compare it to what the resilience literature related to the electricity system is (not) telling us, and (iv) extend it to include vulnerability. In Sect. 10.3 we discuss some problems of the resilience of interacting systems. Section 10.4 discusses some issues of interconnection with other countries which are important in the energy transition when trying to reduce the probability of blackouts. Section 10.5 provides a dynamic model; the dynamic model relates consumption, GDP, KWh of electricity, all measured per capita for German data, and oil prices, and shows that responses to one-time shock on electricity supply and electricity demand variables have permanent level effects in a model, which is stable only in growth rates. Section 10.6 collects resilience policy steps. Section 10.7 summarizes.

---

[1] See Chap. 2 in this book.

[2] A logical pre-condition that is outside the scope of this book is the attribution of climate events to probabilities (see Allen 2003; van Oldenborgh 2007), which is also a scientific area which is just in its early phase, just like resilience research.

[3] Lin et al. (2018) briefly mention a budget problem regarding access to outage information and taking measures after the shock.

10.2 Definitions and Measurement Issues

10.2.1 *Derivation of a Resilience Index*

Summarizing expressions such as resilience, sustainability and vulnerability serve mainly intuitive communication and do not necessarily have to be measured. Measurement can be carried out and may be useful at a lower level such as functional relations, (cap-) abilities, (dis-) advantages, costs and benefits, actions, effects, and reactions. At least we need to distinguish between measurement of resilience from a preventive and an after-crash perspective. The perspectives of *shock prevention* or system reaction after the crash (see Bruneau et al. 2003, Fig. 6 in Kröger 2017, and Fig. 1 in Acatech et al. 2021) are two different perspectives. In the latter, *vulnerability prevention* as preparation for the after-shock phase is the major topic of most of the literature discussed in this chapter, besides preparation for a quick recovery. Brand et al. (2025)[4] point out that there is some literature looking at both phases. However, the discussion of the pre-shock phase focusses typically on reducing the vulnerability of the systems rather than the reduction of the probability of the shock. Bruneau et al. (2003) speaking of 'reduce the chances of a shock' and 'reduced failure probabilities' but actually mean the 'expected degradation in quality' of infrastructure (p.737) in the context of earthquakes.[5] Shen et al. (2020) present the basic formulas of several resilience models. None of all these models has an endogenous probability of shocks.

From the perspective of shock prevention, (lack of) resilience can be defined as the subjective[6] probability p of a crash occurrence in terms of S crashes or outages, in T years, in Z countries. From an after-crash perspective (lack of) resilience can be defined as the loss or damage D received until (im-) perfect re-creation after the crash.[7] Lack of resilience, L, by combining both parts then is

$$L \equiv pD$$

[4] See Chap. 12 in this book.

[5] Bruneau et al. (2003) speaking of 'reduce the chances of a shock' and 'reduced failure probabilities' but actually mean the 'expected degradation in quality' of infrastructure (p. 737) in the context of earthquakes. This is less clear where the authors briefly speak of 'probabilities of occurrences' (p. 742).

[6] We would assume here that probabilities are essentially always subjective. Even if insurance companies try to calculate them from event data of the past, the expectation that there might be social or technological structural change, or none is subjective. In contrast, Brand et al. (2025) (see Chap. 12 in this book) assume that there might be known probabilities. This may be true under conditions of controlled technical experiments if structural change in (mixtures of) materials, technology and other conditions can be excluded.

[7] Moslehi and Agami Reddy (2018) provide a model for losses of K service systems from J sources of shocks.

If the damaging event happens to occur p goes to 1 and $L = D$.[8] This is a redux version of the extensive definitions and descriptions in Chap. 2. Francis and Bekera (2014) provide a sophisticated version of it with distribution functions and stochastic parameters.

Another special case is that of D as energy not supplied (ENS), and p its probability, leading to L as EENS, expected energy not supplied (Espinoza et al. 2016).[9] For the USA, the Department of Energy estimates "...the average annual cost of weather-related power outages to be between \$18 and \$33 billion over the past decade" (Watson et al. 2015), which is expected to increase as a result of climate change.[10] Afgan and Cvetinovic (2011), in their Fig. 10.3, use this expression as a percentage of the maximum damage as a resilience-per-period index.

In order to go from 'lack of resilience' to 'resilience' we go from p to $1 - p$, the probability of no bad event, and from damage to damage avoided, $D_{max} - D$ leading to

$$R(t) = (1 - p)(D_{max} - D)$$

To turn R into an index between zero and one,[11] we divide by D_{max}. The resilience index then is

$$RI(t) = (1 - p)[(D_{max} - D)/D_{max}] = (1 - p)(1 - d),\ d \equiv D/D_{max} \quad (10.1)$$

$D/D_{max} = d$ is the damage as a proportion of the maximum possible damage (called expected degradation in Bruneau et al. 2003), where the latter is sometimes intuitively also called total damage, or $1 - d$ is called sustainability index in Afgan and Cvetinovic (2011). We leave summation over the periods to the reader. The index shows resilience equal to one if $p = d = 0$, meaning that the probability of a risky event has been reduced to zero or the system has a perfect reaction reducing the damage to zero. The index shows resilience zero, if the risk is 100 percent sure and therefore $p = 1$, or if the damage cannot be reduced, $d = 1$.[12] For values of p and d between zero and one we show the values of the index in Fig. 10.1a. Watson et al. (2015) discuss reliability and resilience together, but the literature used here tends to discuss either one or the other. In contrast, our model includes both the resilience cases and the area of reliability or risk analysis. High-impact, rare (low probability or frequency) events (often abbreviated as HIRE or HILP or HILF) defined as low p

---

[8] This idea was inspired by Arnim von Gleich's contribution to the discussion of the Workshop ReMoDigital, March 9, 2021.

[9] If p and D are column vectors, the vector product $p'D$ could be the expected sum of damages from many events.

[10] This compares to a GDP of \$15,854 billion, or 18/15854 = 0.0011 or 33/15854 = 0.0021, which is between 0.11 and 0.21% of the GDP.

[11] This property was suggested by Ortwin Renn during the third video group meeting for the project Resilienz-Monitoring für die Digitalisierung der Energiewende (ReMoDigital), March 10, 2021.

[12] The fear of the global re-insurance system in case of nuclear accidents may come close to this and therefore they do not want to offer insurance.

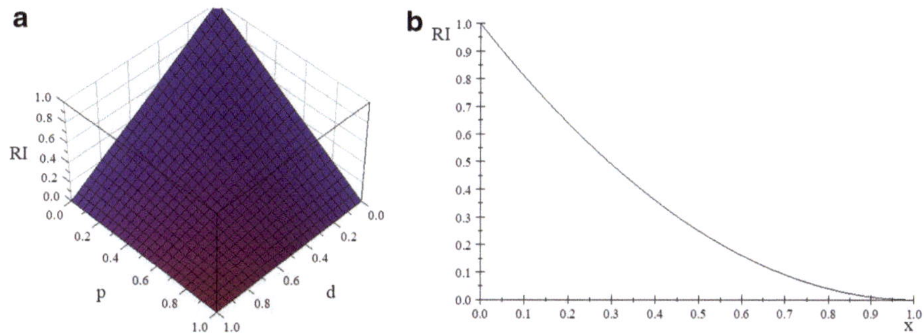

Fig. 10.1 **a** The three-dimensional plot of the resilience index RI. **b** The two-dimensional plot of the resilience index RI when p(robability) and d(amage proportion) are equal, p = d = x

and high *d, D*, D_{max} are located on the left of Fig. 10.1a; this is the area of resilience studies. Low-impact, high-probability events are located on the right of Fig. 10.1a; they are the area of traditional reliability studies (Yu et al. 2021). $p = 0$ would be perfect prevention. However, perfect prevention is not possible (Watson et al. 2015) and therefore we shall consider the possibilities of reducing *p* below.

For equal values *x* of *p* and *d*, the index is $(1-x)(1-x)$ and looks as shown in Fig. 10.1b; bended like a classical Chinese roof.

All aspects described in the extensive version of a definition of resilience and related aspects in Renn et al. (2025) serve the reduction of *p* or the increase of $(1-p)$ and the reduction of damage *D* or the increase of *damage avoidance* = *maximum possible damage* − *actual damage.* Problems of measurement, information and uncertainty, estimation, valuation, modelling (including complexity and ambiguity), assessment, strategies and management are subordinated but possibly important details of this concept when trying to fill the index empirically.

10.2.2 The Economics of Allocating Means Between Damage Prevention and Mitigation

For example, for the case of the electricity system, natural hazards (earthquakes, hurricanes, rock slope failures, ice layers, snow, floods, drought, heat, space weather), human errors (normal fault contingencies like programming or software mistakes), human attack (physical or cyber[13]) and terrorist attack might justify some investment in prevention and resilience (Afgan and Cvetinovic 2011).[14] "…we need to make a compromise between the cost of the strategy deployment and the profit resulting from the resilience enhancement. This is an open research topic which must be addressed in the future studies." (Gholami et al. 2018, p. 32050). One such model minimizes

[13] Gholami et al. (2018) provide a classification of cyber-attacks and its consequences.

[14] Roege et al. (2014) provide a broad table of causes and effects.

the *total cost* = *system costs of reliability and outage costs* through the choice of reliability.[15] The result is that the 'profit' or reliability is worth the system costs. It is by no means natural to minimize costs. When electricity is more than just a cost item, minimization of outage *risk* could also be considered in combination with costs.

We consider next how the cost can be allocated across prevention and mitigation of damage using the resilience index (10.1), providing a basic example for such an investment decision building on the formalization above.

Assumption 1 A certain percentage of the GDP or of the cost of a system under consideration, *b*, is available for resilience measures. Then this budget can be divided into preventive expenditure *x*, and expenditures for damage reduction measures in preparation of damage mitigation after a bad event, y, both per period (as in the case of financing with a no-payback credit).[16] Then the resilience budget is

$$b = x + y \;\; for\, all\, t \tag{10.2}$$

Assumption 2 The (subjective) probability *p* can be reduced with decreasing effects through the percentage x of the cost that is used for prevention such that $p(x)$, with first and second derivative $p' < 0$, $p'' > 0$, as we would see it if projects were ranked according to the strength of the effects.

Assumption 3 The damage proportion *d* can be reduced through the percentage *y* of the GDP that is used for investment in damage reduction or ex-post resilience such that $d(y)$, $d' < 0$, $d'' > 0$.

Maximization of the resilience index RI with respect to the budget (10.2) using the Lagrange function

$$RI = \int (1 - p(x))(1 - d(y)) - \lambda(x + y - b)dt \tag{10.3}$$

yields first-order conditions[17]

$$L_x = -p'(1 - d) - \lambda = 0 \, for\, all\, t \tag{10.4}$$

$$L_y = (1 - p)\left(-d'\right) - \lambda = 0 \, for\, all\, t \tag{10.5}$$

$$L_\lambda = -(x + y - b) = 0 \, for\, all\, t \tag{10.2'}$$

---

[15] Watson et al. (2015) attribute this model to Burns and Cross (1990), but they have no references and google scholar does not bring it up either.

[16] A good example are ice-coatings. If cables crash under the weight of ice layers, putting cables underground may become cost-effective if the avoided damage is large enough to over-compensate the higher cost of underground cables.

[17] For second-order conditions see appendix. $1 - p$ and $1 - d$ are concave in x and y respectively. As the constraint is linear, we should expect second-order conditions to be fulfilled.

Elimination of the Lagrange multiplier λ yields

$$-p'(1-d) = (1-p)\left(-d'\right) = \partial RI/\partial b = \lambda \tag{10.6}$$

The last equation, using the envelop theorem, indicates that the Lagrange multiplier is the value of the change in resilience from extending the budget *b*. Division of (10.6) by RI = (1 − *p*)(1 − *d*) yields

$$-\mathrm{P}'/(1-\mathrm{P}) = -D'/(1-d) = \frac{(\partial RI(t)/\partial b)}{RI(t)} = \frac{\lambda}{(1-p)(1-d)}. \tag{10.6'}$$

Resilience is maximized when the elasticities of 1 − *p* and 1 − *d* with respect to expenditure percentages x andy respectively are equal to each other and equal to the semi-elasticity of the yearly resilience index with respect to the budget share *b*. More intuitively, expenditure should be well balanced between prevention of and preparation for the catastrophe case. Only if the functions have identical properties this would imply $x = y$. We get $(\partial RI/\partial b) = \lambda/RI$, meaning that the interpretation of the Lagrange multiplier is the marginal effect of a budget increase on the maximized resilience index, both also as percentage of the maximized resilience index.

Popular *functional forms* are the log-linear damage functions and the exponential distribution functions for *p* (or more generally the extreme value or Weibull function, normally used for stochastic variables[18] but here for politically chosen expenditures x). Assume (ignoring the indices of the *x* values)

$$p = e^{-h(x)}, h\prime > 0, h'' < 0, log_{x\to\infty}h = \infty, h(0) = 0 \tag{10.7}$$

These assumptions imply that

$$p' = -h'e^{-h(x)} < 0, p'' = -h''e^{-h(x)} + (h')^2e^{-h(x)} > 0 \tag{10.8}$$

For $x = 0$ this implies $p = 1$; for *x* to infinity, we get $p = 0$. A simple example with these properties is

$$h = x^{\alpha}, \alpha < 1$$

Assume

$$d(y),\ d' < 0,\ d'' > 0;\ d(0) = 1,\ d(y_0) = 0 \tag{10.9}$$

[18] The related stochastic variables would be the random determinants of probabilities of extreme weather or software mistakes, which are outside the scope of this book. See also Footnote 1. Integrating both concepts could perhaps be done by multiplying both functions which are both between 1 and zero.

The last two properties of (10.9) are not strictly necessary. The following function is an example fulfilling these requirements.[19]

$$d = (1-y)^2,\ d(0) = 1,\ d(1) = 0,\ d' = 2(1-y)(-1) \le 0\, for\, y \le 1,\ d'' = 2 \quad (10.10)$$

Insertion of (10.7)–(10.10) into the first equation of (10.6) yields the following first-order condition:

$$\alpha x^{\alpha-1} e^{-x^{\alpha}}/(1-e^{-x^{\alpha}}) = 2(1-y)/(1-(1-y)^2) \quad (10.6')$$

with $y = b - x$ from (10.2). For $\alpha = ½$ and $b = 3$, we get the following graph.

The falling curve is the left-hand side of (10.6′). The right-hand side of (10.6′) is shown by the upward-sloping curves. The part for x > 3 is invalid because it is outside the budget b = 3 and has no intersection with the falling curve. The second intersection point around x = 0.6 requires y = 2.4, which would imply d′ > 0 and therefore cannot be a maximum of RI. Under the assumptions made, of the budget b = 3 (percent of the cost) about two thirds is used for prevention, $x = 2.05$, and one third for post-shock damage reduction, $y = 0.95$. Only this latter solution fulfils all requirements.[20]

Climate change may drive the probability elasticity up over time. This could be captured by an exogenous function of a time trend multiplied to $h(x)$, leading to $h(x)/v(\tau)$.[21]

Maximizing (10.1) is not quite the same as minimizing (expected) damages, pd, or losses $L = pd$, as $RI = 1 - d - p + pd$. However, the major point here is that p is not exogenous but rather can be reduced through expenditures at the cost of damage reduction because budgets are given.

Beyond theoretical reasoning, such indices may become useful if someone can estimate values or functions for p(robabilities) and d(amages). For damages there is a software approach call HAZUS.[22] However, this is not the same as measuring

[19] For the alternative formulations of functions $d = 1 - ln(1+y)$, $d' = -(1+y)^{-1} < 0$, $d'' = (1+y)^{-2} > 0$, and $d = -ln(e+y)$, $d' = -(e+y)^{-1} < 0$, $d'' = (e+y)^{-2} > 0$, d can be negative. e is Euler's e = 2,718 281 828 459 045 235, and lne = 1.

[20] In the special case of a given probability for bad events where expenditure has no impact, p' is given as a horizontal line at level zero in Fig. 10.2, intersecting the increasing line at $x = 2$, implying $y = 1$. Equation (10.4) would drop out; the derivative of (10.5) with respect to y would be $(1-p)(-d'')$. Given the functional form, this would give $d'' = 0$, $d = 0$. An interior maximum does not exist at this value. Spending on x makes no sense without impact on p. Spending $y > 1$ for damage reduction also makes no sense as we get $d' > 0$ when assuming (10.10). This serves as a reminder that a not all functional forms lead to standard approaches. A budget of $b = y = 1$ makes sense in this case as neither probability nor damage can be reduced with more money. At $b > 1$ with $x = 0$, $y = 1$, the Lagrange multiplier would be zero.

[21] $V(\tau) = \tau$ is assumed for the exponential function, $v(\tau) = \alpha\tau^{\alpha-1}$ for $\alpha > 0$ yields the Weibull function, and $v(\tau) = \alpha e^{\alpha\tau}$ generates the extreme value function (Kamien and Schwartz 1980). The difference with patent races is that there x and τ increase the probability.

[22] https://www.fema.gov/flood-maps/products-tools/hazus.

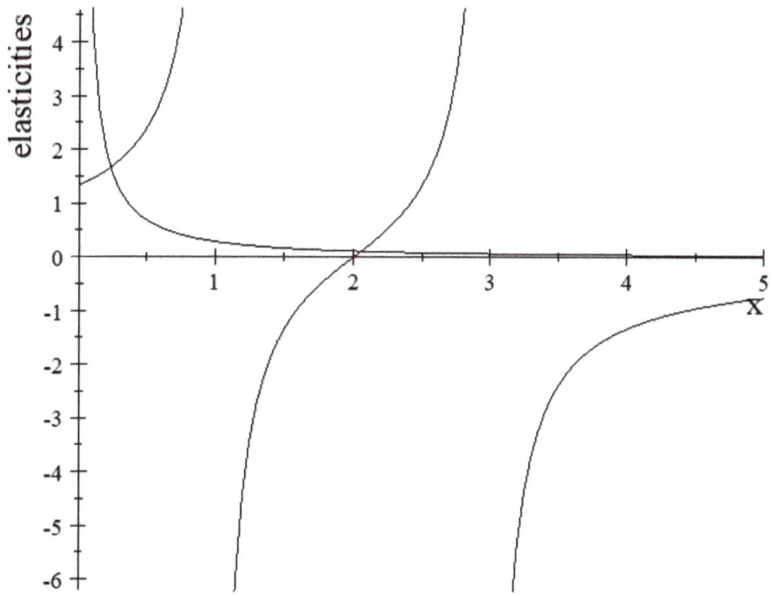

Fig. 10.2 Equal semi-elasticities of prevention probability 1 − p and damage avoidance 1 − d with respect to investment for $p = e^{(-h(x))}$, h(x) = xα, α = 0.5; d = (1 − y)2, b = 3

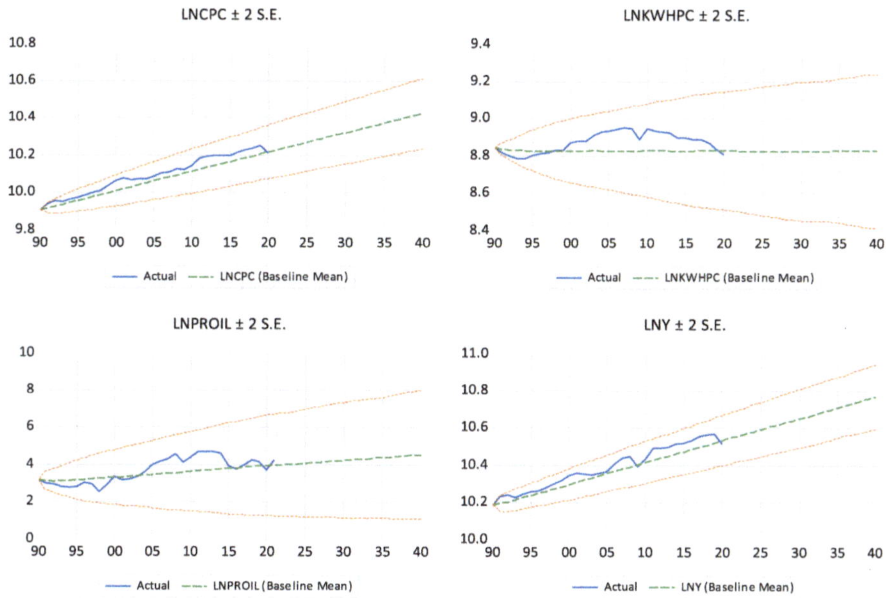

Fig. 10.3 Baseline simulation of model (10.14)–(10.18)

resilience, for which there is currently 'no consistent basis for … measurement' (Watson et al. 2015). The authors speak of 'stochastic optimization will identify … control actions that maximize resilience …', and 'optimization of investment for enhanced resilience given a fixed budget', referring to a '118-bus test case' (besides a petroleum and a gas case). Their mixed integer linear model could be seen as the technical analogue to our non-linear economic model, but probabilities are not endogenous and only consequences of shocks can be reduced through, for example, walls around generators and putting cable underground. This research is just in its beginning phase and much research is needed to find the damages and resilience metrics (Watson et al. 2015).

It follows from the simplicity of the first-principles model that it is set up to indicate the basic line of thought how to allocate money to shock prevention and damage control. It is of course not a real decision model, because we do not have the data and do not know the functions and other circumstances going beyond a first-principles model. Practical steps are suggested by Mayer and Brunekreeft (2021).[23] These practical steps are most important, but in spite of using the concept of resilience as a tool of communication the steps are not merged into a value of resilience, because the concept simply is not yet developed that far.

We are not going to discuss the basics of risk, uncertainty, knowns, unknowns, and unknowables. In practice, often decisions are taken after catastrophes such as great floods, because then there is huge regret, and disaster myopia leads to higher subjective probabilities jointly with the desire not to let it happen again. These decisions are extremely dependent on the institutional context, and they are dynamic in the sense that 'after the flood' is 'before the flood' and regret and disaster myopia lead to reconsideration of probabilities and expected damages.

10.2.3 Generalizing the Resilience Index to Many Systems

If there are n (sub-) systems for each of which we have a resilience index, $(1-p)$ is an $n \times 1$ vector and $(1-d)$ is also an $n \times 1$ vector. A simple form of a resilience index where the expressions for all systems are weighted equally then is

$$\sum_{1}^{n} RI = \frac{1}{n}(1-p)\prime(1-d) = \frac{1}{n}\Big[(1-p_1)(1-d_1) + (1-p_2)(1-d_2) + \cdots + (1-p_n)(1-d_n)\Big]$$

[23] They suggest batteries to bridge blackouts, network-availability subsidies as incentives to care for resilience, and institutional adjustments.

Afgan and Cvetinovic (2011) have a similar formular where the weights are not equal, and the probabilities are dropped. The index then considers only the post-shock or damage phase and not the ex-ante or prevention phase although the abstract of their article supports the idea of the two phases.

10.2.4 Resilience Indices in the Literature on Economics and the Electricity Systems

In the *engineering* literature, Afgan and Cvetinovic (2011) support the two-phase structure in their abstract with probabilities considered in the first phase, but then focus exclusively on the second phase. Espinoza et al. (2016) discuss how to get the probabilities that are related to the expected damages or consequences (Watson et al. 2015). Watson et al. (2015) indicate that there are a lot single-problem indices, but the envisaged integrated system models are far from the stage of practical work. Francis and Bekera (2014), in their survey on definitions of resilience, are most explicit about including the first phase in their definition "Resilience can be understood as the ability of the system to reduce the chances of a shock, to absorb a shock if it occurs (abrupt reduction of performance) and to recover quickly after a shock (re-establish normal performance)."

In the *geographical* literature, for example Maliszewski and Perrings (2012), the basic idea is that most of the damage happens to occur within a radius 1000 feet. The damage then is proportional to the speed of recovery. All other arguments explaining the damage are then subsumed either under what happens roughly within the 1000 feet circle or what determines the speed. Although the emphasis is also here on the second damage phase or its resilience, some of the arguments explaining the duration of outages would also be relevant for prevention regarding probabilities: age of infrastructure, overhead feeders, and near-by vegetation.

In the early papers in *economics*, the literature uses the word resilience only in relation to the second phase with damages after a shock. First-pre-shock phase, in contrast, is considered using the term vulnerability. This would only be a difference in the choice of words if it were not that vulnerability is covering only aspects like dependence on strategic imports, export concentration, which are by assumption policy independent (Briguglio 2004).[24] This implies that policies cannot reduce vulnerability, and probabilities of bad events are therefore not subject to preventive measures (very much like the heal of Achilles at his time and state of the art of medicine). The probabilities of (external) shocks are exogenous, perhaps because the focus of this literature is on small-island countries, public health, tourism, and waste management. Policies can take effect only after the shock. This also precludes any relevance of the principle of precaution regarding the first phase by assumption. By implication, the allocation of economic means for prevention and damage mitigation

[24] In this article Briguglio suggests the development of resilience indices for the macroeconomy on the country level.

suggested in Sect. 2.2 is not a topic under the assumption of policy independence of vulnerability or probability. An exception is 'multidisciplinary climate change education', which could try to build capacity for international negotiation on reductions of probabilities of bad events (Briguglio and Moncada 2019). Of course, it is an empirical, problem-specific question whether or not prevention measures exist. If not, p' and d' in Sect. 2.2 are both zero and no optimum choice exists. However, tree trimming, x-eyes principle in quality control of software programming, investing cyber police, and climate-change policies are all actions that may reduce the probabilities of the causes of bad events besides reducing vulnerability, which is the major perspective (Mahzarnia et al. 2020 and Kandaperumal and Srivastava 2020 provide a longer list). These perspectives may be hard to distinguish if an event happens to occur after storm only if trees are falling or, alternatively, the effects are just getting more severe if trees are falling.

Briguglio et al. (2006) provide resilience indices for 102 countries, emphasizing that the selection of variables and their weights is subjective. They show that their index is correlated with the natural logarithm of the GDP per capita with coefficient 0.94. In economics, it is one of the stylized facts of growth that growth and ratios of GDP and capital stocks are relatively constant thereby also highly correlated over time and countries. But the GDP and the capital stock are the maximum possible damage, D_{max} in the notation of Sect. 2.2, whereas we would associate the second phase of after-shock resilience with the avoided damage $D_{max} - D$ rather than the GDP. In economics, the effects on GDP could be calculated from imposing shocks on models for GDP and calculate the difference between GDP and GDP loss as a measure of ex-post resilience. This method is known under the name impulse response method, which imposes shocks on stochastic difference equations. Section 10.5 demonstrates its working for a shock on electricity supply and other variables representing electricity demand.

Molyneaux et al. (2012) essentially give up the difference between vulnerability and resilience because they define 'the Resilience Index, as a comparative indication of vulnerability to a variety of shocks, …'. Their 'Resilience Index is the geometric mean of the seven normalised indices as follows: Resilience index RI = $\sqrt[7]{abcdefg}$, where a is the non-renewable-fuel-used index; b the generation efficiency weighted for non-renewable fuel dependence index; c the distribution efficiency index; d the carbon intensity index; e the diversity index; f the redundant electricity for use in GDP index; g the reliance on imports index.' (p.193). The seven indices are all items, which reduce the probability of blackout, electricity rationing and/or volatility, or high prices. The second, after-shock phase of damage limitation and getting out of problems is not considered from a post-shock perspective. With exclusive focus on preventive reduction of vulnerability, there is no consideration of allocating means to the improvement for the two phases.

Interdisciplinary literature emphasizes the origin and development of the resilience concept in psychology, ecology, and management and its holistic approach. As such it also includes a first pre-shock phase that tries to prevent adverse events, mentioning explicitly terrorist attacks and other threats to survival (Thoma 2014). Climate change and software problems are discussed in this literature. The possibility to reduce probabilities of adverse events through software mistakes is emphasized by Mayer and Brunekreeft (2021) and Acatech et al. (2021).

10.2.5 Fitting in Vulnerability Without Dropping Prevention for Probabilities

Two problems come to the fore in the literature. First, there is not always a differentiation between the probability of a bad event and the vulnerability where it hits. In case of natural catastrophic events, probabilities cannot be changed through preventions and therefore the focus often goes to vulnerability. In Table 10.1 we look at three examples: Achilles' heal, Turkey's earthquakes, and the interaction of digitalization with the electricity system. In the classical example (Table 10.1, column 2) of Achilles, it is clear that the major vulnerability is his heal (column 2, row 3). The relevant probability is that of an arrow hitting the heal (column 1, row 1). This probability can be reduced through prevention (column 1, row 2), preferably through peace keeping and negotiations, or war strategies, indicating that probabilities may not be just given but under influence of possible actions. Moreover, vulnerability could be reduced through policies (column 1, row 4) such as classical body-protection means or, in the long run, through medical measures and research. The resilience against the consequences of the attack consists of Achilles' health (Table 10.1, column 1, row 5). The policy supporting resilience is the health system or, in modern times, the hospital.

A second instructive example is the case of a recent Turkish earthquake (see Table 10.1, column 2). The probability of an earthquake cannot be reduced unlike those from excessive mining in the Groningen area of the Netherlands and many other

Table 10.1 Instructive examples for shock probabilities, vulnerability, and resilience policies

Example	Achilles	Turkey	Digitalization & electricity system
Probability of shock	Arrow	Earthquake	Human and natural errors
P-prevention	Peace negotiation	–	Computer-safety and network-technology; climate policy
Vulnerability	Heal	Missing esct (a)	Interacting in two systems
Vulnerability policy	Body protection	Introducing esct (a)	Computer safety, cyber police, undergrounding, cross-border cables
Resilience	Achilles' health	Not missing esct (a)	Interacting in two systems
Resilience policy	Hospital system	Rescue service	Target attack, fault, repair

(a) Earthquake-safe construction technique; construction supervision against fraud; liability law effective after fraud areas in the world. Prevention can only reduce the vulnerability through earthquake-safe construction methods. These were present in some houses during the earthquake, but, through fraud of construction firms, not in others. This has unnecessarily increased the vulnerability. Policies improving the vulnerability therefore could be obligations for earthquake-safe construction techniques, construction supervision against fraud, and effective liability law after fraud, perhaps through insurance obligation. This would also improve the resilience because the damage would be less. For the remaining damage there are resilience policies like preparing rescue services.

Third, in the case of two systems (Table 10.1, column 3), the problem may not only go from digital problems to the power system, but also for power outages stopping the digital system. With two systems interacting there are more potential sources of bad events. Under vulnerability, the heal problem of Achilles seems given and cannot be under prevention given the medical knowledge of the time; but earthquake-safe construction of houses is a good example of policy against vulnerability. In other words, in Table 10.1, the row 'vulnerability' can also be split up in 'given' and 'prevention or policy'.

In our case of a combined power and digitalization system (Table 10.1, column 3), the problems may be caused by programming mistakes, cyber-attacks, or technical deficiencies (row 1) in the technique of the power system (Acatech et al. 2021). Policies against malfunctioning can reduce the probabilities of bad events for internal reasons. Cyber-attacks have probabilities which can be reduced through the deterrence effect of the reputation of a cyber police. Climate policies try to change the probabilities of extreme-weather events (row 2). C02 emissions have an effect on the climate after ten years. Reducing emissions therefore can have small effects after ten years. However, visible effects on the weather may take much more time, leaving the probabilities close to exogenous for shorter time horizons and endogenous only for the long term. All aspects of vulnerability then are related to both of the systems because each can set the other out of function (row 3). Cyber-safety[25] against internal failure and external events, and, for example, underground cables are measures trying to reduce vulnerability and improve the probability augmented resilience index (Francis and Bekera 2014). Cross-border cables can be helpful in power crashes to the extent that imports of electricity are not limited as it was the case in a California blackout (row 4).[26] Prevention therefore is related to shocks, vulnerability, and also to resilience.

[25] Nguyen et al. (2020) point to industrial control systems (IDS) and cyber-physical systems (CPS), requiring intrusion detection systems (IDS).

[26] Imports of electric current did slightly mitigate the supply shortage during the California blackout 2001. Also, the interconnectivity is strong enough to transmit price changes to neighbouring states. But strong price difference between states indicate that interconnectivity is far from making prices equal on all sides of the borders of US states (Faruqui et al. 2001). This indicates limited capacity to trade electricity. This is also the case in continental Europe. The prices on the Dutch and Belgium side of the border seem to be equal at the northern border of Belgium (Mulder and Willems 2009) but not at the western border where RWE hopes to get a higher prices when feeding in there. The share of renewables in addition has reduced trade and its ability to equalize prices (Ziesemer 2019). Investment in transborder cables should be a high priority in European climate and competition policies.

One important investment is in algorithms which are able to distinguish between attacks and normal fault contingencies (Nguyen et al. 2020), which conceptually means distinguishing between an event with given probability and one with probability that could have been reduced. As this algorithm works after the attack or fault it belongs to the phases after the shock, and to the policies against vulnerability and support of resilience.

When vulnerability is included, the formal model of Sect. 2.2 needs some adjustment. In case of (i) a given probability for bad event, say π, (ii) the vulnerability can be reduced through some adequate investment z to $v(z) \in [0, 1]$, $v' < 0$, $v'' > 0$, we can re-define $p(z) \equiv \pi v(z)$. Obviously, the optimization model presented above still holds with adequate reinterpretation of vulnerability instead of probability. If instead both, the probability and the vulnerability can be reduced through investment we get lack of resilience or expected damage as

$$L = p(x)v(z)D(y)$$

The problem (10.3) then becomes maximizing in each period

$$RI = \left[1 - p(x)v(z)\right]\left[1 - d(y)\right] - \lambda(x + y + z - b) \tag{10.3'}$$

Investment can now go into probability reduction, vulnerability reduction, and damage reduction. A specification of the function $v(z)$ is necessary to carry out the optimization. The original model (10.3) is the special case of $z = 0$, $v = 1$, which is possibly a corner solution of the larger model. The model of the economic literature is the special case $p(x) = \pi$, meaning that prevention is not possible and good policy can only achieve vulnerability reduction. Ultimately, the model differs from most of the literature by way of giving shock prevention defined as $p'(x) < 0$ a role, which was the view in setting up (10.3). Watson et al. (2015) briefly discuss 'Willingness to pay … to avoid an outage' but it does not become explicit whether this is referring to a reduction in probabilities or in vulnerabilities. Maximizing (3′) yields the following first-order conditions.

$$x : -p'v(z)[1 - d(y)] - \lambda = 0 \tag{10.4'}$$

$$y : [1 - p(x)v(z)][-d'(y)] - \lambda = 0 \tag{10.5'}$$

$$z : -pv'[1 - d(y)] - \lambda = 0 \tag{10.11}$$

$$\lambda : b - x - y - z = 0 \tag{10.12}$$

With marginal effects equal to the Lagrange multiplier, the marginal effects are equal to each other, and equal to the marginal increase of resilience through a budget increase.

$$-p'v(z)[1-d(y)] = [1-p(x)v(z)][-d'(y)] = -pv'[1-d(y)] = \lambda = \frac{\partial RI(t)}{\partial b(t)} \quad (10.13)$$

The last equation follows from the envelop theorem, which allows us to use only the direct effect, because the indirect effects via x, y, and z are equal to the first-order conditions and therefore zero and drop out. Dividing all equations by $(1- d)$ and $(1- pv)$ we get

$$\frac{-p'v(z)}{[1-p(x)v(z)]} = \frac{[-d'(y)]}{[1-d(y)]} = \frac{-pv'}{[1-p(x)v(z)]} = \frac{\lambda}{[1-p(x)v(z)][1-d(y)]} = \frac{\frac{\partial RI(t)}{\partial b(t)}}{RI(t)} \quad (10.13')$$

The elasticities of $[1- p(x)v(z)]$ and $[1 - d(y)]$ with respect to x, y, and z respectively should be equal to each other and equal to the semi-elasticity of the per-period resilience index with respect to the budget b. Francis and Bekera (2014) analyse a case of putting cables underground at cost z, which increases resilience. However, this does not consider the potential alternative ways of spending the money in the budget.

After the shock, $p = 1$, and the resilience formula is

$$RI(p=1) = \int [1 - v(z*)][1 - d(y*)]dt$$

where $z*$ and $y*$ are the values that have been invested before the shock. Lower vulnerability and damage indicate higher resilience. The inclusion of vulnerability is essential here because otherwise the ex-ante RI would be zero when we get $p = 1$. Moreover, the traditional resilience index has $v = 0$ (Bruneau et al. 2003) and is extended here to include vulnerability.

10.2.6 Getting Probability Estimations

$e^{-x^{\alpha}}$, the functional of the distribution function used above for expenditures, is similar to a special case of the Generalized Extreme Value distribution $e^{-(1+\xi(x-\frac{\psi}{\beta}))^{\alpha}}$, used in empirical resilience research, where the GEV is the special case $\alpha = -1/\xi$. For $\psi = 0$, $\xi = 1$, we get $e^{-(1+x)^{\alpha}}$ indicating the difference to a special case. The GEV has been used to estimate probabilities of floods in the UK for hourly data from 33 years (Espinoza et al. 2016) and in the Netherlands (Aerts and Botzen 2008). Moreover, vulnerability is defined as failure of components in the electricity system, depending on threats, wind speed, and accumulated rainfalls (Espinoza et al. 2016). These relations are called fragility curves. This type of empirical application of the resilience concept seemingly is still in its infancy because of the lack of data on systems failure.

10.2.7 No Probabilities: Viability Theory

Even if data are available as in the case of insurance against repeated events, probabilities are not known but rather *estimated* probabilities. Using data, which are always from the past, for decisions for the future requires assumptions about the absence of structural breaks and change. At least in this sense probabilities are always *subjective.* In addition, in first-principles models, probabilities are typically also subjective. If probabilities are too uncertain for alternative scenarios to be informative, as in the case of events happening only every 200 years or even less frequently, the literature speaks of *radical* uncertainty and does not use probabilities and optimization (Karacaoglu and Krawczyk 2021).

Abandoning optimization, Karacaoglu and Krawczyk (2021) suggest to (i) quantify a systemic degree of resilience against shocks, (ii) minimize the potential harm from shocks, and (iii) maximize the opportunities and capabilities to adapt and bounce back after a shock. Their viability theory suggests the 'existence of controls which can keep the system within a viability kernel' where a system operates safely as a solution to a viability problem.

They 'contend that delivering a solution to a viability problem is more relevant to governments than an 'optimal' solution'. This may overlook the case dependence of an effort towards any solution because such comparison makes only sense if both procedures are an alternative. This raises the questions (i) whether relevant cases are really completely in the realm of subjective probabilities or that of radical uncertainty, and (ii) how to handle mixed cases.

Viability theory is modelled for a set of state variables $x(t)$, which change according $dx = f(x(t), u(t))$ under the constraint that $x(t)$ remains in a closed set K. A closed set V is part of K, which is viable and called the viability kernel V. Control variables $u(t)$ are used to keep $x(t)$ in the viable part V and show inertia otherwise, implying indeterminacy of the path of control and state variables inside the viability kernel unless some behavioural idea is added. Whatever the added behaviour is, it is assumed to be costly to stay away from the boundaries. If the system is unstable, $x(t)$ would go to the boundaries of the viable set with ever higher or ever lower externalities like emissions than in a stable model. A regulator with suitable instruments is needed to keep them away through acting in time, which is not guaranteed though because of the absence of optimal behaviour. The environmental problem is similar to neoclassical models, where the natural environment is not stable or regulated, even if they are stable in the market relation. If the system is stable and the steady state is in the viability kernel there is no viability problem and resilience to shocks works as in the standard resilience graph. Whereas these are interesting lines of thought, they also raise questions and comments:

(1) What does viability exactly mean if 'outside the kernel, special crisis control needs be implemented'? Seemingly there is an assumption that we are still alive outside the viability kernel.
(2) Surprisingly, there is no explicit formulation of radical uncertainty or any other form of uncertainty although this was the starting point. One might have expected it in regard to the boundaries of the constraints K and those of the viability kernel V, and also regarding functions f and its derivatives indicating the possibility of marginal steering. The contemporaneous resilience problems do not have exact boundaries in terms of degrees of global warming, biodiversity, use of nitrogen, etc. Therefore, it is tempting for those less inclined to precautionary behaviour to run ever more into situations with unclear limits of viability hoping that this is not very dangerous.
(3) Compared to optimal growth theory, conceptually the viable kernel comes into the place of (maximizing) a utility functional. Still, this is similar to neoclassical models, where the natural environment is not stable or regulated, even if the models are stable in the market relation. Here, the hope is that markets for property rights are stabilizing.
(4) As it is not specified what the regulator is doing to stabilize the system, we cannot compare it to optimization.

In spite of the critical remarks and questions, an interesting conclusion can be drawn from the comparison of the unstable viability model of Karacaoglu and Krawczyk (2021) with stable models potentially underlying the resilience graph: Resilience policy should try to turn an unstable model into a stable one, besides enlarging the viability kernel through public investment. This extends the definition of resilience to unstable models, whereas the standard figure only used stable models with the exception of the case of 'collapsing behaviour' in the graph from Kröger (2017). Resilience policy would aim at excluding collapsing behaviour when trying to turn an unstable model into a stable one. This is conceptually perhaps less ambitious than trying to maximize resilience indices conditional on getting the probabilities formulated in the viability indices, which are underlying the maximization of resilience indices in the model (10.2)–(10.6) and its extension (10.3′)–(10.5′), (10.11), and (10.12).

However, clarifying stability conditions empirically can be very demanding. One way of doing it econometrically is the estimated model of Eqs. (10.14)–(10.18) below, which (i) extracts the probabilities for the residuals (shocks) through estimation (not identical to those of resilience indices), (ii) shows the damages in Figs. 10.3, 10.4, 10.5, 10.6 and 10.7 for an exogenous shock, and (iii) has the possibility to test for stability. This is possible because this type of models deals with one problem at a time, whereas a system of resilience indices would be more comprehensive.

Table 10.2 summarizes this section schematically assuming that there are underlying dynamic economy-environment models with a stable or unstable economy part and index models such as (10.2)–(10.6).

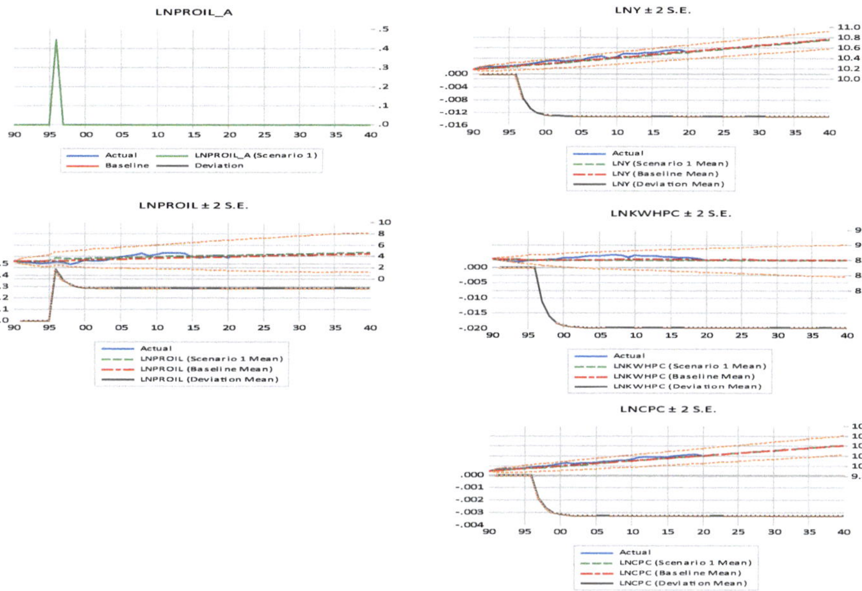

Fig. 10.4 A shock on oil prices and the response of the economy

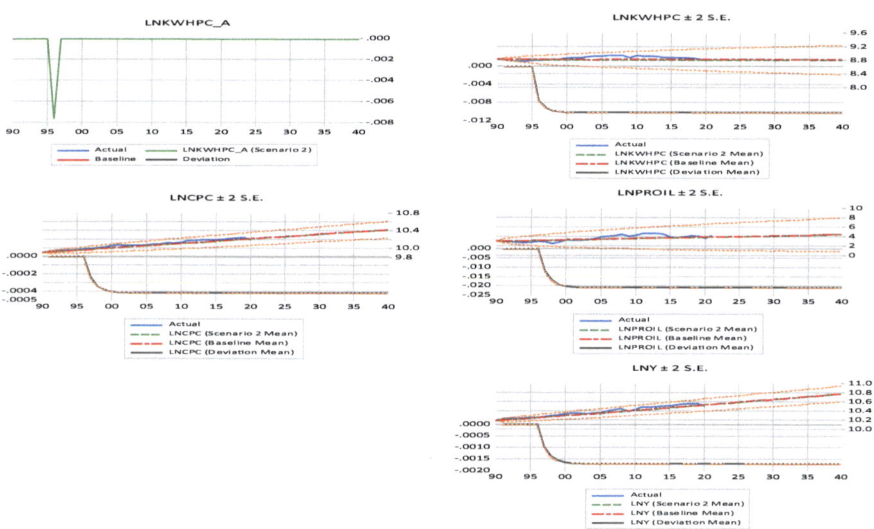

Fig. 10.5 A shock on electricity use and the response of the economy

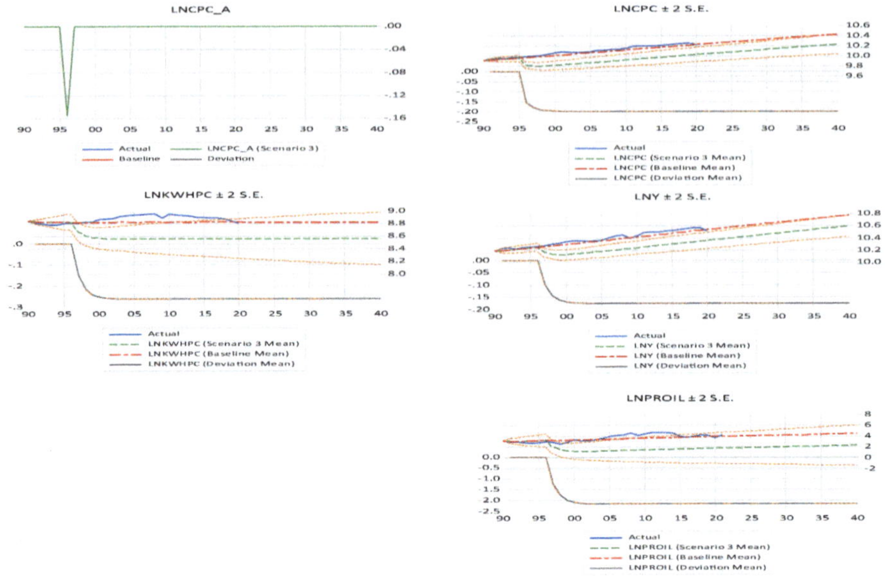

Fig. 10.6 A negative shock on consumption and the response of the economy

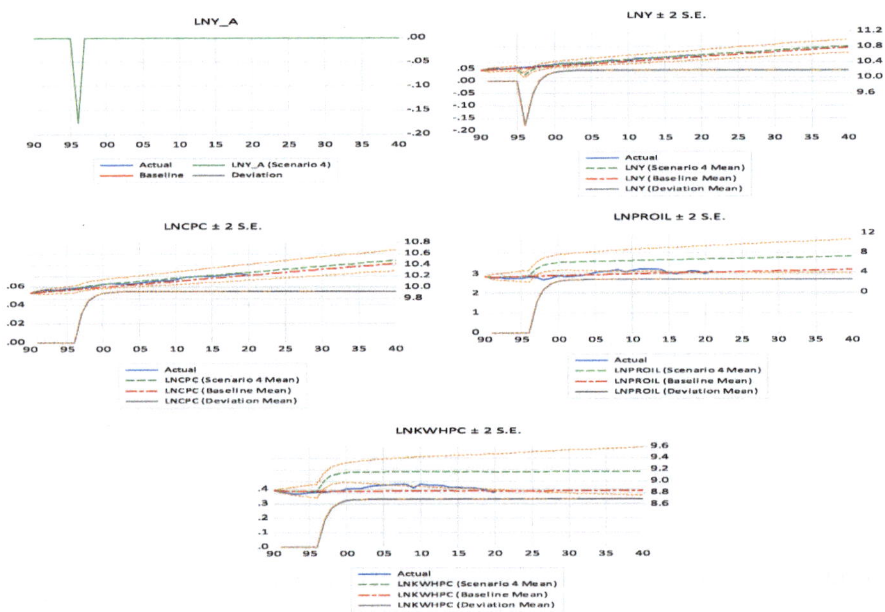

Fig. 10.7 A negative shock on GDP per capita has a positive permanent effect on all variables. GDP per capita is super resilient to its own shock

Table 10.2 Resilience problems, policies, and research for underlying models

Models → Issues↓	Stable dynamic models	Unstable dynamic models	Resilience index models
Environmental control	Tradable permit markets or taxes	Abatement policies	Index observation
Other stabilization through ...	Market prices, decreasing marginal product of capital	Policies	Jump into optimum (a)
Resilience problem	Vulnerability, recovery	Collapsing behaviour	Multiple reduction of damages and their probabilities
Resilience policy	Reduce damage	Turn unstable into stable model	Quasi maximization of resilience index
Empirical problem	Getting probabilities and damage functions	Measure stability	Getting probabilities; balancing marginal impact of costs on probabilities, vulnerability, damages, and budget
Empirical method	Estimate residuals for (10.14)–(10.18); test model for (in-) stability	Testing model for (in-) stability	Test optimality conditions?

(a) Absence of adjustment costs assumed

10.3 Resilience of Interacting Systems

10.3.1 Literature on Interacting Systems

Whereas the early resilience literature discussed above looks at the power system alone, the more recent literature looks at the power system in connection with the digital communication system, the natural and car gas system, the heating system, the transportation system (Manshadi and Khodayar 2015; Nguyen et al. 2020), and the food, water and wastewater systems, health, telephone, and police (Acatech et al. 2021). The presumption there is that there is a two-way direction of causality with electricity outages and communication failure, implying that under repair both parts have to be (re-)implemented simultaneously, because none of the parts is separable (Liu et al. 2020; Mayer and Brunekreeft 2021). Table 10.3 indicates the interaction between several systems delivering services to each other. Electricity and ICT services are delivering to all other systems (indicated by a 'x') whereas other systems have a much more limited role (indicated by a '–').[27] Health and food are important simply because all systems need labour power. We leave the interpretation of the single symbols to the reader.

[27] Lin and Bie (2016) provide a nicely systematic scheme of the interactions within an integrated energy system.

Table 10.3 Delivery matrix of interdependent systems

Service delivery from ↓ to →	Car gas	Electricity	Heating	ICT	Natural gas	Transport	(Waste) water	Health	Phone	Police	Food
Car gas, oil	–	–	x	–	–	x	–	(x)	–	x	x
Electricity	x	x	x	x	x	x	x	x	x	x	x
Heating	–	–	–	–	–	–	–	x	–	x	x
ICT	x	x	x	x	x	x	x	x	x	x	x
Nature. gas	–	–	x	–	–	x	–	–	–	–	?
Transport	x	–	–	–	x	x	–	x		x	x
Water	–	–	x	–	–	x	x	x	–	x	x
Health	x	x	x	x	x	x	x	x	x	x	x
Phone	–	x	–	x	–	x	–	–	x	x	x
Police	–	–	–	x	–	x	x	–	–	x	x
Food	x	x	x	x	x	x	x	x	x	x	x

This literature links to the idea of probabilities for consequences (Watson et al. 2015; Mahzarnia et al. 2020) in the form of, for example, loss-of-loading probability (lolp), which increases whenever the supply of electricity grows more slowly than the demand. However, these probabilities are not the same as the *p* for a damaging shock event introduced above but rather those for a normal blackout. No part of the literature does consider a maximization of the expected avoided damage as we have introduced it above.[28]

10.3.2 Some Problems of Three Systems Meeting at the Highway

The systems of greatest interest to this project are:

- Traffic (with highway)
- Energy system with car gas and electricity stations
- Digital information for traffic and energy.

10.3.2.1 Electricity and Communication Have a Special Interaction Via Satellites (Acatech et al. 2021)

'…Satellitenkommunikation, die zunehmend kostengünstig wird und unbeeinflusst von Stromausfällen ist …'. Communication with satellites requires an electricity network, perhaps not at the end of the satellite, which may have a battery or solar module, but at the other end on earth the ICT requires electricity. Are there any services that cannot be delivered without electricity on the earth's end of communication? If the earth's end of communication also has its own islands of battery or solar/wind modules, what are the probabilities and consequences of their crashes?

10.3.2.2 Investment Policies

The new plans for investment in electricity loading at German highways uses techniques where loading time of 20 min allows for driving 200 km afterwards. This compares to the current practices for car gas loading of 5 min for 400 (or 500) km or even more. The duration of loading therefore is eight times as long in the new system compared to the old. Keeping loading times constant requires eight times as many loading points for the future compared to today. As loading stations are crowded in holiday periods already now, a delay may yield to even more crowded loading and parking places and more traffic trying to circumvent this with some traffic leaving the highways. This will result in more activity of GPS systems in cars. This will

[28] Zhou et al. (2012) discuss statistical finance methods for related problems of reducing vulnerability, not shock prevention.

require a high speed of carrying out the investments in loading equipment in order to avoid more traffic jam and overburdening of the electricity system during the holiday period and perhaps outside.

10.3.2.3 Public Goods and Lack of Incentives?

A blackout reduces the revenue of providers of network services. When deciding on their investment they consider these potential losses. However, it is unlikely that they will consider the losses of electricity users. This suggests that there is a negative network externality leading to under-investment in network safety measures. Similarly, software mistakes reduce the reputation of software suppliers, which they consider when developing products and managing quality control. Also, here it is unlikely that they will consider their clients' costs of blackouts from software mistakes unless it is easy to switch to better suppliers. In both cases resilience policies should think about policies to reduce these externalities (Mayer and Brunekreeft 2021; Acatech et al. 2021).

10.4 Relations with Neighbouring Countries and the EU

10.4.1 Background

The transition to higher shares of renewable energy implies increasing volatility of the electricity system. Storage capacity is the major means to mitigate volatility (Droste-Franke et al. 2012). However, the cost of storage and kWh of electric current can be reduced through international trade in electric currents (Bussar et al. 2014; Falcan 2021). Besides the technical link, international trade in electric current also connects the ICT system to the international electricity system. Resilience monitoring then has to take care of international relations.

Acatech et al. (2021) raises the question whether the international relations should be taken care of early on in the process of organizing resilience monitoring or at the end of this process, without giving an answer. At the same time Acatech et al. (2021) speaks of important path dependences. If there are several national path dependent processes, international coordination of research and implementation must happen to occur right from the beginning. Otherwise, there are no possibilities to adjust to international aspects and problems later on because path dependence has a national perspective. Fortunately, there is some international cooperation in the areas of grid control and standardization of techniques and information exchange (Mayer and Brunekreeft 2021). These aspects gain in importance when there is more international trade in electric current.

10.4.2 Consequences of Neglecting the International Dimension

Currently electricity prices for industrial users are higher in the Netherlands than in Germany. Therefore, the Dutch aluminium industry has requested and received permission to put a cable from their plants in the Netherlands to Germany in order to get the lower electricity prices. The mirror image of this is that a lack of international trade in electric current does not allow other firms or household to get the low German prices as well. The reason is the lack of transborder transmission because of limited cable capacity.

Similarly, the German electricity provider RWE owns a gas fired power station in Maasbracht in the Netherlands. It is now unpacking this station because gas has become more competitive relative to coal. RWE wants to link that station to the Belgium network because in Belgium they obtain higher prices. The Dutch government has not given permission to link to Belgium.

In both cases, having more cross-border transmission capacity would help having the same prices on both sides of the border, lower prices for consumers on the high-price side and higher prices for producers at the current low-price side, if prices settle between the current ones.[29]

If a lack of international coordination in resilience monitoring of digitalization would lead to a further separation of countries' electricity systems from each other, the integration of electricity markets, which is weak anyway (Barteková and Ziesemer 2015), would be hindered.

10.5 Alternatives to Resilience as Transitory Shock Effects in Macroeconomic Growth

In this section we demonstrate that empirical economic models in the form of difference equations react to one-time shocks, for example on electricity supply, in a way that may be similar to that of the standard resilience figure but there may also be considerable differences in the economic reactions. In that figure (Bruneau et al. 2003), and Fig. 10.1 in Acatech et al. (2021) also used in Brand et al. (2025) a system works at 100 percent of its possible performance, then gets a shock that reduces the performance, and afterwards returns to its 100% level. Such a result would hold by definition for a difference equation system that is stable in the levels of (stationary) variables. Compared to the general conceptualisation,[30] Fig. 10.6, is a bit richer

[29] Conversely, for households the opposite holds true: households in the Netherlands pay much less per kWh: 31 cents in Germany versus 12 (17) cent in the Netherlands on a yearly basis (not) taking into account the energy subsistence minimum per household. The reason for this gap is the difference in taxes, subsidies (EEG and other taxes or subsidies) and differences in electricity pricing.

[30] See Fig. 2.1 in Chap. 2 in this book; from Kröger (2017).

in that it allows for higher or lower than initial performance after recovery (to be distinguished from higher or lower adjusted *resilience*, which can be seen only after the next shock) and also no recovery. Vector-error-correction models may be stable not in the levels of the variables but only in the growth rates of variables because they combine levels of non-stationary variables and their differences. The *maximum* number of transitory, resilient variables defined as returning exactly to their base-line levels is equal to the number of long-term relations between variables in levels. As this leans on advanced econometrics, we illustrate this by way of constructing a simple example. We obtain both cases: higher and lower performance of the system after recovery compared to the pre-shock phase.

We consider the variables GDP per capita, *y*, consumption per capita, *cpc*, electricity per capita, *KWhpc*, all for Germany, and US oil prices, *proil*,[31] as proxy for closely related German oil prices, all in natural logarithms indicated by *ln*. Data for consumption and GDP are taken from World Bank (2022), oil price data from US EIA at www.eia.gov, data for electricity use from the BMWK website at www.bmwk.de.

Statistical considerations explained below lead to the following model with four equations. The long-run relation estimated for the period 1991–2020 is[32]

$$\begin{aligned} CE &= LNCPC(-1) - 1.1LNY(-1) + 0.196LNKWHPC(-1) \\ &\quad - 0.026LNPROIL(-1) + 0.00344t - 0.33 = u_{-1} \end{aligned} \tag{10.14}$$

The four simultaneously estimated equations including (10.14) abbreviated as *CE* are

$$D(LNCPC) = 0.159CE + 0.01 + \varepsilon_{1,t} \tag{10.15}$$

$$D(LNY) = 0.65CE + 0.011 + \varepsilon_{2,t} \tag{10.16}$$

$$D(LNKWHPC) = 0.97CE - 0.00137 + \varepsilon_{3,t} \tag{10.17}$$

$$D(LNPROIL) = 7.97CE + 0.019 + \varepsilon_{4,t} \tag{10.18}$$

CE (cointegrating equation) is an abbreviation for the first equation with all terms on the left-hand side except for a residual if it deviates from its expected value of zero, $Eu_{-1} = 0$. *CE*, which is zero in equilibrium, then is a part of all the other four equations. The log differences on the left-hand side of the four equations are the

[31] Europe Brent Spot FOB (free on board). Oil price movements of many countries are closely linked to each other.

[32] T-values are for (14): [−6.67], [2.255], [−2.8], [1.6]; for (15): [0.98], [3.87]; *for* (16): [2.73], [2.8]; *for* (17): [5.2], [−0.43]; *for* (18): [2.99]; [0.434].

growth rates of the variables appearing in the form of levels in *CE*. The variables in (10.14) are allowed to have any form of two-way causality.

One interpretation of *CE* is that per capita consumption depends positively on GDP per capita, negatively on electricity use and positively on oil prices (indicating limited substitution), where coefficients are elasticities because variables are in logs. Moreover, there is a small positive time trend compensating for the high coefficient of *lny*. In equilibrium, $CE = 0$, and therefore the intercepts of Eqs. (10.15)–(10.18) are the long-run growth rates, which are negative only for electricity, and statistically insignificant for electricity and real oil prices because of the high volatility (Droste-Franke et al. 2015, Chap. 5.3), implying that it makes no sense to put the latter coefficient to zero. The coefficients of *CE* in (10.15)–(10.18) are adjustment coefficients indicating how strongly disequilibrium $u_{-1} \neq 0$ affects the growth rates.

10.5.1 Statistical Considerations

Estimation of an equation like (10.14) independent of (10.15)–(10.18) may be spurious because the variables may have a unit root, defined as unit coefficients in a regression of a variable on its own lag, leading to accumulation of residuals. If such accumulation happens to occur in two variables in a similar way, they may falsely appear as correlated. In Table 10.1 we summarize the results of Augmented Dickey-Fuller unit root tests with breakpoint. All variables have a probability of a unit root larger than 10 percent, and therefore the unit root hypothesis cannot be rejected. Alternatively, using the argument of low power of the test (a low probability of rejecting a false hypothesis) and in view of the low coefficients (far away from unity) we could declare all variables as stationary and estimate a VAR in log-levels. P-values then indicate the error probability of our assumption. We call this

Assumption 1 All variables in log-levels are I(0) and a VAR in log-levels could be estimated.

Because of the disequilibrium effect, the type of model used is called Vector Error Correction Model (VECM). Adj. R-squared four the four equations are −0.0012, 0.182, 0.47, 0.215. Low adjusted R-squared values are an invitation to add more

Table 10.4 Results for ADF unit root tests with breakpoints

Variable	Ln y	Ln cpc	Ln KWhpc	Ln proil
Break type: outlier	Innovational	Innovational	Innovational	Additive
Lags, break date, specific., selection	1, 2016, c,t, F-statistic	0, 2010, c,t, F-statistic	0, t, 2008, Dickey-Fuller t-st	0, 2005, c, t, F-statistic
P of unit root	0.1015	0.8218	0.1983	0.36
Coefficient (a)	0.87	0.374	0.4876	0.418

(a) Comparable to unity

regressors. However, we keep the model simple because it serves only illustrative purposes for the communication of the basic logic and sometimes is called 'minimum necessary model' in the theoretical literature (Krugman 1993) or '(very) small model' in the econometric literature Brillet (no year).

In regressions we should not use variables with unit roots, but rather their difference, or combinations of variables, which have no unit root problem. In order to test for combinations, we use the vector-error-correction model (VECM), which uses differences and combinations of variables in levels.[33] The underlying vector-autoregressive model (VAR), which explains each variable by lags of all variables and a time trend, has optimal lag length $p = 1$ according to all lag length criteria (LR, FPE, AIC, SC, HQ) and is stable. The trace test and the maximum-eigenvalue tests to determine the number r of long-term, cointegrating relations would suggest a p-value of 0.34 and 0.22 for the hypothesis $r = 0$, respectively. This would normally lead to a rejection of the hypothesis of cointegration. However, as we have a very low number of observations and the test has low power, a reasonably looking cointegration graph (not shown) and a reasonable regression (10.14), we assume that (10.14) is a cointegrating equation.

Assumption 2 There is one cointegrating equation, which is Eq. (10.14) and the VECM (10.15)–(10.18) is valid.

The VECM under Assumption 2 is stable.

10.5.2 Shocks and Resilience

For a stable VAR in log-levels (under Assumption 1), after a shock all variables return to the long-run equilibrium. Thus, all variables show resilient behaviour. The only interesting aspect that can be seen from such a shock analysis is then how long the deviations of the variables from equilibrium are statistically significant (not shown).

For the VECM, it follows from the Beveridge-Nelson decomposition of the variables into stationary and non-stationary stochastic processes that the *maximum* number of possibly transitory shocks in terms of effects from a one-time shock is equal to the number of long-term relations (Lütkepohl 2005, Chap. 9). As the number of long-term relations is $r = 1$ here, at most one variable can have transitory effects after a shock and thereby can show resilient behaviour in terms of levels. In Figs. 10.3, 10.4, 10.5, 10.6 and 10.7 we show the baseline simulation of the model with Eqs. (10.14)–(10.18) and adjustment processes after a shock of one standard deviation of the baseline simulation value.

In the baseline simulation shown in Fig. 10.3 all data points are inside the confidence interval. The graph for the GDP per capita, and also the other graphs, show

[33] The general version of a VECM for a $(K, 1)$ vector of variables y is $dy = \alpha\beta' y(-1) + \Sigma_j \Gamma_j dy(-j)$, where d indicates a differences, α and β are (K, r) matrices, j indicates the lags and the number of long-term relations r in the $(r, 1)$ vector $\beta' y(-1) = u_{-1}$ is determined through the trace and maximum eigenvalue test.

dips in the data for the COVID crisis in 2020, the financial crisis in 2009, and the ICT bust in 2003.

Figure 10.4 shows the effects of a one-time shock on oil prices by one standard deviation, 0.45, in 1996. The upper left graph shows the shock. GDP per capita is reduced by 1.3% and does not return. Electricity use and consumption per capita also decrease permanently. Oil prices return partly to their baseline value. Resilience is strongly limited for oil prices essentially absent for the other variables.

Figures 10.5 and 10.6 show a negative shock on electricity consumption and consumption per capita by one standard deviation, −0.0077 and −0.155 respectively, in 1996. All variables go down by a constant percentage from baseline. There is no resilience at all.

Figure 10.7 shows a negative shock in 1996 on GDP per capita of one standard deviation, −0.178, which is again high because of the included trend (it is not taken from the residuals but rather from the baseline simulation values). GDP per capita is low for 1996 and 1997 and is back and above its baseline value in 1998 and remains four percent beyond baseline. This may be called *super resilient*; it is similar to politicians' idea to get out of the crisis stronger than before. It also reveals that the one variable that may have a transitory shock (according to the Beveridge-Nelson decomposition because of having one long-term relation) at first sight is the GDP per capita. However, the shock is not only returning to zero but rather to permanently positive values (the maximum number of *one* variable having transitory shocks is not realized, but it is zero). All other variables do not react negatively at all from 1997 onwards and follow the long-run increase of the GDP per capita.

10.6 Summary of Resilience Policies

10.6.1 Shock Prevention Policies

Climate change policies can reduce the speed of global warming and reduce the intensity of heavy rains and the ensuing floods.

The x-eyes principle and externality policies, such as network-availability subsidies, can perhaps reduce the probability of programming and other software mistakes (Acatech 2021). International cooperation in the areas of grid control and standardization of techniques and information exchange is necessary and will gain in importance (Mayer and Brunekreeft 2021). Other cases are discussed in Brand et al. (2025).

National and international cyber police and policy can deter later national and international ransom attacks by acting tough on current attacks or make them less likely. For example, Ukrainian hackers have paid back money from the 2019 attack on the University of Maastricht after the EU had promised to support EU integration of Ukraine.

10.6.2 Vulnerability Reduction Policies

Tree trimming, walls around generators, cross-border cables, and underground cables can reduce the vulnerability of the electricity system.

Earthquake-safe construction techniques, construction supervision against fraud, liability laws effective after fraud can help to reduce the vulnerability of buildings.

Cyber- and computer safety measures can reduce the vulnerability of the digital parts of the electricity systems. Industrial control systems (IDS), cyber-physical systems (CPS), requiring intrusion detection systems (IDS) are prominent means in the engineering literature.

10.6.3 Recovery and Damage Reduction Policies

Locally available batteries may help to bridge blackouts. Rescue and repair services and their required inputs and budgets should be available.

10.7 Conclusion

Resilience literature focusses strongly on the post-shock phase. The high-probability-low-impact events are discussed in older reliability literature. The endogeneity of the low probabilities for high-impact events is a gap in the literature. The resilience literature should get broader again and include shock prevention in resilience indices and avoid unstable dynamics in line with viability theory.

We provide a first-principles model defining a resilience index, which includes low-impact high probability reliability risk analysis and high-impact low probability as special cases. Probabilities for bad events are endogenous. A budget can be used for three resilience purposes: reducing shock probabilities, reducing vulnerabilities, and reducing damage. The resilience index is maximized under the budget constraint when the elasticities of no-event probabilities, expected vulnerability reduction, and damages with respect to expenditures are equal to each other and equal to the elasticity of the resilience index with respect to a marginal budget increase. The message is that the budget for the reduction of probabilities, vulnerabilities, and damages should be allocated in a well-balanced manner.

The electricity system interacts with many other domestic systems, electricity systems in foreign countries, and, of course, the whole economy. The essentials of these interactions can be summarized as follows.

Where the three systems for electricity, digitalization, and traffic meet at the highway, a sufficient speed of investment is required in order to avoid congestion of electricity loading because the frequency and duration of loading is becoming a multiple of the car-gas system. E-car loading is already bothered by waiting queues in the Netherlands. Satellite systems may have their own risks in their electricity provisions endangering satellite services.

International trade allows consumers and firms to get electricity from the cheapest suppliers and improves EU competitiveness through integration. Moreover, international trade of electric current provides some insurance against lack of national supply and reduces the cost of storage. Instability problems of national, simultaneous, uncoordinated action (Acatech 2021) can be mitigated through open borders for electric current. International cooperation in the areas of grid control and standardization of techniques and information exchange is necessary (Mayer and Brunekreeft 2021). Also, international computer trade in electric current links ICT to the electricity systems. International coordination of research and implementation must happen to occur right from the beginning. These international aspects are very important for the resilience of the electricity system.

A dynamic empirical model using German data shows that a negative shock on electricity has permanent negative effects for GDP per capita, consumption per capita, and oil prices, and therefore should be prevented. A negative shock on GDP per capita has negative effects on all mentioned variables in the short run and positive effects in the long run. A positive shock on oil prices and a negative shock on consumption per capita both have a negative impact on all other variables. In sum, the economic system is only (super) resilient against GDP shocks but has long-run percentage deviations from baseline from all other shocks.

Finally, we have derived a list of specific recommend policy measures: 1. Prevention measures against programming and other software mistakes; 2. Cyber- and computer safety measures: Industrial control systems (IDS), cyber-physical systems (CPS), intrusion detection systems (IDS); 3. Climate change policies; 4. Strengthening the national and international cyber police; 5. Vulnerability measures: Tree trimming, walls around generators, underground cables, earthquake-safe construction techniques, construction supervision against fraud, and liability laws; 6. Strengthen international trade in electric current through investment in cross-border cables. 7. International cooperation in the areas of grid control and standardization of techniques and information exchange. 8. Provision and maintenance of batteries, as well as rescue and repair services.

Appendix

Second-order conditions

The second derivatives in the Hessian matrix of the Lagrange function (10.3) obtained from differentiating (10.4), (10.5), and (10.2′) are as follows.

$$L_{xx} = -p''(x)(1-d(y)) < 0$$

$$L_{xy} = -p'(x)(-d'(y)) > 0$$

$$L_{x\lambda} = -1 < 0$$

$$L_{yx} = \left(-p'(x)\right)(-d'(y)) > 0$$

$$L_{yy} = (1-p(x))\left(-d''(y)\right) < 0$$

$$L_{y\lambda} = -1 < 0$$

$$L_{\lambda x} = -1 < 0$$

$$L_{\lambda y} = -1 < 0$$

$$L_{\lambda\lambda} = 0 \tag{10.19}$$

The determinant is positive and therefore the second-order conditions for the basic model of Sect. 10.2 are fulfilled (see Henderson and Quandt 1980):

$$\begin{array}{l} 0 + L_{xy}L_{y\lambda}L_{\lambda x} + L_{x\lambda}L_{\lambda y}L_{yx} - L_{\lambda x}L_{yy}L_{x\lambda} - L_{\lambda y}L_{y\lambda}L_{xx} - 0 > 0 \\ + - - - - + - - - - - - - - - \end{array}$$

For the vulnerability-extended model second-order conditions remain unclear. The substitution method yields:

$$Max\, RI = (1 - p(x)v(b - x - y)))(1 - d(y))$$

$$Foc\, x: \ -p'v(1-d) + pv'(1-d) = 0\, implies\, p'/p = v'/v$$

$$Foc\, y: +pv'(1-d) - (1-pv)d' = 0\, implies\, pv'/(1-pv) = d'/(1-d)$$

2nd derivatives

$$RI_{xx}: \ -p''v(1-d) + p'v'(1-d) + p'v'(1-d) - pv''(1-d)$$

$$\begin{aligned} RI_{xy}: \ & p'v'(1-d) + p'vd' - pv''(1-d) - pv'd' = (1-d)[p'v' + p'vd'/(1-d) \\ & - pv'' - pv'd'/(1-d)] > (<) 0 \end{aligned}$$

$$RI_{yx}: \ p'v'(1-d) - pv''(1-d) + p'vd' - pv'd' = RI_{xy}.$$

$$RI_{yy}: -pv''(1-d) - pv'd' - pv'd' - (1-pv)d'' < 0$$

Second-order conditions require

$$RI_{xx} < 0;\ RI_{xx}RI_{yy} - RI_{yx}RI_{xy} > 0$$

$$\begin{aligned} RI_{xx} &= -p''v(1-d) + p'v'(1-d) + p'v'(1-d) - pv''(1-d) \\ &= (1-d)[(-p''/p' + v'/v)vp' + (p'/p - v''/v')v'p]' < 0; \\ &\textit{if second derivatives dominate.} \end{aligned}$$

$$\begin{aligned} &RI_{xx}RI_{yy} - RI_{yx}RI_{xy} = [-p''v(1-d) + 2p'v'(1-d) - pv''(1-d)] \\ &[-pv''(1-d) - 2pv'd' + (1-pv)d''] \\ &- [p'v'(1-d) - pv''(1-d) + p'vd' - pv'd'][p'v'(1-d) \\ &+ p'vd' - pv''(1-d) - pv'd'] \\ &= -p''v(1-d)[-pv''(1-d) - 2pv'd' + (1-pv)d''] \\ &+ 2p'v'(1-d)\left[-pv''(1-d)\right] + 2p'v'(1-d)\left(-2pv'd'\right) \\ &+ \left[-pv''(1-d)\right]^2 - pv''(1-d)\left(-2pv'd'\right) + [-p''v(1-d) \\ &+ 2p'v'(1-d) - pv''(1-d)](1-pv)d'' \\ &- \{p'v'(1-d)[p'v'(1-d) + p'vd' - pv''(1-d) - pv'd'] \\ &- pv''(1-d)[p'v'(1-d) + p'vd' - pv''(1-d) - pv'd'] \\ &+ p'vd'[p'v'(1-d) + p'vd' - pv''(1-d) - pv'd'] \\ &- pv'd'[p'v'(1-d) + p'vd' - pv''(1-d) - pv'd']\} = \end{aligned}$$

$$\begin{aligned} &-p''v(1-d)[-pv''(1-d) - 2pv'd' + (1-pv)d''] \\ &+ [-p''v(1-d) + 2p'v'(1-d) \\ &- pv''(1-d)](1-pv)d'' + p'v'(1-d)(-2pv'd') \end{aligned}$$

$$\begin{aligned} &-\{p'v'(1-d)[p'v'(1-d) + p'vd'] - pv''(1-d)\left[p'vd'\right] + p'vd'[p'v'(1-d) \\ &+ p'vd' - pv''(1-d) - pv'd'\right] - pv'd'\left[p'vd' - pv'd'\right]\} > 0? \end{aligned}$$

Without substitution method, the second derivatives in the Hessian matrix of the Lagrange function (3′) obtained from differentiating (10.4′), (10.5′), (10.11) and (10.12) are as follows.

Repro foc:

$$x: -p'v(z)(1-d(y)) - \lambda = 0; \tag{10.4′}$$

$$y: (1-p(x)v(z))\left(-d'(y)\right) - \lambda = 0; \tag{10.5′}$$

$$z:\ -pv'(1-d(y))-\lambda=0 \tag{10.11}$$

$$\lambda:\ b\ -x-y-z=0 \tag{10.12}$$

$$L_{xx}=-p''(x)v(z)(1-d(y))<0,\ L_{xy}=-p'(x)v(z)(-d'(y))>0,$$
$$L_{xz}=-p'v'(1-d)<0,\ L_{x\lambda}=-1<0$$

$$L_{yx}=\big(-p'(x)\big)v(z)(-d'(y))>0,\ L_{yy}=(1-p(x))v(z)\big(-d''(y)\big)<0,$$
$$L_{yz}=-pv'(-d')>0\,L_{y\lambda}=-1<0$$

$$L_{zx}=-p'v'(1-d)<0,\ L_{zy}=-pv'(-d')>0,$$
$$L_{zz}=-pv''(1-d)<0\,L_{z\lambda}=-1<0$$

$$L_{\lambda x}=-1<0,\ L_{\lambda y}=-1<0,\ L_{\lambda z}=-1<0,\ L_{\lambda\lambda}=0$$

Second-order conditions require that (i) the determinant of the Hessian matrix of the Lagrange function must be negative for a maximum, and (ii) that the determinant without the third row and column must be positive. Condition (ii) is fulfilled because the Hessian without third row and column differs from that in (10.19) only through the multiplicative, positive v-terms. The determinant of the Hessian matrix developed with respect to the first row is

$$L_{xx}(L_{yy}L_{zz}0+L_{yz}L_{z\lambda}L_{\lambda y}+L_{y\lambda}L_{\lambda z}L_{zy}-L_{\lambda y}L_{zz}L_{y\lambda}-L_{\lambda z}L_{z\lambda}L_{yy}-0\,L_{yz}L_{zy})$$
$$-\ 0+-----+-------0\ \textit{negative sign for this line}$$

$$-\ L_{xy}(L_{yx}L_{zz}0+L_{yz}L_{z\lambda}L_{\lambda x}+L_{y\lambda}L_{\lambda z}L_{zx}-L_{\lambda x}L_{zz}L_{y\lambda}-L_{\lambda z}L_{z\lambda}L_{yx}-0\,L_{yz}L_{zx})$$
$$+0+------------+0\ \textit{unclear sign for this line}$$

$$+\,L_{xz}(L_{yx}L_{zy}0+L_{yy}L_{z\lambda}L_{\lambda x}+L_{y\lambda}L_{\lambda y}L_{zx}-L_{\lambda x}L_{zy}L_{y\lambda}-L_{\lambda y}L_{z\lambda}L_{yx}-0\,L_{yy}L_{zx})$$
$$-\ 0--------+----+0\ \textit{positive sign for this line}$$

$$-\ L_{x\lambda}(L_{yx}L_{zy}L_{\lambda z}+L_{yy}L_{zz}L_{\lambda x}+L_{yz}L_{\lambda y}L_{zx}-L_{\lambda x}L_{zy}L_{yz}-L_{\lambda y}L_{zz}L_{yx}-L_{\lambda z}L_{yy}L_{zx})$$
$$-++----+---++--+---\ \ \textit{unclear sign for this line}$$

References

Acatech/Leopoldina/Akademienunion (Hrsg.): Resilienz digitalisierter Energiesysteme. Wie können Blackout-Risiken begrenzt werden? Stellungahme. (Schriftenreihe zur wissenschafts-basierten Politikberatung) (2021). ISBN: 978-3-8047-4224-6

Aerts, J.C.J.H., Botzen, W., van der Veen, J., Krywkow, J., Werners, S.: Ecol. Soc. **13**(1), article 41 (2008)

Afgan, N.H., Cvetinovic, D.B.: Resilience of high voltage transmission system. Energy Power Eng. **2011**(3), 600–606 (2011)

Allen, M.: Liability for climate change. Nature **421**, 891–892 (2003). https://doi.org/10.1038/421891a

Barteková, E., Ziesemer, T.H.W.: The impact of electricity prices on foreign direct investment: evidence from the European Union. Appl. Econ. **51**(11), 1183–1198 (2015)

Briguglio, L.: Economic vulnerability and resilience: concepts and measurements. In: Briguglio, L., Kisanga, E.J. (eds.) Economic Vulnerability and Resilience of Small States (Malta: Islands and Small States Institute of the University of Malta; London: Commonwealth Secretariat) (2004)

Briguglio, L., Cordina, G., Farrugia, N., Vella, S.: Conceptualising and measuring economic resilience (Chap. 3). In: Chand, S. (ed.) Pacific Islands Regional Integration and Governance. ANU Press, pp.26–49 (2006)

Briguglio, L., Moncada, S.: The benefits and downsides of multidisciplinary education relating to climate change. In: Climate Change and the Role of Education, pp. 169–187 (2019)

Brillet, J.-L.: Structural econometric modelling methodology and tools with applications under reviews. https://www.eviews.com/StructModel/structmodel.html

Bruneau, M., Chang, S., Eguchi, R., Lee, G., O'Rourke, T., Reinhorn, A., Shinozuka, M., Tierney, K., Wallace, W., von Winterfelt, D.: A framework to quantitatively assess and enhance the seismic resilience of communities. EERI Spectra J. **19**(4), 733–752 (2003)

Bussar, C., Moos, M., Alvarez, R., Wolf, P., Thien, T., Chen, H., Cai, Z., Leuthold, M., Sauer, D.U., Moser, A.: Optimal allocation and capacity of energy storage systems in a future European power system with 100% renewable energy generation. Energy Procedia **46**, 40–47 (2014)

Droste-Franke, B., Paal, B.P., Rehtanz, C., Sauer, D.U., Schneider, J.-P., Schreurs, M., Ziesemer, T.: Balancing Renewable Electricity. Ethics of Science and Technology Assessment, vol. 40. Springer, Berlin, Heidelberg (2012)

Droste-Franke, B., Carrier, M., Kaiser, M., Schreurs, M., Weber, C., Ziesemer, T.: Improving Energy Decisions. Springer International, Berlin (2015)

Espinoza, S., Panteli, M., Mancarella, P., Rudnick, H.: Multi-phase assessment and adaptation of power systems resilience to natural hazards. Electr. Power Syst. Res. **136**, 352–361 (2016)

Falcan, I.: Transition towards a renewable European electricity system. What are the implications for the power technology mix, pan-European power trade and the electricity market? Maastricht University (2021)

Faruqui, A., Chao, H.-P., Niemeyer, V., Platt, J., Stahlkopf, K.: Analyzing California's power crisis. Energy J. **22**(4), 29–52 (2001)

Francis, R., Bekera, B.: A metric and frameworks for resilience analysis of engineered and infrastructure systems. Reliab. Eng. Syst. Saf. **121**, 90–103 (2014)

Gholami, A., Shekari, T., Amirioun, M.H., Farrokh Aminifar, M., Amini, H., Sargolzaei, A.: Toward a consensus on the definition and taxonomy of power system resilience. IEEE Access **6**, 32035–32053 (2018)

Henderson, J.E., Quandt, H.: Microeconomic Theory. A Mathematical Approach. McGraw-Hill Kogakusha, Tokyo; German 5th edn., Verlag Franz Vahlen, Munich (1980)

Kamien, M.I., Schwartz, N.L.: A generalized hazard rate. Econ. Lett. **5**, 245–249 (1980)

Kandaperumal, G., Srivastava, A.K.: Resilience of the electric distribution systems: concepts, classification, assessment, challenges, and research needs. IET Smart Grid **3**(2), 133–143 (2020)

Karacaoglu, G., Krawczyk, J.B.: Public policy, systemic resilience, and viability theory. Metroeconomica **72**, 826–848 (2021)

Kröger, W.: Securing the operation of socially critical systems from an engineering perspective: new challenges, enhanced tools, and novel concepts. Eur. J. Secur. Res. **2**, 39–55 (2017)

Krugman, P.: How I work. Am. Econ. **37**(2), 25–31 (1993)

Lin, Y., Bie, Z.: Study on the resilience of the integrated energy system. Energy Procedia **103**, 171–176 (2016)

Lin, Y., Bie, Z., Qiu, A.: A review of key strategies in realizing power system resilience. Glob. Energy Interconnect. **1**(1), 70–78 (2018)

Liu, X., Chen, B., Chen, C., Jin, D.: Electric power grid resilience with interdependencies between power and communication networks – a review. IET Smart Grid **3**(2), 182–193 (2020). https://doi.org/10.1049/iet-stg.2019.0202, www.ietdl.org

Lütkepohl, H.: New Introduction to Multiple Time Series Analysis. Springer, Heidelberg (2005)

Maliszewski, P.J., Perrings, C.: Factors in the resilience of electrical power distribution infrastructures. Appl. Geogr. **32**(2), 668–679 (2012)

Manshadi, S.D., Khodayar, M.E.: Resilient operation of multiple energy carrier microgrids. IEEE Trans. Smart Grid **6**(5), 11 pp. (2015)

Mayer, C., Brunekreeft, G. (ed.): Resilienz digitalisierter Energiesysteme. Blackout-Risiken verstehen, Stromversorgung sicher gestalten (Schriftenreihe Energiesysteme der Zukunft), München (2021). ISBN: 978-3-9820053-3-1

Mahzarnia, M., Moghaddam, M.P., Baboli, P.T., Siano, P.: A review of the measures to enhance power systems resilience. IEEE Syst. J. **14**(3), 4059–4070 (2020)

Molyneaux, L., Wagner, L., Froome, C., Foster, J.: Resilience and electricity systems: a comparative analysis. Energy Policy **47**, 188–201 (2012)

Moslehi, S., Agami Reddy, T.: Sustainability of integrated energy systems: a performance-based resilience assessment methodology. Appl. Energy **228**, 487–498 (2018)

Mulder, M., Willems, B.: Borging van Publieke Belangen in de Energievoorziening. In: van Damme, E., Schinkel, M.P. (eds.) Marktwerking en Publieke Belangen. Preadviezen van de Koninklijke Vereniging voor de Staatshuishoudskunde, pp. 147–178. KVS, Amsterdam (2009)

Nguyen, T., Wang, S., Alhazmi, M., Nazemi, M., Estebsari, A., Dehghanian, P.: Electric power grid resilience to cyber adversaries: state of the art. IEEE Access **8**, 87592–87608 (2020)

van Oldenborgh, G.J.: How unusual was autumn 2006 in Europe? Clim. past **3**, 659–668 (2007). https://doi.org/10.5194/cp-3-659-2007

Roege, P.E., Collier, Z.A., Mancillas, J., McDonagh, J.A., Linkov, I.: Metrics for energy resilience. Energy Policy **72**, 249–256 (2014)

Shen, L., Cassottana, B., Heinimann, H.R., Tang, L.C.: Large-scale systems resilience: a survey and unifying framework. Qual. Reliab. Eng. Int. **36**(4), 1386–1401 (2020)

Thoma, K. (Hrsg.): Resilien-Tech – "Resilience by Design": a strategy for the technology issues of the future. In: German as Acatech STUDY. Reihenherausgeber: acatech – Deutsche Akademie der Technikwissenschaften, München. In English: Herbert Utz Verlag, München (2014)

Watson, J.-P., Guttromson, R., Silva-Monroy, C., Jeffers, R., Jones, K., Ellison, J., Rath, C., Gearhart, J., Jones, D., Corbet, T., Hanley, C., Walker, L.T.: Conceptual framework for developing resilience metrics for the electricity, oil, and gas sectors in the United States SANDIA REPORT SAND2014-18019, September 2015, Sandia National Laboratories, Albuquerque (2015)

World Bank: World Development Indicators|DataBank. Washington (2022). www.databank.worldbank.org

Yu, Z.Y., Li, Z.P., Qian, T., Huang, K.C., Chen, X.Y., Tang, W.H.: Optimal restoration strategy based on resilience improvement for power transmission systems under extreme weather events. IOP Conf. Ser. Earth Environ. Sci. **645**, 012042; IOP Publishing (2021)

Zhou, Q., Lambert, J.H., Karvetski, C.W., Keisler, J.M., Linkov, I.: Flood protection diversification to reduce probabilities of extreme losses. Risk Anal. **32**(11), 1873–1887 (2012)

Ziesemer, T.: Renewable electricity supply, infrastructure, and gains from international trade in electric current. Appl. Econ. Lett. **26**(17), 1439–1443 (2019)

Part IV
Operationalisation of Resilience

Chapter 11
Indicators for the Resilience Monitoring of Energy Systems

Henning Wigger, Gandhi Pragada, Henrik Netz, and Urte Brand-Daniels

Abstract Indicators are crucial for any management as they allow for monitoring, reviewing and adapting strategies towards a previously planned objective. The key challenge for fulfilling this task is to identify and define indicators that enable the assessment of the system under consideration. Regarding energy systems, if understood as complex socio-technical systems, a multitude of indicators for resilience assessments have been proposed, which vary according to the different systems analysed as well as indicators and stakeholder requirements. Hence, it seems that universal indicator frameworks are not applicable for resilience assessments and thus specific indicators are needed. However, a crucial step towards developing a general resilience assessment and monitoring framework is to understand on which aspects indicators depend, what "good" indicators can cover and what the limits of these indicators are. Therefore, this chapter gives an insight into the general characteristics of indicators. Then, it describes the indicator requirements particularly with respect to the resilience (monitoring) for energy systems. Furthermore, the chapter will address some specifics for the resilience assessment and management of energy systems based on some exemplary indicators.

Keywords Indicators · Assessment · Energy system · Resilience · Reliability · Indicator selection · Monitoring

11.1 Introduction

Indicators have become crucial elements for any management related processes and were initially introduced by economics and management disciplines (Linser 1999; Merry 2011). Since then, the use of indicators spread into other disciplines such as governance and politics, medicine and engineering, as well as sustainability related

H. Wigger (✉) · G. Pragada · H. Netz · U. Brand-Daniels
Institute of Networked Energy Systems—German Aerospace Center (DLR)—Department of Energy Systems Analysis, Oldenburg, Germany
e-mail: henning.wigger@dlr.de

D. van Doren et al. (eds.), *Towards Energy System Resilience*, Ethics of Science and Technology Assessment 55, https://doi.org/10.1007/978-3-032-03187-7_11

aspects. Indicators facilitate the monitoring and evaluation of systems, tasks or objectives. Further, they allow the comparison of different systems, technologies or products regarding their performance and the effectiveness of implemented measures (ILO 2015). For example, based on previously defined indicators, situations can be analysed regarding the performance and if indicators surpass certain thresholds potential counter measures can be initiated in time. Consequently, indicators are relevant for any assessment and evaluation, however, potential users should consider certain relevant aspects of indicators and their selection. Therefore, this chapter not only focuses on the overview of key indicator characteristics, but also give indicator examples for the resilience monitoring of energy systems.

11.2 General Aspects of Indicators and Their Requirements

Although a general unified theory of indicators does not yet exist, indicators share common aspects throughout various disciplines. If the sole meaning of the term is considered, it can be traced back to the latin verb *indicō*, which means to point out or to show at/to/on something. The German dictionary "Duden" defines indicators as *"something (fact, feature), which relates to a (statistically applicable) sign/evidence for a certain progress or a condition entered*" (Duden 2023, translated by the authors). Hence, indicators describe certain aspects of the object in question and thus they should be related to the overall purpose of the analysis.

Before indicators can be defined, several steps of abstraction have to be done. The term "indicator" is often synonymously used for different levels of the data pyramid with data, information, indicators and indices as shown in Fig. 11.1 with an increasing degree of abstraction used during indicator definitions. Firstly, the initial point is the real-world object that can be either analysed by experiments (e.g. lab-scale) or computational models. Secondly, from these tools, data can be obtained focusing on the model/experiment purpose. Thirdly, data are processed (interpreted) and thereby transformed into information. Fourthly, the indicators are defined according to a specific purpose and can be aggregated at process information level. Fifthly, indices can aggregate and weight several indicators, which are also named composite indicators.

When analysing systems, the measuring point and basis for indicators can be assigned to different stages of the system model ranging from inputs, processes (throughput), outputs, outcomes and impacts, which are shown in Fig. 11.2. At each of these metering points data, information, indicators and indices can be ascertained following different purposes of the indicators obtaining data and information. Whilst, the input and output can refer to the efficiency and quality of the system service (i.e. purpose), respectively, the process of the system is mostly related to the system behaviour and its state during an observation period. The system outputs directly contribute to the results causing additional outcomes in mid-term and at higher

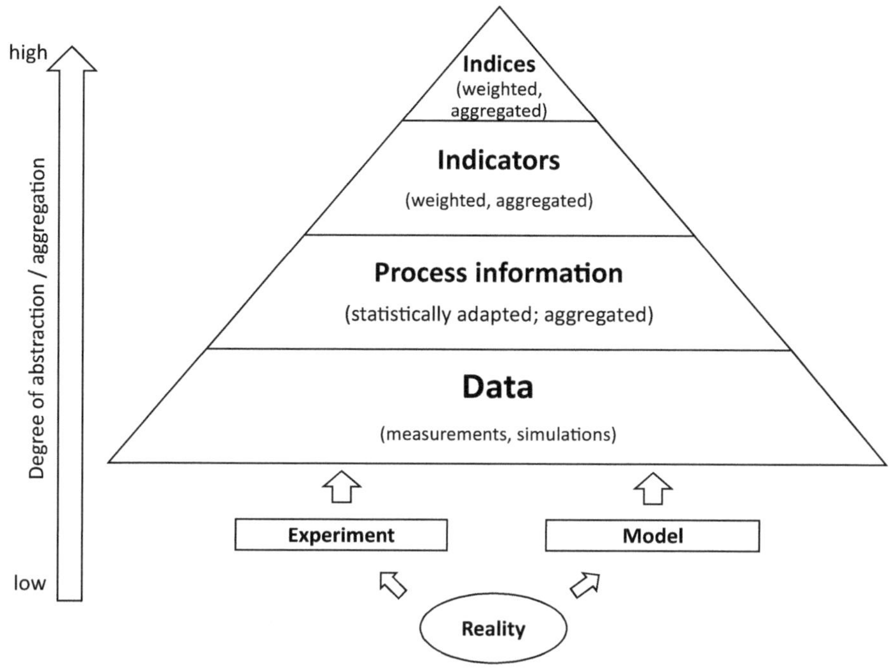

Fig. 11.1 Abstractions steps considered in modeling and evaluation processes for indicator and indices definitions. (adapted and extensively extended from Adriaanse (1995) cited in Linser (1999))

geographical scale (e.g. nationally and not only local). In contrast, outcomes and their impacts are less tangible because they are not directly traceable from system outputs and results (Zwart 2017). For example, if the output of a system (or a certain management measure) creates higher productive employment, the outcome of this system output would be the increase of the national state's welfare. Nevertheless, the global impact might be also positive in the mid- or long-term, but it is often challenging to identify direct linkages of outcomes and impacts. Depending on study objective, the indicator definition has to consider this differentiation since particularly the outcomes and impacts as reference points for the success of a management action often are not related to the direct system output and thus may have to include other aspects avoiding misinterpretations.

11.2.1 Types of Indicators

Due to the wide range of indicator applications, it is worthwhile to examine various indicator classifications addressing different aspects. The most common distinction is between **quantitative or qualitative** indicators. Presumably, the former is preferred over the latter, as numbers (often in cardinal scales) enable an easier comparison

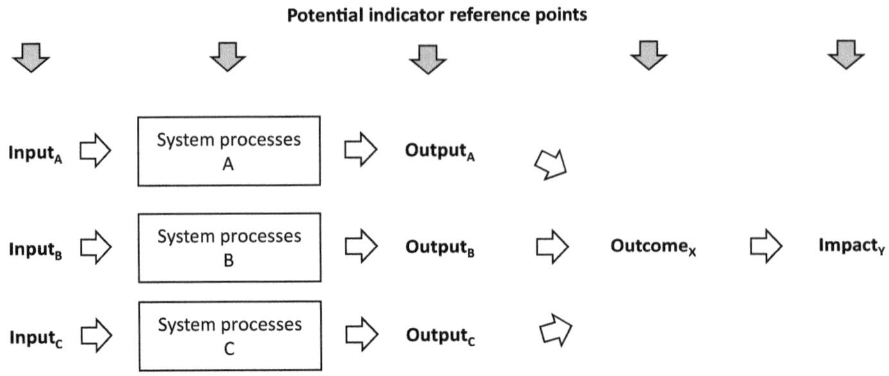

Fig. 11.2 Potential reference points for indicators for system inputs, processes, outputs, outcomes to impacts for the assessment of systems (own illustration based on Zwart 2017)

and understanding. Therefore, qualitative indicators are often converted into semi-quantitative numbers by using ordinal scales (Merry 2011).

The subcategory **direct and indirect (i.e. proxy)** indicators is another relevant differentiation. Direct indicators directly measure the phenomenon of interest. Proxy indicators are used when direct measurements are not possible, which often applies for complex systems. For instance, in medical contexts, the high or low blood pressure can indicate some severe problems related to the heart itself or blood vessels, without measuring directly the phenomenon and the actual cause (e.g. heart activity).

From a management perspective, indicators can be used not only for **descriptive** but also for **normative** purposes. Whilst, the former neutrally depicts the system state and behaviour as well as impacts, the latter focuses on either the **evaluative** or **prescriptive** (management) aspects by comparing to strictly set goals. A special type of indicator are **composite indicators**, also called indices, that are used to aggregate several indicators to describe the intended purpose. By doing so, complex issues can be easier communicated. Furthermore, the temporal perspective with the focus on past, present or future developments is a relevant characteristic of indicator to consider for differentiations.

11.2.2 Pros and Cons of Indicators

The use of indicators has benefits and detriments about which a vast amount of literature is available. Thus, only a brief overview will be given in this subsection.

Indicators have the advantage of facilitating the communication of complex issues to both non-experts and decision-makers (Meadows 1998; Mair et al. 2018). Hence, indicators should be appropriately aligned to the decision objectives (i.e. the purpose). As it is shown in Fig. 11.2, indicators generally generate knowledge at an abstract level and thereby, when applied to comparable objects, enable the scoring, ranking, comparing as well as monitoring and evaluating of the objects in question. By doing so, indicators support the decision-making for experts and non-experts.

Although the "simplification" of complex matters appears to be appealing for communication and decision purposes, every abstraction and aggregation process involves the loss of information. In the abstraction process, the relevant aspects for the specific purpose have to be identified and accordingly considered in the indicator definition. Hence, the abstraction process and with that the indicator definition strongly depends on the individual's expertise and experience who develops the indicator. Consequently, indicators are not an objective description of the research object (Mair et al. 2018) and thus are value-laden (Bossel 1996; Turnhout et al. 2007).

Another critical aspect of indicators is their interpretation. Even though indicators ease the evaluation and communication of complex matters, it is of high relevance to report the actual indicator purpose and to consider the corresponding limitations for the interpretation. This is particularly relevant when indicators are used in different contexts in which the original purpose does not fit to the decision context (Copeland et al. 2020). Furthermore, the often-desired quantitative illustration of indicators can be misleading. The crucial point of criticism is that those indicators may suggest a wrong precision, because particularly quantitative indicators are associated to the claim of (mathematical) correctness (Porter 1995). Consequently, the limits of indicators are discussed in different disciplines, and it is generally recommended to illustrate the uncertainty enabling the transparent interpretation of the results.

Furthermore, it has to be noted that cause and effect relations of specific measures cannot necessarily be traced back, because particularly in complex systems non-linear relationships may dominate and hamper to the identification of cause-effect-chains with the risk of oversimplifications (Mair et al. 2018; Merry 2011).

11.2.3 Quality Criteria for Selecting Indicators

In order to avoid or minimize the disadvantages mentioned in the previous subsection, it is important to ensure a transparent indicator selection. The process of developing and selecting indicators should be well documented and methodologically guided, particularly when stakeholders and experts are involved in the decision preparation. Thus far, a multitude of assessment frameworks were proposed to guide through the process of the decision preparation for various contexts (e.g. sustainability, environmental protection, etc.) (Guo et al. 2021; Nguyen and Akerkar 2020; Bill et al. 2020).

The need for transparency even becomes more crucial when multiple indicators have to be considered in the assessment, because the indicators have not only to be defined and distinct but also have to be weighed against each other. For this, the method of multi-criteria decision analysis (MCDA) is established as a transparent approach to guide through complex decision situations. Baumann et al. (2019)

summarised the MCDA approach in the following steps: goal and problem definition, selection of alternatives for comparison, identification of criteria (i.e. indicators), choice of suitable MCDA method, aggregation and weighing of criteria/indicators, interpretation of the results, if possible with stakeholder involvement. In the following subsection steps of goal and scope together with the indicator selection will be discussed in more detail due to their relevance for the indicator setup.

11.2.3.1 Indicator Selection

In the first step, the goal and scope of the study have to be determined, if possible by the decision-makers who have information on the decision situation and object. Once the goal and purpose are defined, indicators can be collected and defined that are fitting to the agreed decision objectives. Basically, two major approaches exist to identify indicators for any assessment that usually complement each other (Mangoyana et al. 2013). Firstly, the top-down strategy attempts to identify already existing official indicator frameworks (e.g. sustainable development goals), reports or other published scientific literature that deal with the objective intended. Secondly, the bottom-up strategy accesses the knowledge of experts and other stakeholders that are familiar with the decision context or are affected by the decisions, respectively (Quinlan et al. 2016). This knowledge and indicator needs can be obtained by conducting workshops, interviews or questionnaires, for instance. The top-down strategy is more often applied due to time and resources constraints, although stakeholder and expert workshops give a practical insight into ranking and selection of indicators due to the perceived relevance. This additionally support the trust into the indicators and the acceptance by decision-makers. The actual indicator selection is often not explicitly documented in the literature, which hamper the transparency and potential support of decision-makers for the selected indicators.

11.2.3.2 Criteria for Selecting Indicators

So far, there is no consensus in scientific literature on which quality criteria should be fulfilled by indicators and thus each study has to define their criteria, as no universally accepted indicator set can be defined. The main reason is that the decision objective and participating stakeholders finally determine the indicator selection in bottom-up studies. However, Hirschberg et al. (2008) summarized several criteria that can be applied to select indicators, which are depicted in Table 11.1. The different criteria for indicator selection are assigned to the characteristics scientific, pragmatic and functional aspects determining the decision situation. It has to be noted that some of these criteria cannot undoubtedly be assigned to one category, because several characteristics can be attributed to one criterion. The following Table 11.1 is mainly based on Hirschberg et al. (2008).

The selection of indicators is particularly challenging for the assessment of complex systems, because such systems own features such as being dynamic, having

Table 11.1 Summary of potential criteria for selecting indicators sorted by scientific, pragmatic and functional characteristics. (based on Hirschberg et al. (2008))

Scientific	Pragmatic	Functional
Measurable	***Manageable***[1]	***Relevant***+
Qualitatively and quantitatively	Adequate number of indicators	For most stakeholders involved
Meaningful*+	***Understandable****	***Compelling***
Reflects the users' needs	Possible to comprehend	Interesting and suggestive for actions
Clear in value*	***Feasible***	***Leading***
Clear positive or negative denotation	Possible to collect within time and cost constraints	Basis for the decision about actions
Clear in content*	***Timely***	***Influenceable***
Understandable units	Easy and quick to calculate	Indicators are able to influence actions
Appropriate in scale#1	***Coverage of the intended aspects***#*+	***Comparable***
Suitable dis-/aggregation scale	Purpose is covered	If applied in different systems, the comparison should be possible
No redundancy		***Comprehensive***+
Avoid double counting		Describes the relevant aspects of the system under study
Robust / reproducible		
Transparent and sound indicator calculation method		
Sensitive / specific		
Sensitive to changes during the time scale considered		
Verifiable		
Verifiable by third persons		
Hierarchical*		
Adjustable levels of detail		

*, +, 1 and # marked criteria that are linked to other criteria of different characteristics

non-linear or emergent properties that aggravate the identification of suitable indicators. Consequently, it is generally recommended to involve experts and stakeholders to gather in-depth knowledge, which is required for the assessments. Stakeholders and experts can contribute to both the identification of the long list (i.e. all potential indicators) and the final short-listed indicators which can be selected based on the reasons mentioned in Table 11.1.

11.2.3.3 Selection of Indicator Sets

When selecting indicators for monitoring systems, the question automatically arises how many indicators are needed for a sound assessment and monitoring process. There generally is no definite answer to the correct number of indicators or indices as it depends on the study purpose, system complexity and stakeholder needs. For example, when analysing systems with a low complexity, simple data points (cf. Fig. 11.2) can be sufficient describing the necessary systemic aspects that allow for the assessment. In contrast, for complex systems it can be challenging to identify and select suitable indicators (Quinlan et al. 2016) and thus more pragmatic aspects have to be considered for the selection of indicators. Describing complex phenomena with the help of a single indicator or indices (i.e. aggregated indicators) is in general critically discussed (Martišauskas et al. 2022), since information is lost because of omitting indicators or by the indicator aggregation, respectively. However, when defining indicators sets with non-aggregated indicators, different indicator sets can cover the same study purpose, because the system behaviour originates from different system characteristics and depending on the stakeholder and expert perspectives can lead to different selections. Therefore, it is recommended to include stakeholder and experts in the participatory definition and selection process to gain the trust of decision-makers as final users. Furthermore, the weighing of different indicator domains in the sustainability assessments should be targeted to avoid any biases favouring either the economic, environmental and social dimension, for instance (Gillespie-Marthaler et al. 2019).

11.3 Challenges for Resilience Assessment and Monitoring Frameworks for Energy Systems

The subsection will focus on resilience and its challenges regarding indicator definition as well as give a short overview on frameworks and exemplary indicators for resilience assessments. The assessment and monitoring of energy systems regarding their resilience is a multidimensional challenge due to several reasons. Firstly, energy infrastructures are considered as complex systems, which are characterized by their spatially and temporally dependent parameters that aggravate the resilience assessment and monitoring (Kaloti and Chowdhury 2023). Secondly, energy systems cover large geographical areas with diverse system states depending on time and location that need to be considered for the assessments. Thirdly, resilience depends on the stress cases with their various stressors (i.e. resilience against what) and also on the stakeholders (i.e. resilience for whom) who can be affected by the loss of energy supply (Wied et al. 2020). Fourthly, resilience is an emergent system property and thus it is not directly measurable (Cains and Henshel 2021). Resilience depends not only on the stress case, but also on system design (e.g. redundancy, diversity), recovery, mitigation and adaptation strategies that determine the system capacity and capability to (re)act resilient (Watson et al. 2022). Consequently, resilience assessments usually have to establish several proxy indicators for determining or monitoring the energy system's resilience (Martišauskas et al. 2022).

These combined challenges aggravate the selection of indicators and development of frameworks applicable for energy systems. For example, when assessing a specific part of the energy system, potential stress cases will affect the physical infrastructure and its operation at different geographical areas. In such cases, indicators and indices at low geographical resolution (i.e. high aggregation) may show a positive resilience performance, indicators and indices at higher temporal and spatial resolution can point to the opposite. The main reason is that the adaptive capacity, which reduces the system vulnerability, has to be available and accessible at the affected region to cope with the local stress case impacts. Even if the energy system may have high adaptive capacities in general, it is crucial to consider if these can be accessed at the right point of time during the stress case to minimise or even mitigate the stressor impacts. Thus, the potential disruptive events (and failures) as well as the available capacity and capability to react and recover are always context-dependent (Watson et al. 2022). This challenge will become even more severe when the resilience assessment is conducted as permanent monitoring. Resilience monitoring is understood as the continuous observation of the system state (Mentges et al. 2023), the aforementioned challenges have to be considered for the interpretation and/or when used for performance checks, the indicators should be selected only for the agreed objectives. By doing so, monitoring processes can address both real-time (i.e. operational) and strategical aspects and allow the identification of necessary adaptation strategies. Also, in the context of resilience monitoring, the decision makers are interested in how potential resilience measures may affect the resilience level of the system so that already taken decisions can be revised or a priori defined mitigation strategies initiated, if necessary.

11.3.1 Resilience Assessment Frameworks

Numerous resilience assessment frameworks were developed in recent years that differ in purposes and covered aspects (Guo et al. 2021; Nguyen and Akerkar 2020; Bill et al. 2020). Likewise, the considered types and number of indicators vary among the studies (Mujjuni et al. 2021; Gillespie-Marthaler et al. 2019) and thus aggravates their comparison and the identification of standard indicator sets (Martišauskas et al. 2022). Resilience assessment frameworks can employ quantitative, semi-quantitative, qualitative or combinations of these indicator types. Resilience assessments are usually not limited to one domain, but cover several domains as for example technical, organizational/management, social, safety, security and sustainability (Osei-Kyei et al. 2023; Ahmadi et al. 2021). Each of these domains comes with its own set of requirements not only regarding the indicator definitions and purposes, but also the data required for the assessment, setting a cumbersome task

(Cantelmi et al. 2021). Although the comprehensive view on all domains is becoming more relevant (Poulin and Kane 2021), the main focal point for energy systems was laid on the technical domain, so far.

The technical domain is most often considered quantitatively by applying the resilience curve approach, therefore focusing on the system performance (Gasser et al. 2020; Poulin and Kane 2021; Bill et al. 2020). The resilience curve, or also named bath tube curve, shows the development of the system performance depending on time, where the former is impacted by disruptive events decreasing the performance.

Generally, the resilience curve can be distinguished by the temporal stages of the event. This includes the system performance before, during and after event (Resilens 2016; Bill et al. 2020). During each of these stages different actions can be established to improve the resilience and with that the system performance. The *pre-event stage* is characterized by proactive strategies such as to prepare, prevent and protect the system against potential stressors. It thus focuses on questions about system design and/or on decreasing potential exposures to the stressors. The *during event* stage includes mitigation, absorption and mobilisation of adaptive capacities (in short-term) for buffering or minimizing potential impacts on energy infrastructures (Resilens 2016). The *post-event stage* represents the respond, recovery and learning capabilities (Resilens 2016). Similarly, to the event-related differentiation, the *pre-*, *during* and *post-event stage* can be differentiated according to the main actions that a system should perform. In these stages, the system should resist, re-stabilize, rebuild and reconfigure to deal with the impacts of the stressors (Gasser et al. 2021). Likewise, for each of these four stages different indicators can be defined. Consequently, for each domain and stage of the resilience curve, several indicators are required to describe complex socio-technical systems (Jovanović et al. 2020; Martišauskas et al. 2022),

The described multidimensionality of resilience assessments and monitoring is also reflected in the various frameworks that have been developed so far. This is also reflected by several reviews on frameworks, metrics, models and general assessment methods that were published in the last years (Hosseini et al. 2016; Kaloti and Chowdhury 2023; Cassottana et al. 2023; Wang et al. 2022; Wied et al. 2020; Stanković et al. 2023). Moreover, professional associations such as the Institute of Electrical and Electronical Engineers (IEEE) worked out a specific resilience assessment framework applicable for power systems (Bill et al. 2020). This IEEE framework considers potential hazards of any kind and their impacts on the power system with five focus areas. Several technical and operational metrics were defined to assess critical energy infrastructures and prioritize investment decisions (Bill et al. 2020). Besides the technical domain, other researchers emphasized the need also to include other resilience curve stages such as recovery/repair (Kottmann et al. 2021) as well as other domains such as social-economic aspects (Ottenburger et al. 2020; Quinlan et al. 2016; Meerow and Newell 2019; Copeland et al. 2020). Since it is not the aim of this chapter to provide a comprehensive overview on all potential indicators and frameworks (cf. e.g. Martišauskas et al. 2022), the following sections will provide some examples of commonly used indicators.

Table 11.2 Overview of the exemplary indicators presented in this subsection

Indicator name	Structural/Performance
Loss of Load Probability (LOLP)	Performance
Loss of Load Expectation (LOLE)	Performance
Storage capacity	Structural
Diversity—Primary fuels/energy carrier	Structural
Diversity—Installed capacity	Structural
Share of renewable energy in generation	Structural
Import dependency	Structural
Change of network connection of the system	Structural

11.3.2 Indicators for the Resilience Assessment and Monitoring of Energy Systems

The assessment and monitoring of energy systems has a long tradition as it was always the main objective to provide energy to different stakeholders at highest quality with minimized interruptions. For this reason, various tools and indicators are already well-established for determining the status of power systems, which are related to reliability and robustness. Although the boundaries of reliability and resilience are inconclusive, both share common aspects so that they can be used complementarily (Gasser et al. 2020).

In the following subsections some exemplary indicators are described, which can be used for the monitoring of the energy system during disruptive events. An overview of the discussed indicators is available in Table 11.2. It has to be noted that this list does not aim for comprehensiveness. The indicators were selected as relevant examples since they can be used in most of the assessments or are quantifiable with power system models used in the ReMoDigital project. It also has to be emphasized that these indicators mainly cover the technical aspects of power system and does not include additional aspects such as sustainability perspective which should be included in future assessments (Netz et al. 2023).

Before the indicators are briefly discussed, firstly the distinction of reliability and resilience in the context of performance indicators is described and some selected reliability indicators are introduced. Secondly, the difference between structural and performance indicators is given. Thirdly, exemplary structural indicators will be introduced, which will be integrated in the resilience curve.

11.3.2.1 Reliability Indicators for the Performance Assessment of Energy Systems

In long term planning of energy systems, reliability of the generation system plays a key role for the expansion of installed capacities to meet the future demand (Nikzad et al. 2012). Reliability is defined as the **ability to meet customers energy demand** needs despite unexpected equipment breakdowns reducing the supply of energy (Felder and Petitet 2022). On the other hand, the ability to **quickly bounce back** from an undesirable situation was described as resilience. Thus, reliability focuses on certain aspects of resilience, but resilience includes additional aspects regarding the adaptation and recovery capabilities (Chi et al. 2018). Consequently, a higher reliability can increase the resilience and vice versa, but there can be certain situations in which this does not pertain (Kaloti and Chowdhury 2023).

During the planning process, reliability indicators aid to take decisions on investment opportunities of forthcoming installed capacities (Nikzad et al. 2012). The reliability of the energy system can be evaluated by following standard indicators such as 'Loss of Load Probability (LOLP)' and 'Loss of Load Expectation (LOLE)', which usually depicts the interruptions of the annual average. The definitions of these indicators are:

- **Loss of Load Probability (LOLP):** It refers to probability of an occurrence of event where loss of one generation unit '*i*' causing risk of the power supply (Ferro Ferdinant et al. 2020).

$$LOLP = \sum_{i=1}^{n} p_i \times t_i$$

  where, n is the number of units that supply energy to the system, p_i is the probability of loss of capacity, and t_i is the duration of loss of capacity
- **Loss of Load Expectation (LOLE):** It refers to the expected number of hours per year where the occurrence of an outage might possibly occur in an energy system (Ferro Ferdinant et al. 2020).

$$LOLE = \sum_{i=1}^{n} p_i \times t_i$$

  where, n is the number of outages in a year, p_i is the probability of loss of capacity, and t_i is the duration of loss of capacity.

Especially the LOLE is a useful indicator for performance measurement, as energy systems are usually benchmarked by their outage hours for end-users. Methodologically, these indicators need to be distinguished from structural indicators.

11.3.2.2 Structural Indicators for the Resilience Assessment of Energy Systems

Indicators in quantitative resilience frameworks are helpful for comparing the degrees of resilience in various systems or for assessing the efficacy of certain resilience strategies (Bill et al. 2020). When following the standard resilience curve approach, indicators are often based on quantifications for performance assessment in a technical meaning. However, regarding resilience assessments of energy systems, indicators can be categorized in three different categories: performance-, structure- and hybrid-based approaches (Hosseini et al. 2016; Wied et al. 2020; Martišauskas et al. 2022; Biringer et al. 2013).

Performance-based indicators focus on the benefits (i.e. services) that a system provides for stakeholders as the main concern of resilience (Wied et al. 2020; Jasiūnas et al. 2021). However, the sole focus on performance indicators cannot contribute to an understanding why certain system designs perform better than others against disruptive events. Hence, it is crucial to consider also the system's structure, even though relevant system characteristics can be difficult to identify (Biringer et al. 2013). Thus, structural indicators are required for assessing how the internal characteristics (i.e. architecture and design) of a system affect the system's resilience. Those characteristics are attributes that can decrease or increase the energy system's resilience depending on the stress case, such as the availability of redundant power lines in the grid (Martišauskas et al. 2022). Structural resilience indicators have the disadvantage, though, to be complex, more time-consuming to obtain and require a high level of expertise, as the system's inner structure needs to be modelled (Hosseini et al. 2016; Mottahedi et al. 2021). On the other hand, performance indicators are more validated, but as all systems have certain dynamics, it can be difficult to determine what the base performance of a system was before an event, and when exactly that event started. This may lead to uncertainty when the base performance is compared to the performance within an adverse event (Biringer et al. 2013; Wied et al. 2020).

In a hybrid-based approach, that employs both structural and performance indicators, the structural indicators provide potential explanations for the system behaviour that are measured and evaluated with performance indicators. In the best case, the relationship between structural (i.e. system design) changes and the performance can be explained (Biringer et al. 2013; Martišauskas et al. 2022). This knowledge can be used for improving the system's resilience, because resilience assessment and monitoring is a continuous process.

11.3.2.3 Structural Indicators: Examples for the Energy Sector

The structural indicators for the energy sector try to measure properties of a power system related to resilience enhancing system characteristics such as redundancy or diversity. In the following an exemplary overview about structural indicators is given, which addresses some system characteristics. It has to be noted that these structural indicators are not comprehensive but are relevant in most frameworks.

Storage Capacity ($\mathbf{I_{SCA}}$):

This indicator $\mathbf{I_{SCA}}$ represents the amount of available reserves that can be used for the energy supply. Higher storage capacities can show the potential characteristics of stronger resilience. However, the impact of an uncertain event on storage capacity relies on the available and accessible reserves at the time of disruption.

The advantages of the storage capacity indicator are that the indicator itself and the principle behind it is approachable and easy to assess for most systems. In some cases, with higher demand of stored energy it has to be considered that storage systems can only discharge to certain power system specific limits and there not the capacity, but the power output of the system is the limiting factor.

Inputs needed to evaluate the storage capacity includes capacity of storage/reserves, used generation capacity, available generation capacity, and usable stored energy.

$$I_{SCA} = \frac{SC}{EP}$$

where *SC* is the capacity of storage/reserves in MWh and *EP* is the total energy consumption in MWh (Martišauskas et al. 2022).

Diversity—Primary fuels/energy carrier ($\mathbf{I_{PF}}$):

The indicator $\mathbf{I_{PF}}$ demonstrates the diversification of installed capacity through various fuel sources. A diversified energy system with multiple fuels or energy carriers for energy production can result in higher resilience during the disruption than less diversified systems, because energy carriers can be differently affected by the same disruptive event. Moreover, domestic and imported fuels can be combined in this indicator to consider the diversified sourcing strategies, which points at the supply chain of energy carriers. Thereby, this indicator gives insights into the dependence on single energy carriers, this is especially relevant for fossil fuels that are imported from abroad, as these resources can be easier restricted for political reasons than nationally sourced energy carriers. A strong dependency on one renewable energy carrier can be especially relevant for abnormal climate patterns, for example, droughts can limit the supply of hydroelectric energy. (Martišauskas et al. 2022; Gasser et al. 2020).

The following equation is based on the Herfindahl–Hirschman index (HHI) and Shannon–Wiener index (SWI) that applies to different types, shares and number of primary fuels.

$$I_{PF} = -\sum_{i=1}^{N} p_i * \ln(p_i)$$

where $\boldsymbol{p_i}$ is the share of single energy carrier 'i', **N** is the number of primary fuels in energy supply (Martišauskas et al. 2022).

Diversity—Installed generation capacity ($\mathbf{I_{IGC}}$):

The indicator $\mathbf{I_{IGC}}$ represents the diversified installed capacity for energy production through various energy carriers. Whilst, the previous I_{PF} indicator only represents the supply-side of specific energy carriers, the I_{IGC} indicator focuses on the capacity to make use of these energy carriers. A restricted generation capacity because of a disruptive event would endanger the system's resilience, for instance, even if energy carriers are available and accessible. Thus, a diversified installed capacity for energy production can increase the resilience in face of disruptive events. This will always allow one type of the installed capacities to compensate the damaged one. Like the primary fuel indicator, this indicator gives guidance for overdependence. However, the diversity of the installed generation capacity focuses on constraints in energy production facilities, not only the fuel supply as the diversity of primary fuel/energy carrier indicator. A good example for overdependence on one type of energy production facility is the French nuclear power fleet. The reactors were mostly built in the same period and therefore needed repair and renovation works at a similar point in time, which lead to concerns about the security of electricity supply in France (Müller-Lancé 2022).

Inputs needed to evaluate the installed capacity include types of energy production technologies, shares of energy production technologies, number of energy production technologies. The following equation reflects the diversity of generation capacities,

$$I_{IC} = -\sum\nolimits_{i=1}^{N} s_i * \ln(s_i)$$

where s_i is the share of individual installed capacity, **N** is the number of energy production technologies 'i' (Martišauskas et al. 2022; Jewell et al. 2014).

Share of renewable energy in generation (I_{RES}):

The indicator I_{RES} can evaluate the share of the renewable power generation over the total generation. Renewable power with independent generation from primary fuels has a capability to influence the system's resilience. However, renewable power generation is completely dependent on the weather and day/night patterns. This indicator is similar to the diversity indicators but gives a special insight about the renewable energy sources and the constraints and risks that may be associated with them. Further, is this indicator relevant for sustainability assessment.

Inputs such as total generation, and renewable power generation are needed to evaluate the indicator.

$$I_{RES} = \frac{TG}{RPG} * 100\%$$

where **TG** is the total generation in MWh and **RPG** is the total renewable power generation in MWh (Martišauskas et al. 2022).

Import dependency (I_{ID}):

This indicator represents the share of energy and electricity import in the total energy and electricity demand. This indicator can be difficult to interpret, because supply risk may vary widely between the regions of origin and are dependent on unpredictable events in the future. A higher value indicates the higher dependency on energy imports and thus reflects the capability of energy system with less resilience. For example, the energy system can export energy if the value of indicator is less than 1 (energy production is higher than consumption) and further indicates that the energy system is robust and resilient. On the other hand, the energy system could depend on energy imports if the value of indicator is more than 1 (energy consumption dominates the production). This makes the energy system less robust and more susceptible to interruptions (Martišauskas et al. 2022).

Inputs needed to evaluate the indicator include electricity demand, used generation capacity, imported electricity, type of primary fuels, and imported primary fuels.

$$I_{ID} = (1 - \frac{PE}{DE}) * 100\%$$

where **PE** is the amount of produced energy regardless of imports, **DE** is the amount of energy demand (Martišauskas et al. 2022).

Change of network connections ($\mathbf{I_{CNC}}$):

Connectedness and redundancy are important aspects of resilience (Zhang et al. 2015). To analyse the interconnectedness and redundancy of the network, the method developed by Omer et al. (2014) can be used. The method was originally used to show the resilience of communication networks, but has been also considered in engineering contexts (Patriarca et al. 2021). It is based on the graph theory describing a network with nodes and edges. Similarly, the nodes represent the technical components (i.e. generators, transformers, households etc.) and the links represent the power lines in the power system. The indicator "closeness to centrality" is used as a measure of structural resilience, which considers the fact that the more connections a network node has, the more redundancies exist if the edges are impacted by disruptive events. Consequently, a higher resilience can be reached. This indicator is especially important to include, as it is the only one that measures the graphical structure of the system. Since the indicator is measured as the closeness to centrality before and after the disruptive event, it is thus named as change of network connections. For all nodes n the equation is defined as (Omer et al. 2014):

$$Changeofnetworkconnections = \frac{\sum_{i=1}^{n} C_C(v)_{before\_event}}{\sum_{i=1}^{n} C_C(v)_{after\_event}}$$

Therefore, the change of network connection is a measure of the structural resilience for a node. The closeness of a single node v to all other nodes is defined as $C_C(v)$. The highest value is 1, the lowest 0. To allow for the inclusion of disconnected nodes, the following alternative formulas are used. In case all nodes are reachable, the closeness is described with the nodes of the network V, where A_v is the number of reachable nodes (Mathworks 2023):

$$C_c(v) = \left(\frac{A_v}{n-1}\right)^2 \frac{1}{\sum_{\substack{s \in V \\ s \neq v}}^{n} d_G(v,s)}$$

11.3.2.4 Structural and Performance Indicators for the Resilience Assessment and Monitoring

The resilience monitoring of energy systems can pursue different objectives, which are centred at the resilience curve stages system (design) planning (i.e. *before* disruptive events), system operations *during* disruptive events and feedback loops *after* disruptive events, which corresponds to the well-established resilience curve approach (Poulin and Kane 2021). Thereby, the resilience monitoring process can either focus on the operational status (Mentges et al. 2023) or on assessing the effectiveness of initiated measures by comparing planned and actual results (ILO 2015).

The former aims for continuously monitoring the system status in order to identify any deviations from the operational defaults. If the status does not match the required parameter range, predefined or improvised fallback and recovery strategies can be initiated for stabilizing the system performance. For this purpose, performance indicators are used, because they cover the system performance with an optional temporal resolution that can be relevant for monitoring purposes.

The latter aspect focuses on the success or failure of designed system capability/capacities in face of disruptive events. If predefined measures (e.g. redundant generators, transformers) are not effective during disruptive events, these measures have to be reviewed in the continuous resilience management process to derive recommendations for future improvements. Consequently, the focus on performance indicators only cannot provide explanations for a certain system behaviour and hence structural aspects also become relevant for the decision preparation regarding the system's resilience. In each stage of the resilience curve, structural and performance indicators can be different significant and thus are relevant to consider for certain decision situations (Jackson and Ferris 2013).

Structural indicators can be allocated to different stages of the resilience curve, which consists of plan & prepare, absorb & recover as well as adapt to the performance loss. For instance, on the one hand, the storage capacity, which provides insights about buffer and storage of electricity, heat as well as gas storage systems, aids to the preparedness before an uncertain event. Although the storage capacity is basically determined in the plan & prepare stage, the resilience measure will be only effective once the system is exposed to the disruptive event, which is why this indicator would also be representative for the absorb and recover stage. Similarly, diversity indicators related to the installed capacity, primary fuels and share of renewables with diversified nature of production and transmission, enables to absorb and recover from disruptive events (Brand et al. 2017), but can also affect the plan & prepare stage.

Hence, the quest for linking structural system characteristics with their performance has been difficult (Watson et al. 2022). The reasons are that the system performance depends not only on the system structure at a given location, but also on the (type of) disruptive event, failure duration and recovery strategy (Watson et al. 2022) as well as temporal progress of the disruptive event and its impacts. Moreover, various terms are used for resilience design principles in the literature, which is why a commonly accepted framework would contribute to the progress towards proactively designing energy systems more resilient (Specking et al. 2022). However, this kind of feedback would be required from the continuous resilience assessment and monitoring to recommend successful design for improving of the system's resilience. According to Poulin and Kane (2021), analysts, operators, designers and stakeholders should carefully choose indicators metrics. Based on the situation which should be evaluated, different metrics can yield significantly different recommendations.

11.4 Conclusion

Indicator theory is a commonly applied concept in any management contexts, which was successfully used throughout the last decades in several research disciplines. Indicators have the advantage to enable the communication of complex matters to informed or uninformed decision-makers by abstraction. Resilience (i.e. the continuous supply of a service) became a relevant aspect to consider in systemic decision situations, particularly in the context of critical infrastructures such as energy systems. This book chapter described the different types of indicators used for resilience assessment together with the pros and cons of with each indicator definition is accompanied. Thereby, it was emphasized that the indicator purpose crucially is to be defined together with experts and stakeholders to gain the trust of the users particularly in resilience monitoring applications, where different perspectives unite in the assessment of complex socio-technical systems. Thus, a comprehensive criteria list was proposed to guide the indicator selection. Furthermore, the book chapter shed light on the complexity to assess and monitor the resilience of energy systems. Thereby, several exemplary indicators were briefly introduced, which are relevant and easily obtained in most assessments. Finally, the relation of structural and performance indicators was discussed. It was emphasized that both indicator types are required to be considered in resilience monitoring processes to derive successful resilience design principles for future system designs.

References

Adriaanse, A.: In search of balance. A conceptual framework for sustainable development indicators. In: MacGillivray, A. (ed.) Accounting for Change: Papers from An International Seminar, pp. 3–10. Toynbee Hall, London (1995)

Ahmadi, S., Saboohi, Y., Vakili, A.: Frameworks, quantitative indicators, characters, and modeling approaches to analysis of energy system resilience: a review. Renew. Sustain. Energy Rev. **144**, 110988 (2021). https://doi.org/10.1016/j.rser.2021.110988

Baumann, M., Weil, M., Peters, J.F., Chibeles-Martins, N., Moniz, A.B.: A review of multi-criteria decision making approaches for evaluating energy storage systems for grid applications. Renew. Sustain. Energy Rev. **107**, 516–534 (2019). https://doi.org/10.1016/j.rser.2019.02.016

Bill, C., Bose, A., Brown, S., Chalamala, B., Immermann, D., Khodaei, A., et al.: Resilience framework, methods, and metrics for the electricity sector. The Institute of Electrical and Electronic Engineers - Power & Energy Society (2020). https://opcom.velco.com/library/document/download/7134/DOE-IEEE_Resilience%20Paper_10-30-2020%20for%20publication%20%28002%29.pdf. Accessed 17th Jan 2024

Biringer, B., Vugrin, E., Warren, D. (eds.): Critical Infrastructure System Security and Resiliency, 1st edn. CRC Press, Boca Raton (2013)

Bossel, H.: Deriving indicators of sustainable development. Environ. Model. Assess. **1**, 193–218 (1996)

Brand, U., Giese, B., von Gleich, A., Heinbach, K., Petschow, U.: Resilient design of energy systems: "RE-methane-system" and "regional self-supply" as options for transformation. University of Bremen. Bremen, Germany (Executive Summary of the final report of the project RESYSTRA) (2017). http://www.resystra.de/files/publikationen/resystra-executive-summary-final.master.pdf, Accessed 31 Aug 2023

Cains, M.G., Henshel, D.: Parameterization framework and quantification approach for integrated risk and resilience assessments. Integr. Environ. Assess. Manag. **17**(1), 131–146 (2021). https://doi.org/10.1002/ieam.4331

Cantelmi, R., Di Gravio, G., Patriarca, R.: Reviewing qualitative research approaches in the context of critical infrastructure resilience. Environ. Syst. Decis. **41**(3), 341–376 (2021). https://doi.org/10.1007/s10669-020-09795-8

Cassottana, B., Roomi, M.M., Mashima, D., Sansavini, G.: Resilience analysis of cyber-physical systems: a review of models and methods. Risk Anal. **43**, 2359–2379 (2023)

Chi, Y., Xu, Y., Hu, C., & Feng, S. (eds.): A state-of-the-art literature survey of power distribution system resilience assessment, 5–10 August 2018. Institute of Electrical and Electronics Engineers; IEEE Power & Energy Society. IEEE, Piscataway, NJ (2018). http://ieeexplore.ieee.org/servlet/opac?punumber=8540807

Copeland, S., Comes, T., Bach, S., Nagenborg, M., Schulte, Y., Doorn, N.: Measuring social resilience: trade-offs, challenges and opportunities for indicator models in trans-forming societies. Int. J. Disaster Risk Reduct. **51**, 101799 (2020). https://doi.org/10.1016/j.ijdrr.2020.101799

Duden: Keyword "Indikator" (2023). https://www.duden.de/rechtschreibung/Indikator. Accessed 31 Aug 2023

Felder, F.A., Petitet, M.: Extending the reliability framework for electric power systems to include resiliency and adaptability. Electr. J. **35**(8), 107186 (2022). https://doi.org/10.1016/j.tej.2022.107186

Ferdinant, P.F., Nurdiana, L., Irman, A.: Evaluation of power plant reliability using index loss of load in the Suralaya power plant. IOP Conf. Ser. Mater. Sci. Eng. **909**(1), 12070 (2020). https://doi.org/10.1088/1757-899X/909/1/012070

Gasser, P., Lustenberger, P., Cinelli, M., Kim, W., Spada, M., Burgherr, P., et al.: A review on resilience assessment of energy systems. Sustain. Resilient Infrastruct. **6**(5), 273–299 (2021). https://doi.org/10.1080/23789689.2019.1610600

Gasser, P., Suter, J., Cinelli, M., Spada, M., Burgherr, P., Hirschberg, S., et al.: Comprehensive resilience assessment of electricity supply security for 140 countries. Ecol. Indic. **110**, 105731 (2020). https://doi.org/10.1016/j.ecolind.2019.105731

Gillespie-Marthaler, L., Nelson, K., Baroud, H., Abkowitz, M.: Selecting indicators for assessing community sustainable resilience. Risk Anal. off. Publ. Soc. Risk Anal. **39**(11), 2479–2498 (2019). https://doi.org/10.1111/risa.13344

Guo, D., Shan, M., Owusu, E.: Resilience assessment frameworks of critical infrastructures: state-of-the-art review. Buildings **11**(10), 464 (2021). https://doi.org/10.3390/buildings11100464

Hirschberg, S., Bauer, C., Burgherr, P., Dones, R., Simons, A., et al.: Deliverable D3.2 – RS 2b "Final set of sustainability criteria and indicators for assessment of electricity supply options". Brussel, Belgium (2008)

Hosseini, S., Barker, K., Ramirez-Marquez, J.E.: A review of definitions and measures of system resilience. Reliab. Eng. Syst. Saf. **145**, 47–61 (2016). https://doi.org/10.1016/j.ress.2015.08.006

ILO: Basic principles of monitoring and evaluation. International Labour Organization. Geneve, Switzerland (2015)

Jackson, S., Ferris, T.L.J.: Resilience principles for engineered systems. Syst. Eng. **16**(2), 152–164 (2013). https://doi.org/10.1002/sys

Jasiūnas, J., Lund, P.D., Mikkola, J.: Energy system resilience – a review. Renew. Sustain. Energy Rev. **150**, 111476 (2021). https://doi.org/10.1016/j.rser.2021.111476

Jewell, J., Cherp, A., Riahi, K.: Energy security under de-carbonization scenarios: an assessment framework and evaluation under different technology and policy choices. Energy Policy **65**, 743–760 (2014). https://doi.org/10.1016/j.enpol.2013.10.051

Jovanović, A., Klimek, P., Renn, O., Schneider, R., Øien, K., Brown, J., et al.: Assessing resilience of healthcare infrastructure exposed to COVID-19: emerging risks, resilience indicators, interdependencies and international standards. Environ. Syst. Decis. **40**(2), 252–286 (2020). https://doi.org/10.1007/s10669-020-09779-8

Kaloti, S.A., Chowdhury, B.H.: Toward reaching a consensus on the concept of power system resilience: definitions, assessment frameworks, and metrics. IEEE Access **11**, 81401–81418 (2023). https://doi.org/10.1109/ACCESS.2023.3300292

Kottmann, F., Kyriakidis, M., Dang, V.N., Sansavini, G.: Enhancing infrastructure resilience by using dynamically updated damage estimates in optimal repair planning: the power grid case. J. Risk Uncertainty Eng. Syst. Part A Civ. Eng. **7**(4), 4021048 (2021)

Linser, S.: Theoretical background of indicators and indicator system for the assessment of sustainable development. In: Niskanen, A., Väyrynen, J. (eds.) Regional Forest Programmes: A Participatory Approach to Support Forest Based Regional Development. EFI Proceedings, vol. 32, pp. 205–221 (1999)

Mair, S., Jones, A., Ward, J., Christie, I., Druckman, A., Lyon, F.: A critical review of the role of indicators in implementing the sustainable development goals. In: Filho, W.L. (ed.) Handbook of Sustainability Science and Research. World Sustainability Series, pp. 41–56. Springer International Publishing, Cham(2018)

Mangoyana, R.B., Smith, T.F., Simpson, R.: A systems approach to evaluating sustainability of biofuel systems. Renew. Sustain. Energy Rev. **25**, 371–380 (2013). https://doi.org/10.1016/j.rser.2013.05.003

Martišauskas, L., Augutis, J., Krikštolaitis, R., Urbonas, R., Šarūnienė, I., Kopustinskas, V.: A framework to assess the resilience of energy systems based on quantitative indicators. Energies **15**(11), 4040 (2022). https://doi.org/10.3390/en15114040

Mathworks: Centrality. Measure node importance. MathWorks, Inc. Natick, MA, U.S. (2023). https://de.mathworks.com/help/matlab/ref/graph.centrality.html. Accessed 31 Aug 2023

Meadows, D.: Indicators and information systems for sustainable development. The Sustainability Institute. Hartland four corners VT, USA (1998)

Meerow, S., Newell, J.P.: Urban resilience for whom, what, when, where, and why? Urban Geogr. **40**(3), 309–329 (2019). https://doi.org/10.1080/02723638.2016.1206395

Mentges, A., Halekotte, L., Schneider, M., Demmer, T., Lichte, D.: A resilience glossary shaped by context: reviewing resilience-related terms for critical infrastructures. Int. J. Disaster Risk Reduct. **96**, 103893 (2023). https://doi.org/10.1016/j.ijdrr.2023.103893

Merry, S.E.: Measuring the world: indicators, human rights, and global. Governance **52**(3), S83–S95 (2011)

Mottahedi, A., Sereshki, F., Ataei, M., Nouri Qarahasanlou, A., Barabadi, A.: The resilience of critical infrastructure systems: a systematic literature review. Energies **14**(6), 1571 (2021). https://doi.org/10.3390/en14061571

Mujjuni, F., Betts, T., To, L.S., Blanchard, R.E.: Resilience a means to development: a resilience assessment framework and a catalogue of indicators. Renew. Sustain. Energy Rev. **152**, 111684 (2021). https://doi.org/10.1016/j.rser.2021.111684

Müller-Lancé, K.: Energiekrise: Wieso die Atomenergie in Frankreich zum Sorgenkind wird. (engl. Energy crisis: Energy crisis: Why nuclear energy is becoming a problem child in France. In Sueddeutsche Zeitung GmbH, 10/15/2022 (2022). https://www.sueddeutsche.de/politik/frankreich-kernkraft-macron-1.5656779. Accessed 23 Jan 2024

Netz, H., Schönwandt, I., Wigger, H., Brand-Daniels, U., Lichte, D., Vogt, T.: Integrating sustainability and resilience aspects into power system technology assessments. In: 2023 IEEE PES Innovative Smart Grid Technologies Europe (ISGT EUROPE), Grenoble, France, pp. 1–5 (2023). https://doi.org/10.1109/ISGTEUROPE56780.2023.10407277.

Nguyen, H.L., Akerkar, R.: Modelling, measuring, and visualising community resilience: a systematic review. Sustainability **12**(19), 7896 (2020). https://doi.org/10.3390/su12197896

Nikzad, M., Farahani, S.S.S., Tabar, M.B., Tourang, H., Yousefpour, B.: Calculation of generation system reliability index: loss of load expectation. Life Sci. J. **9**, 3595–3599 (2012)

Omer, M., Mostashari, A., Lindemann, U.: Resilience analysis of soft infrastructure systems. Procedia Comput. Sci. **28**, 565–574 (2014). https://doi.org/10.1016/j.procs.2014.03.069

Osei-Kyei, R., Almeida, L.M., Ampratwum, G., Tam, V.: Systematic review of critical infrastructure resilience indicators. Constr. Innov. **23**(5), 1210–1231 (2023). https://doi.org/10.1108/CI-03-2021-0047

Ottenburger, S.S., Çakmak, H.K., Jakob, W., Blattmann, A., Trybushnyi, D., Raskob, W., et al.: A novel optimization method for urban resilient and fair power distribution preventing critical network states. Int. J. Crit. Infrastruct. Prot. **29**, 100354 (2020). https://doi.org/10.1016/j.ijcip.2020.100354

Patriarca, R., de Paolis, A., Costantino, F., Di Gravio, G.: Simulation model for simple yet robust resilience assessment metrics for engineered systems. Reliab. Eng. Syst. Saf. **209**, 107467 (2021). https://doi.org/10.1016/j.ress.2021.107467

Porter, T.M.: Trust in numbers. The pursuit of objectivity in science and public life. Princeton University Press, Princeton N.J. (1995)

Poulin, C., Kane, M.B.: Infrastructure resilience curves: performance measures and summary metrics. Reliab. Eng. Syst. Saf. **216**, 107926 (2021). https://doi.org/10.1016/j.ress.2021.107926

Quinlan, A.E., Berbés-Blázquez, M., Haider, L.J., Peterson, G.D.: Measuring and assessing resilience: broadening understanding through multiple disciplinary perspectives. J. Appl. Ecol. **53**(3), 677–687 (2016). https://doi.org/10.1111/1365-2664.12550

Resilens: Deliverable D2.2: qualitative, semi-quantitative and quantitative methods and measures for resilience assessment and enhancement. Brussel, Belgium (2016)

Specking, E., Parnell, G.S., Pohl, E., Buchanan, R.: Engineering resilient systems: achieving stakeholder value through design principles and system operations. IEEE Trans. Eng. Manag. **69**(6), 3982–3993 (2022). https://doi.org/10.1109/TEM.2021.3056934

Stanković, A.M., Tomsovic, K.L., de Caro, F., Braun, M., Chow, J.H., Čukalevski, N., et al.: Methods for analysis and quantification of power system resilience. IEEE Trans. Power Syst. **38**(5), 4774–4787 (2023). https://doi.org/10.1109/TPWRS.2022.3212688

Turnhout, E., Hisschemöller, M., Eijsackers, H.: Ecological indicators: between the two fires of science and policy. Ecol. Indic. **7**(2), 215–228 (2007). https://doi.org/10.1016/j.ecolind.2005.12.003

Wang, C., Ju, P., Wu, F., Pan, X., Wang, Z.: A systematic review on power system resilience from the perspective of generation, network, and load. Renew. Sustain. Energy Rev. **167**, 112567 (2022). https://doi.org/10.1016/j.rser.2022.112567

Watson, B.C., Chowdhry, A., Weissburg, M.J., Bras, B.: A new resilience metric to compare system of systems architecture. IEEE Syst. J. **16**(2), 2056–2067 (2022). https://doi.org/10.1109/JSYST.2021.3062444

Wied, M., Oehmen, J., Welo, T.: Conceptualizing resilience in engineering systems: an analysis of the literature. Syst. Eng. **23**(1), 3–13 (2020). https://doi.org/10.1002/sys.21491

Zhang, X., Miller-Hooks, E., Denny, K.: Assessing the role of network topology in transportation network resilience. J. Transp. Geogr. **46**, 35–45 (2015). https://doi.org/10.1016/j.jtrangeo.2015.05.006

Zwart, R.: Strengthening the results chain: synthesis of case studies of results-based management by providers. OECD (OECD development policy papers, No.7) (2017)

Chapter 12
Resilience of Digitalised Power Systems—Challenges and Solutions

Michael Brand, Sanja Stark, Stefanie Holly, Jirapa Kamsamrong, Christoph Mayer, and Sebastian Lehnhoff

Abstract Power systems are undergoing a transition from centralised generation to distributed renewable generation. That calls for more flexibility in balancing generation and consumption because distributed energy resources in most cases rely on weather-dependent resources and can, therefore, only be controlled to a certain degree. The necessary flexibility can be achieved by introducing digitalisation. However, digitalisation also leads to new vulnerabilities of the power system and increases its complexity. Additionally, the often rapid and unforeseeable development of information and communication technology introduces a new level of uncertainty in the system design. These circumstances make it necessary to shift from a classical design of robust to the design of resilient power systems that are able to anticipate, react to, and recover from them. In this paper, these real-time attributes of a resilient digitalised power system are analysed with respect to potential measures: an enhanced situational awareness, virtualisation, flexibilisation, and a distributed black start. The measures tackle different challenges for resilient digitalised power systems and cover different phases of the resilience process, visualised in the so-called resilience bathtub curve. The implementation of these measures can increase the resilience of a digitalised power system and are meant to be combined with further, also non-real-time measures.

M. Brand (✉) · S. Stark · S. Holly · J. Kamsamrong · C. Mayer · S. Lehnhoff
Energy Division, OFFIS—Institute for Information Technology, Oldenburg, Germany
e-mail: michael.brand@offis.de

S. Stark
e-mail: sanja.stark@offis.de

S. Holly
e-mail: stefanie.holly@offis.de

J. Kamsamrong
e-mail: jirapa.kamsamrong@offis.de

C. Mayer
e-mail: christoph.mayer@offis.de

S. Lehnhoff
e-mail: sebastian.lehnhoff@offis.de

D. van Doren et al. (eds.), *Towards Energy System Resilience*, Ethics of Science and Technology Assessment 55, https://doi.org/10.1007/978-3-032-03187-7_12

Keywords Digitalised power systems · Resilience · Anticipate · React · Recover

12.1 Introduction

Power systems are a critical infrastructure and therefore have a huge impact on safety, health, and economy of modern society (Rinaldi et al. 2001). Incidents can lead to failures and failures to blackouts, which cascade and impact other critical sectors like banking, emergency, and government services. To prevent this, power systems must fulfil strict requirements regarding safety, security, and reliability (Cortekar and Groth 2015). Currently, power systems undergo a transition to create a more environmentally friendly energy supply. Conventional generators like coal-fired or nuclear power plants are replaced by smaller generators that rely mostly on renewable resources such as wind and solar. Unlike traditional power plants, these generators are widely distributed geographically and are most prominent in medium and low voltage grids. They are referred to as distributed energy resources (DERs) (Akorede et al. 2010). In addition, the consumers are changing as well and are no longer passive by definition. New roles like pro-active consumers and prosumers emerge. Pro-active consumers can manage their electricity demand corresponding to electricity tariffs (demand response), while prosumers are consumers and producers at the same time (e.g., private rooftop photovoltaic (PV) (Janssen and Dankbaar 2008, Gautier et al. 2018].

This transition not only increases the complexity of the power system but also demands more flexibility in balancing generation and consumption, due to the fluctuating behaviour of renewable resources (Dena and G.E.A. 2024). Digitalisation can help to handle the higher system complexity as well as efficiently utilise the aforementioned flexibility by providing extended monitoring and control, especially in medium and low voltage grids. In this paper, digitalisation of the power system is defined as an "ongoing process of networking applications, processes, actors and devices or objects using information and communication technology. It involves the collection, processing, exchange and analysis of information and data and supports the generation of knowledge, the making of decisions, and the derivation of control actions. The resulting new processes are automated, supported by AI [artificial intelligence] mechanisms and learning systems" (Acatech et al. 2021). In a digitalised power system, all devices, systems, and applications can potentially exchange information, such as the current power feed-in of a DER, weather data, and market data. This supports the system operator in controlling the power system and ensuring its stability. However, not only generators are affected by the digitalisation. In addition, pro-active consumers and prosumers can also support with their flexibility in shifting generation and consumption of power over time. Examples are combined heat and power plants, heat pumps, and electric vehicles (EVs) (Knezovic et al. 2017; Huang et al. 2019; Brange et al. 2016).

While digitalisation is necessary to support and accelerate the transition of power systems, it also introduces more uncertainty about the future into the system design, since developments in the ICT domain can be fast and unforeseeable. Moreover, digitalisation introduces new interdependencies between the power and the ICT system (Pillitteri and Brewer 2014). A failure of the ICT system that handles the communication between the field devices and the control room can impact the performance of the power system and, in worst case, interrupt the power supply. Furthermore, the digitalised system has a large attack surface and is more vulnerable for cyber-attacks compared to a non-digitalised. One example is the attack on the Ukraine power system in 2015, where attackers got access to the supervisory, control, and data acquisition systems and disconnected several substations (Case 2016). When designing a digitalised power system, the goal is, therefore, to use the potential for supporting the energy transition on the one hand while managing the aforementioned risks on the other.

Classical risk assessment analyses the robustness of systems against known incidents and known occurrence probabilities (Shiwen et al. 2017). This approach is not sufficient for digitalised power systems due to the possibility of unpredictable events, developments and incidents. Alternatively, the concept of resilience can be used. "A resilient power system has the ability to absorb the adverse effects of a spontaneous or creeping disruptive event, with possible declines in quality of supply, but without the system collapsing with a major blackout, and then rapidly, affordably and with a minimum of damage returning to normal operating status" (Acatech et al. 2021). A resilient system should have the abilities to anticipate, withstand, recover, and evolve. This requires a resilience strategy, that combines different measures to achieve the desired system behaviour. It is important to note that the power system not only influences society, and the way people live but at the same time society shapes the technological developments in the power system through social innovations and trends, resulting in a complex, socio-technical system (Acatech et al. 2021). Therefore, it is important that a thorough resilience strategy should not only include technical measures but also measures in all other relevant domains, such as social or political. However, the focus of this paper is on technical resilience measures, specifically those that benefit from digitalisation and are applied at run-time, i.e. not in planning phase. The contributions are the following:

- Challenges for Digitalised Power Systems: A summary of basic causes for incidents in digitalised power systems is presented.
- Resilience of Power Systems: Resilience is defined, described, and delimited from risk analysis in more detail.
- Classification of Resilience Measures: An overview of existing classifications of resilience measures is provided.
- Real-Time Measures for Resilient Digitalised Power Systems: Main real-time attributes of a resilient digitalised power system and concrete measures for the attributes are presented.
- Discussion: The presented measures are discussed in terms of challenges they tackle, resilience phases they contribute in, and how they can be classified.

12.2 Challenges for Digitalised Power Systems

The decentralisation and digitalisation in the power system result in new kinds of incidents that the system must face. To get an overview over the different types of incidents, they can be assigned to different *basic causes*. A basic cause describes one aspect—or a group of similar aspects—of the system or the system environment that is not problematic on itself but can be a risk factor for disruptive events. Examples include climate change, which can cause more frequent extreme weather events and therefore more natural disasters, or the fluctuating power feed-in from PV or wind power plants, which, if not handled adequately, can cause instabilities in the system.

This paper focuses specifically on basic causes that result from the process of digitalisation. In the context of the project ESYS,[1] a working group has identified four new basic causes based on possible future scenarios for the year 2040, which can cause potentially large blackouts brought about by new types of incidents (Acatech et al. 2021).

12.2.1 Basic Cause 1: Simultaneous Behaviour of Small Producers and Consumers

Digitalisation enables the sending of triggering signals to many small electricity producers (e.g., rooftop PVs or small wind plants) or consumers (e.g., heat pumps, charging EVs, or home electricity storage systems) at the same time and with very little effort. The signal can cause several small producers and consumers to behave simultaneously, either by directly controlling the device itself or by encouraging consumers' behaviour change, for example with an attractive electricity tariff scheme (demand-response).

The aggregated power of many small producers or consumers can impact power system stability if operated simultaneously. In Acatech et al. (2021) it is shown, for example, that only 9.5\% of PVs units in the state of Baden-Württemberg in southern Germany already equals an amount of 420 megawatts, which is considered critical infrastructure under German law.[2]

Simultaneous behaviour can either be a one-time event, which creates one large peak in generation or consumption in the distribution grid (e.g., the simultaneous charging of many EVs) or several events over a period (e.g., disconnecting and reconnecting rooftop PVs), thereby creating an oscillating behaviour.

When used in a grid-supporting way and under supervision of the grid operators, this simultaneous behaviour may be desired. However, in a digitalised power system, this triggering signal could also be sent by a third party (e.g., software updates from the manufacturers or market signals) or even be part of a cyber-attack.

---

[1] https://energiesysteme-zukunft.de/en/.

[2] https://www.gesetze-im-internet.de/bsi-kritisv/anhang_1.html.

12.2.2 Basic Cause 2: Mutual Interdependence Between ICT-Related and Energy Technical Resources

All ICT systems are dependent on power supply and, therefore, would be heavily affected by a blackout (unless they are equipped with an uninterruptible power source such as battery backup). However, the dependence of the power system on ICT systems gets more significant in the digitalised power system. While transmission system operators (TSOs) and large distribution system operators (DSOs) traditionally only use dedicated and blackout-safe communication infrastructure, ICT will be a necessary and inseparable part of digitalised power systems, especially in the smaller DSO areas. As explained in the Introduction, this is a necessary step to handle the future system complexity.

Nevertheless, because of the mutual interdependence, failures in the ICT system can affect the entire power system as well. Even small subsystem failures can result in a partial or complete blackout due to cascading failures (Wirtz and Monti 2020). Defective or maliciously compromised ICT systems cannot simply be switched off since it would no longer be possible to control the power grid.

12.2.3 Basic Cause 3: Technical Complexity in System Operation

It is challenging to balance power supply from weather dependent DERs such as wind and PV plants to always meet the demand. Therefore, replacing the large power plants with DERs requires more flexibility to ensure the same power quality as today. This growing number of decentralised power generation and storage units also means that the interdependencies between the different system components are becoming more and more complex. Fast reactions are required by the system operators, while at the same time it can be difficult to oversee the results of certain control actions, since they might have unforeseen and mutually reinforcing effects. ICT helps system operators to deal with this system complexity on the one hand but introduces new complexity on the other hand. The use of artificial intelligence to autonomously control decentralised units, for example, can result in emergent behaviour that makes it harder to predict the behaviour of the system. Similarly, the incidents themselves get more complex. It is not possible to separate ICT and power system incidents completely anymore due to the previously mentioned mutual interdependencies.

While the optimisation of system design and adequate training of the operational staff are important to handle the system complexity as best as possible, it is still to be expected that failures can happen, certain aspects are overlooked, or unforeseen effects take place.

12.2.4 Basic Cause 4: Uncertainty About Future Developments

The design of the power system is based on explicit or implicit assumptions about the future. Once a technology has been implemented or an infrastructure built, it is usually part of the system for a long period of time, making it hard to adapt the system to new developments at a later point in time. This can lead to certain path-dependencies. Decisions that have been made in the past about the system design are still followed in the present, even though they are not the best choice for the current system anymore. In a digitalised power system, however, two very different rates of development then exist side-by-side. While power system components remain in use for several decades, software is often updated several times a year and ICT innovation cycles are just a few years long.

Given the uncertainty of future developments, an inflexible system design may not be prepared for unknown future incidents. Moreover, it might also not be capable of keeping up with the development of ICT, which is critical for integrating large numbers of DERs.

12.3 Resilience of Power Systems

The reliability of power system across the world varies significantly. An international comparison for selected power systems can be seen in Fig. 12.1, showing the system average interruption duration index (SAIDI) value in minutes for 2016. The figure shows that especially the European power system is very reliable with a median value of 76 min for SAIDI across all countries, including exceptional events (see CEER 2018 for a detailed comparison between the different European countries). In comparison, the SAIDI values for USA (250 min Lindstrom and Hoff 2017) or Australia (152, 59 and 410 min for the three grids (Economic Regulation Authority Western Australia 2019)) are much higher.

This level of stability has been achieved by using a robust design approach based on risk analysis. The risk for known incidents, e.g., incidents that have already occurred in the past, is measured by evaluating the damage that an incident could cause as well as the probability of this incident to happen. The goal is to find weak points in the system design and then "harden" these parts, thereby increasing the system's robustness against a set of specific incidents. Robustness means that an incident has no impact on the system performance. A classic example for a known incident is the failure of one large component, e.g., a transmission line, a transformer, or a large generation unit. To make the system robust against this specific kind of incidents, the n-1 criterion is implemented. It states that a system should be able to always withstand the failure or outage of one system component, which can be achieved by redundancy.

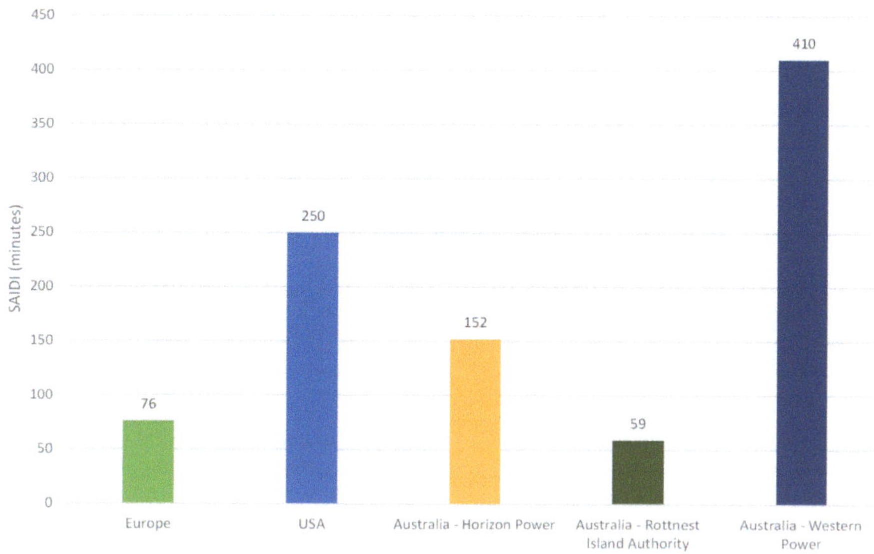

Fig. 12.1 SAIDI comparison in 2016 of selected electric power networks worldwide

While this concept has been very successful in the past, it might not be sufficient for future systems anymore. As described in the Sect. 12.2, the power system faces new challenges, making a robust design approach solely based on risk analysis infeasible. On the one hand, due to the uncertainty about future technical, social, or political developments, it is not possible anymore to measure the probability of incidents—even the probability of already known incidents can change. On the other hand, the heterogeneous components of the digitalised power system can have non-linear interdependencies that result in unforeseen events. One example is the integration of many EVs into the grid, which can lead to voltage instabilities due to the charging behaviour of the batteries (Habib et al. 2018). These incidents would not be caught by the currently installed robustness measures and cannot be evaluated with a risk analysis.

Here, the concept of resilience can help to tackle those types of incidents that cannot be derived from experience or current knowledge. Resilient system design shifts the focus from the actual incident to the time after it has happened. The primary goal is no longer only to harden the system against known incidents (although robustness is a part of resilience) but rather to absorb the effects of incidents as much as possible and then return the system to a normal state as fast as possible.

Figure 12.2 shows a so-called *bathtub curve* (as described in Acatech et al. 2021), with which the desired resilient system behaviour can be described and shows the difference in behaviour of a non-resilient (pink) and a resilient (blue) system. While an incident in a non-resilient system can lead to a major blackout, it has less effect in a resilient system. The following five phases of a resilient system behaviour can be distinguished:

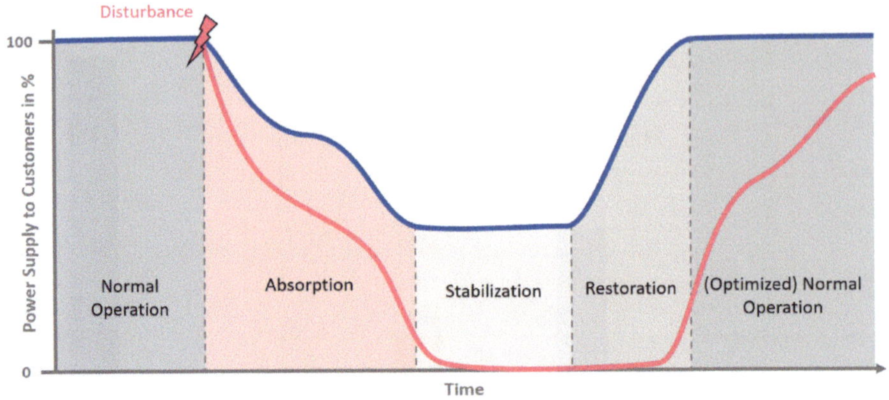

Fig. 12.2 Restoring system functionality: responding appropriately to an incident (absorption of disruption), reverting to critical functions and stabilising the system (stabilisation), followed by a controlled return to normal system operation (restoration) [7]

(1) Normal operation: All loads are supplied, and all system parameters are within their operational limits. Robustness measures like the n-1 criterion are in place. If an incident happens in this phase, it might be (partly) handled by the system robustness.
(2) Absorption of disruption: The system robustness was not enough to handle the incident and the system performance decreases, i.e. not all loads are supplied anymore. The goal of this phase is to stop the degradation. It is especially critical to prevent cascading and escalating failures, which might result from the initial incident.
(3) Stabilisation: The next goal is to stabilise the system on a lower performance level and gain full control over the system again. Stabilisation here refers to frequency and voltage stability, which is necessary to prevent a (further) triggering of protection equipment.
(4) Restoration: Starting from a stable but degraded state, the system should now be brought back to full performance with the necessary restoration measures. This means to reconnect all loads that have been disconnected, restart generators, reconnect separated grid areas, etc.
(5) (Optimised) Normal operation: The final goal is not only to bring the system back into its normal operation but also to improve the system's resilience. One central part of resilient systems is the ability to learn from past events and prepare for similar future events. This requires a detailed (automated) analysis of the incident as well as of the second, third, and fourth phase. Based on the results, new or improved resilience measures (e.g., replacing of components, adaption of technical processes or improved information exchange) can be developed. Therefore, after a successfully handled incident, the system becomes more resilient than before.

Current power systems already have several measures that enable resilient system behaviour in case an incident could not be absorbed by the system robustness. These measures are part of the ancillary services, which are divided into four categories: frequency control, voltage control, operation management, and restoration (dena 2014). Each of these categories has several measures at disposal to fulfil its respective goal. Examples for measures are balancing power and load-shedding for frequency control, reactive power management for voltage control, redispatch and switching operations for operational management, and detailed plans for several black start scenarios for restoration. These measures can be used to react to incidents and return the system back to normal operation, an example of which can be seen from the incident of the Continental Europe Synchronous Area Separation on 8 January 2021 (ENTSO-E 2021). The system split into two areas as a result to the cascading trips of several transmission network elements (busbar couplers, transformers, transmission lines). While the south-eastern area had an increased frequency from a surplus of generation, the north-western area had a surplus of load and thus a decreased frequency. The absorption of this disruption was possible due to the automated defence actions in the individual areas, mainly primary frequency reserve in both areas and automated load shedding in the north-western area with underfrequency. Next, the TSOs in both areas performed manual control actions to further stabilise their respective area. For the north-western area, this included the starting of additional turbines and then reconnecting the interrupted load as soon as there was enough generation available. In the south-eastern area, power production was reduced and loads, which have been disconnected automatically during the cascading events, were resupplied. Finally, both areas were resynchronised, thereby fully restoring the European power system. The entire process was completed in just over an hour.

This example shows, how currently installed resilience measures are already capable of creating a resilient system behaviour in reaction to an incident. However, these current resilience measures need to be adapted and extended to ensure resilience in digitalised power systems.

12.4 Classification of Resilience Measures

The digitalisation of power systems, along with the decentralisation of energy supply, offers opportunities for new kinds of resilience measures, which can handle the new challenges. To develop resilience measures in a structured way, it can help to split up the concept of "system resilience" into different categories and classify the resilience measures accordingly. The use of such a classification allows to evaluate the system resilience based on how well different categories are covered by the resilience measures.

So far, there is no standard classification for system resilience or resilience measures. The following subsections will introduce and discuss different criteria for classification, categorised into time-based, threat-based, and attribute-based approaches.

12.4.1 *Time-Based*

Time-based classification is the most common approach for resilience measures that distinguishes between power system planning and operation (Kandaperumal and Srivastava 2020, Ciapessoni et al. 2020, Mahzarnia et al. 2020). A similar classification approach can be achieved by defining passive (i.e. redundancies, component hardening) and active resilience measures (i.e. threat assessment, restoration procedures) (Ciapessoni et al. 2020). The active resilience measures can be further classified into proactive (i.e. measures that provide preventive functions) and reactive (i.e. measures that provide restorative functions) (Kandaperumal and Srivastava 2020). Resilience measures are typically designed by time horizon in the planning phase to respond to incidents in the operational phase as described in Mahzarnia et al. (2020) and as shown in Fig. 12.3.

To emphasise the focus of resilient power systems on the dynamic reaction to incidents, measures can also be classified with respect to the occurrence of an incident: pre-event/before threat—measures that can predict, prevent, and/or mitigate an incident before it occurs, during event/during threat—measures to withstand, absorb and survive the incident, and post-event/after threat—measures including recovery, repair, reconfigurability and learning capability of the system (Kandaperumal and Srivastava2020; Mahzarnia et al. 2020; Shen et al. 2020). Following the structure of the resilience bathtub curve, resilience measures can also be mapped into the four categories plan, absorb, recover and adapt (Lin et al. 2018).

The time-based classification gives the impression that resilience measures must be applied sequentially and are building on each other, while they might be overlapping; for example, restorative measures may already be used concurrently with system stabilisation. The same resilience measure can also be used to prevent an incident and—if that fails—to help absorbing the effects.

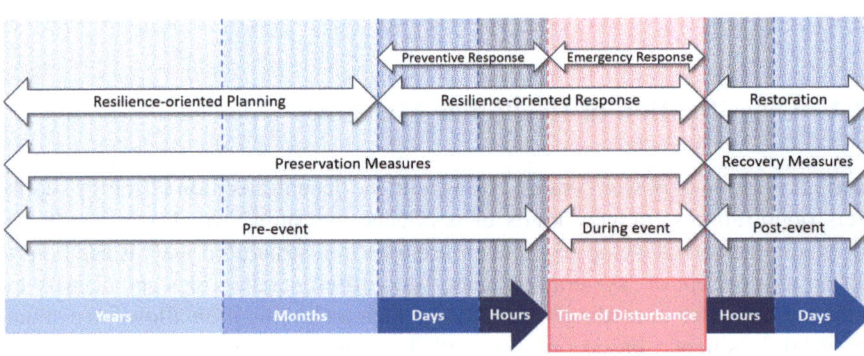

Fig. 12.3 Classification of power system resilience in terms of the time of the event's occurrence

12.4.2 Threat-Based

Threat-based classification orders measure according to the type of incident they are supposed to mitigate. The authors in Kandaperumal and Srivastava (2020) separate between physical and cyber resilience. Physical resilience encompasses all measures that increase the system's resilience against any (man-made or physical) threats, while cyber resilience covers all measures related to cyber threats. Of the examples given, the focus is mostly on preventive hardening measures, such as pole hardening or elevated substations for physical resilience and encryption or redundant communication links for cyber resilience.

A threat-based classification focuses on measures that prevent specific threats without considering the resilient system behaviour after a threat has happened. Separating between different threat types also does not consider potential interdependencies between the different threats as well as cascading or escalating failures. Moreover, as described in Sect. 12.2, the developments of the power system can also lead to new unknown threats, which would not be covered.

12.4.3 Attribute-Based

In an attribute-based classification, the first step is to define abstract attributes a resilient system needs to have and then develop concrete measures to strengthen these attributes.

In Petit et al. (2020), the authors propose to define a resilience index by considering the preparedness, mitigation measures, response capabilities, and recovery mechanisms of a system. Preparedness includes on the one hand measures to be performed long before any incidents take place, such as emergency operation plans, cyber security plans, or preventive maintenance. On the other hand, it also covers measures that enhance the awareness to potential incidents of any kind.

Mitigation measures can either help the system to resist an incident or at least absorb its consequences. This includes, but is not limited to, system hardening measures, such as redundancy, which focus on robust design.

Response capabilities is the category for all those measures, that directly respond to any persistent negative effect of an incident—except for measures, that explicitly handle the recovery of the system, which would be classified under recovery mechanisms.

An attribute-based classification allows a more general view on resilience, where the different attributes can be relevant in several phases of the resilience curve as well as tackling different types of threats. The authors in Shen et al. (2020) show how different attributes can be mapped to time phases by classifying the same resilience measures in a time-based as well as an attribute-based structure. The attributes include resistance, robustness, and attentiveness for the pre-event phase, recovery and coping capability for the during-event phase, and reconfigurability and learning capability for the post-event phase. While this comparison of time-based and attribute-based classes can help to define which attributes are most relevant for each phase, it should be noted that nearly all attributes are in some way necessary—or at least supporting—in every phase.

12.5 Real-Time Measures for Resilient Digitalised Power Systems

The previously described classification approaches not only focus on system resilience but also include a significant amount of robustness measures as well. Considering the new challenges (cf. Sect. 12.2), robustness measures will become less and less relevant and the key aspect of resilience—dynamic real-time reaction to incidents—should be more emphasised. Moreover, except for the threat-based classification, the digitalisation of the future power system is not explicitly considered in any of the approaches.

Therefore, this paper analyses specifically those resilience measures, that can be used in real-time and benefit from digitalisation, using the attribute-based classification scheme as a base. To emphasise the focus on real-time elements, instead of preparedness, mitigation measures, response capabilities, and recovery mechanisms, the three categories *anticipation*, *reaction*, and *recovery* are defined.

Anticipation is the real-time part of preparedness, which should take an integrative perspective regarding the physical power system and the ICT system. Past blackouts have shown the importance of considering the state of the ICT system for anticipation. For example, in the 2003 North American blackout, an alarm system failure led to loss of situational awareness of the operators (NERC 2004). Since the alarm system failure itself was unnoticed, no measures were taken to regain situational awareness, and failures in the power system—tripping of lines due to tree contact—were hidden from the operator's view, leading to cascading failures and finally to a blackout that affected 55 million people.

Reaction describes the real-time part of mitigation and response. Here an individual perspective is taken, because the react measures depend on the concrete kind of incident and in which domain the impact takes place. Therefore, power and ICT domain are viewed separately. A key requirement to be able to react in real-time to incidents are flexibilities in both domains. Flexibilities describe all parameters that can be changed as a response to an incident. In the power grid, this could be either the ability from consumers or producers to modify their consumption or production, respectively, flexibility in the grid itself (e.g. remote-controllable switches) or energy storage. An example for flexibilities in the interconnected ICT system is the use of virtualisation to maintain ICT service functionality.

Recovery should, as for the anticipation, take an integrative perspective. The restoration of a power system is highly complex and requires careful coordination. Traditionally, the responsibility for a black start lies with the TSOs, who have developed a detailed restoration plan and perform regular simulation trainings to prepare for a potential black start (ENTSO-E 2017). To have full control over the restoration process, distribution grids are considered only as passive loads and all DERs in the distribution grid are automatically disconnected during the restoration process. Only after the restoration from top-down is completed, DERs can proceed to supply active or reactive power to the grid. The increasing number of DERs, especially on the distribution level, enables new restoration concepts, where distribution grids take a more active role. Digitalisation can help the system operator to handle the increased complexity of the restoration process by collecting and analysing data from the distribution grid, support decision making or even perform a fully automated restoration. However, especially in case of large-scale blackouts, the ICT system is very likely to be affected as well, potentially limiting the communication from field to the control room. Therefore, not only the power system needs to be recovered but also the interconnected ICT system to regain full performance.

In the following subsections, different exemplary approaches in the three mentioned classes of anticipate, react, and recover are presented and discussed: Anticipate: An Integrative Perspective, React: An ICT Perspective, React: A Power System Perspective, and Recover: An Integrative Perspective.

12.5.1 Anticipate: An Integrative Perspective

Situational awareness and, especially, the estimation of the physical system state, i.e. the complex voltages in the power system, is of high importance for a reliable and resilient operation of the system. State estimation is the foundation for many grid services like, for example, voltage control. It is, therefore, crucial for anticipation to be able to estimate the current system state and to rely on it. Without a reliable situational awareness, incidents are most probably recognised too late and suitable resilience measures cannot be conducted in time or at all.

For the anticipation in digitalised power systems, traditional aspects for situational awareness like functional correctness and reliability are still important but need to be extended by additional aspects or perspectives like security and credibility. This becomes intuitively clear by raising the following questions. Are the field measurements that are collected in the control room correct? Can they be manipulated? Are the data sources credible? An example for incidents that cannot be anticipated so far are coordinated cyberattacks (Pillitteri and Brewer 2014). Liu et al. (2011) have shown that attackers performing a so-called coordinated false data injection attack can influence the estimated system state without being detectable by state-of-the-art detection schemes. Potential goals can be harmful control actions by the system operator or hiding harmful manipulations.

This is because state of the art for monitoring energy and especially power systems rely only on physical measurements like produced and consumed power, voltage, and current values acquired by, e.g., remote terminal units (RTUs). Many grid operators have information security management systems and intrusion detection systems

(IDSs) in place, however, an inclusion of information from these systems into the situational awareness is not state of the art. The approaches described in this subsection aim at closing this gap.

The first approach is called abnormal traffic-indexed state estimation (ATSE) and focuses on the integration of information from IDSs into a state estimation (Liu et al. 2015). The IDS record suspicious activities in the ICT system and the corresponding alerts from these systems are allocated to affected field devices. A network impact factor matrix is built with that information, reflecting how occurring alerts affect which measurements. The matrix is then integrated into the state estimation as a matrix of weighting factors. The more alerts and the higher the priority of alerts, the lower the measurement's weight in the state estimation process (Liu et al. 2015). The benefit for situational awareness is that the estimated system state relies, if possible, more on measurements, to which no alerts from IDSs are allocated. Measurement redundancy is, therefore, a requirement for this approach. Another limitation is the focus on the integrity of the measurements. The digitalisation of a power system and, especially, the consideration of an ICT system, however, introduces more potential risks than integrity issues. Issues in the topics of functional correctness and credibility, for example, are not covered by this approach.

Another approach is proposed by Basciftci and Ozguner (Basciftci and Ozguner 2012). Unlike the ATSE, no state estimator with weighted least squares but a particle filter is used. Accordingly, Basciftci and Ozguner propose a new trust-aware particle filter in which, based on the measured values, a confidence value is calculated for each sensor. The authors define two confidence classes for each sensor, trustworthy and untrustworthy, and work with probabilities for these two classes. On this basis, a confidence transition matrix is constructed, which consists of the probabilities with which a sensor is trustworthy or untrustworthy under the condition that another sensor is trustworthy or untrustworthy, respectively. The calculation of the conditional probabilities is based largely essentially on the sample variance of the sensor values of the two sensors involved (Basciftci and Ozguner 2012). As for the ATSE, the calculated confidence probabilities are integrated as weights into the objective function. Therefore, the same limitations hold as for the ATSE.

A third approach to be presented is called anomaly sensitive state estimation (ASSE)[3] (Brand et al. 2021a, 2023). The solution uses trust as an umbrella term to augment state estimation with aspects like functional correctness, security, or credibility. Based on a trust model, the credibility of power system components, data, and services is assessed. This provides new insights for the situational awareness and can improve anticipation and decision making in control rooms. Instead of singularly investigating the integrity of ICT devices and the ICT system (as the aforementioned approaches do), additional ways of compromising components, data, and services are considered and combined to a trust assessment (Brand et al. 2020, 2023; 2021a,

[3] Demonstration of ASSE: https://youtu.be/3hwi49sflIQ.

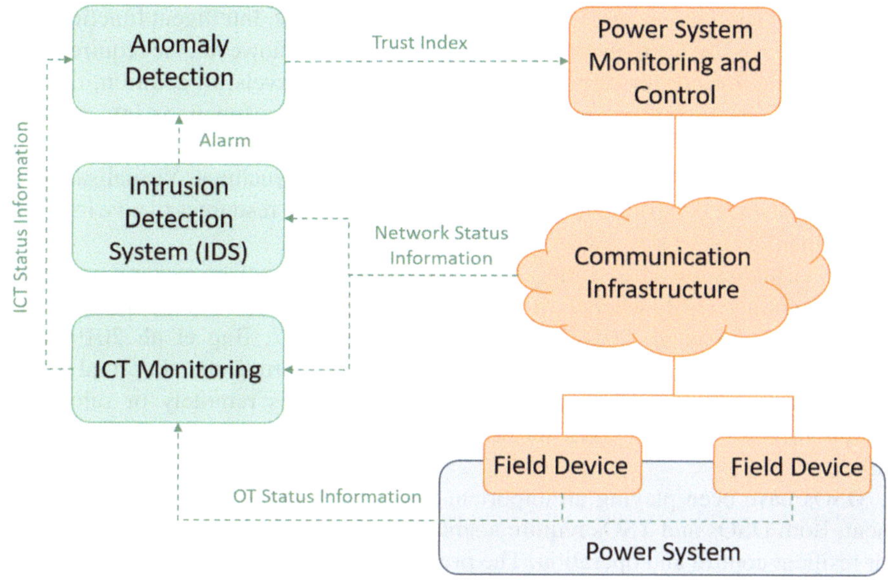

Fig. 12.4 An ICT monitoring framework for trust assessment

b). Trust is defined in this context as "a subjective, context-dependent, and multivariate sense about an entity with respect to its functional correctness, safety, security, reliability, credibility, and usability'"[4] (Brand et al. 2020).

Figure 12.4 shows the methodology of trust assessment as describes in Brand et al. (2019). The trust in field components like RTUs or routers is estimated in real-time based on information from monitoring systems such as ICT health monitoring and, as well as for the ATSE, IDSs. The trust in measurements is derived from the trust in the field devices involved in the data acquisition and the trust in the system state is derived from the trust in the measurements (Brand et al. 2021b). The benefit for the situational awareness is that all measurements and state variables are annotated with compound trust values reflecting probabilities for the different facets (e.g., functional correctness and security) and for different trust estimators (e.g., one based on the performance of devices and one based on alerts from an IDS) (Brand et al. 2020, 2021b). Because of the framework character, it does not provide concrete algorithms for the trust estimation but existing algorithms like the ATSE can be included.

12.5.2 React: An ICT Perspective

Integrating DERs in electrical network requires the intelligent protection and control functions of electrical grids to maintain system stability. All actors connected to the power system must comply with technical requirements defined by the grid code to

[4] This definition of trust is based on trust in organic computing (Steghöfer et al. 2010).

upgrade and configure control systems and/or field devices. Intelligent functions are mostly deployed on the high-voltage level of the network, however, the requirements for these functions differ significantly at lower voltage levels. In addition, there is still inflexibility for on-site testing and integrating of new functions into existing networks. Virtualisation can improve the resilience of power system by reacting to new integration and configuration into existing infrastructures. Virtualisation in computer science is "abstraction and sharing of computing resources to give multiple parties the illusion of sole ownership" (Wang et al. 2013).

The virtualisation is a well-known technique at the power system' information technology level, whereas it is slowly being applied at the operation technology level (Li et al. 2016; Kariyawasam et al. 2017; Bag et al. 2019). For example, virtualisation can help power systems during normal and abnormal events by executing control functions with allocated resources remotely or relocating computing resources at a given substation caused by communication infrastructure failure.

DSOs have been playing an important role with TSOs for congestion management. Both DSOs and TSOs require a smart substation automation system (SSAS) for resilient control and operation. The project SEAPATCH[5] provides software that enables an automation platform for deploying virtualised automation and protection applications. The software platform allows users to develop open-source code for integrating new specific functions and services according to hardware requirements. SEAPATCH also provides guidelines and best practices for testing a proof-of-concept in a realistic software platform. In addition, it also addresses requirements according to standards, such as IEC61850, IEC80870, IEC60255, and IEC61829.

Similarly, the i-Automate project[6] provides a modular configurable and testable approach that can map automation functions of a digitalised power system on hardware. For example, DSOs can monitor their power network by integrating network status detection and topology optimisation functions at the given substation to identify possible network bottlenecks and avoid network reinforcement. The control function can be configured flexibly from the function pool, e.g., a change of an operational setpoint of reactive power, as shown in Fig. 12.5.

Although the function pool platform is designed to provide flexibility in protection and control configurations, the engineering process for integrating these functions into infrastructure is still missing. This challenge is addressed in the i-Autonomous[7] project, where a holistic engineering process is developed to enable flexibility over the entire lifecycle including project planning, installation, and maintenance.

The SSAS concept has been implemented in several pilot projects using the virtualisation concept such as a 230 kV substation in Philadelphia as well as substations in South Africa, France, and Germany (Vardhan et al. 2018; Dayabhai and Prestwich 2018; I.S.G.A.N. 2016). The virtualised SSAS provides a fast reaction by selecting

[5] https://www.offis.de/offis/projekt/i-automate.html.

[6] https://www.offis.de/offis/projekt/i-automate.html.

[7] https://www.offis.de/offis/projekt/i-autonomous.html.

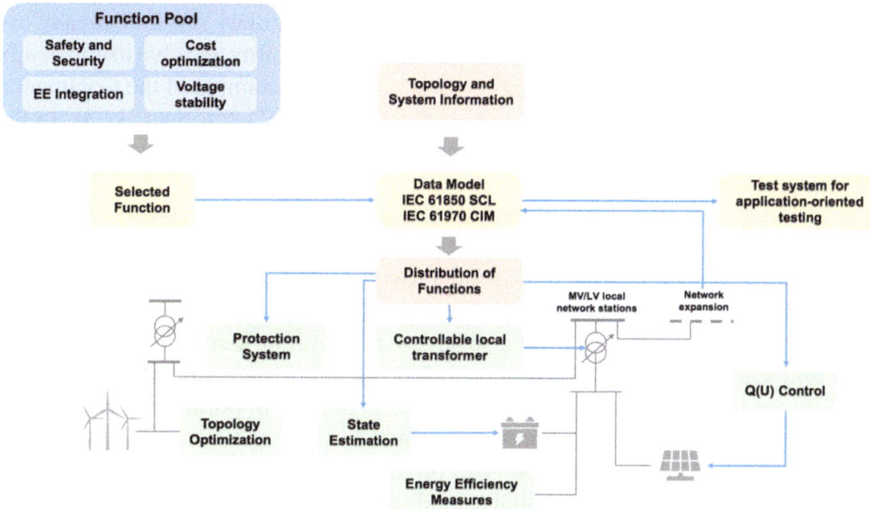

Fig. 12.5 A modular configurable and testable automation architecture

protection and control functions from a function pool and deploying at a given substation (Ma et al. 2010). Development of cloud technologies accelerates SSAS adoption by enabling flexible deployment and configuration in container-based architectures in the cloud application (Asif et al. 2020; Hunt et al. 2019). Each container consists of the code for a control function such as outage detection, state estimation, wide-area monitoring, or management of DERs, and can be deployed anywhere regardless of the target environment, e.g., cloud, classical information technology infrastructure, or desktop computer. Docker and Kubernetes are well-known containerisation and orchestration platforms, where each control function is packaged in a Docker container and Kubernetes reacts as a tool for managing infrastructure resources for multiple containers.

Virtualisation enhance resilience of digitalised power systems by reacting to new technical requirements quickly, reliably, and economically throughout the entire system. System operators can upgrade and incorporate new features into existing power systems with lower costs and less time-consuming by minimising the testing time before the real deployment.

12.5.3 React: A Power System Perspective

The goal of reaction measures in the power system is to mitigate the effects of an incident in such a way that the performance loss is minimised, and the power system can continue its primary task of supplying all connected loads. On a general level, this means to use the system's flexibilities in a coordinated—potentially automated—way to solve the specific problems caused by the incident. Ancillary services already use these flexibilities. They are a core part of power system stability and able to effectively reduce the effects of incidents, as has been shown in the example of the Continental Europe Synchronous Area Separation on 8th January 2021 (cf. Sect. 12.3). In the following, the focus is on the provision of generation flexibility, specifically for balancing power.

It is likely that the demand for balancing power will rise as a result of the ever-increasing share of fluctuating power supply by renewables (dena and G.E.A. 2024). Usually, generation flexibilities are provided by large conventional generators connected to the transmission grid. As the share of DERs at the distribution grid level continues to grow, these plants will also have to take over an increasing proportion of these grid-stabilising tasks (Hillberg et al. 2019). Unlocking the flexibility potential of DERs for the provision of balancing power is thus an important cornerstone for the robustness and resilience of power systems.

Digitalisation of the distribution grids provides the technical basis for leveraging the flexibility of DERs and controllable loads while, at the same time, the conditions for the tendering of balancing power have recently been fundamentally changed by a daily tendering procedure, the day before the provision, and a substantial reduction of product duration from one week to 4-h blocks (TSOs of Austria, Belgium, Denmark, France, Germany, the Netherlands and Switzerland 2018). This makes it possible to participate in a tender for balancing power and especially frequency containment reserve (FCR) with such plants. One of the main challenges is to meet the strict availability and reliability requirements that apply for the provision of balancing services considering the fluctuating nature of most DERs.

Considerable research has already been conducted on how DERs can coordinate or aggregate their flexibility to enable the distributed provision of ancillary services, e.g., (Dominguez-Garcia and Hadjicostis 2010, Hadjicostis and Charalambous 2011, Müller et al. 2017]. There are also approaches for using demand side flexibility for ancillary services (Zhang et al. 2017).

Not only stationary power generators can be used for the provision of balancing power. Battery storages with sufficient state of charge can absorb and release energy. Consequently, they are well suited for the symmetrical provision of balancing services. The authors in Angenendt et al. (2020) introduce a model of a residential home storage system participating in the FCR market. The integrated home includes a PV, a battery energy storage system (BESS), and a heating system with heat pump and thermal storages. The combination with a thermal storage provides additional flexibility since, for example, negative FCR can also be transferred to the heat storage. The maximum FCR power that can be offered by the integrated home is calculated using a worst-case scenario for the 30-min criterion, which ensures that the BESS can provide the maximum FCR for at least 30 min. The results show that participating in the FCR market generates additional incomes for integrated homes while, at the same time, it causes additional costs due to higher battery aging and lower self-consumption.

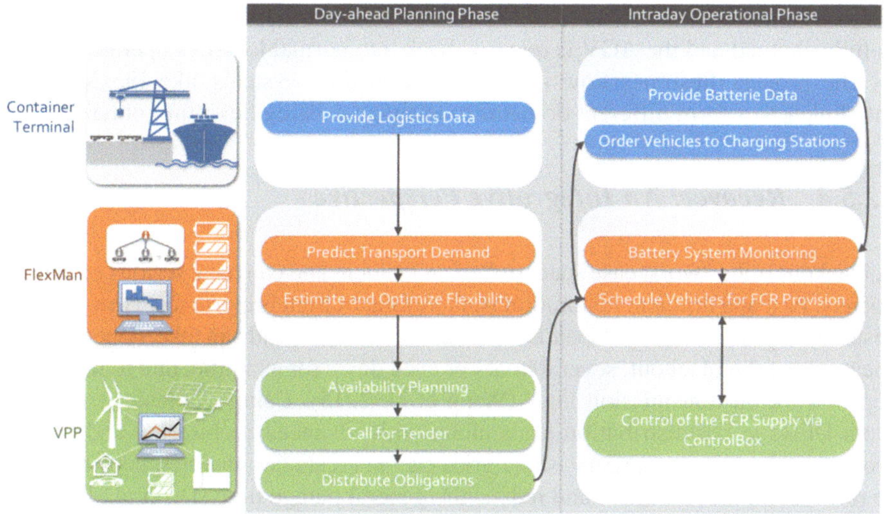

Fig. 12.6 The asks and phases for FCR provision with a fleet of battery-electric vehicles

In the FRESH[8] project ("Flexibility management and balancing power supply of heavy-duty vehicles in a port"), a flexibility management system (FlexMan) has been developed for a fleet of heavy-duty vehicles performing automated container handling at the container terminal Altenwerder in the port of Hamburg. Idle automated guided vehicles (AGVs) can be ordered to idle charging stations, where the flexibility of the battery storage can be leveraged. The FlexMan acts as a pre-aggregator and offers the aggregated flexibility of the fleet to a virtual power plant (VPP), which integrates the fleet into its asset pool and participates in the call for tender for FCR. Emphasis is put on ensuring that the logistics process of the terminal is not impaired and that the offered flexibility can be provided reliably (Holly et al. 2020).

Figure 12.6 shows the tasks that the three actors, i.e. the container terminal, FlexMan, and the VPP must perform. In the day-ahead planning phase, a transport demand forecast for the following day is created based on the terminal's current logistics planning data. The flexibility is determined by an agent-based optimisation heuristic that relies on the forecast and submitted to the VPP. The VPP then conducts availability planning with its entire asset pool, participates in the tender for FCR, and subsequently allocates the resulting obligations to its assets—with one of those being the AGV fleet.

These obligations set the target for AGV scheduling in the next day's operational phase. During this phase, FlexMan continuously monitors the states of the battery systems. In the manner of model predictive control, individual AGVs are selected for FCR provision and the time at which they must be sent to a charging station is scheduled. When an AGV is ready, control for FCR provision is transferred to the

[8] https://hhla.de/en/company/innovation/research-and-funding/fresh.

control boxes of the VPP. After the scheduled time has elapsed, the battery storage is fully charged and the AGV is released back into normal logistics operation.

The project illustrates how intelligent automation systems like FlexMan can interlink actors from different domains to leverage existing flexibility potential.

12.5.4 Recover: An Integrative Perspective

A blackout is the "worst case scenario" of an incident and can have serious effects on society. Fast and efficient restoration is, therefore, of high importance. However, the restoration process highly depends on the actual situation and can take, depending on the cause of the blackout, several hours or even days, for example, due to necessary repair work on transmission lines (Wang et al. 2015). Using available generation from DERs in the distribution grids along with advanced monitoring and control from digitalisation, temporary island grids can be created as an emergency measure to restore critical customers and reduce the general outage time (Du et al. 2021).

Forming stable island grids is difficult from both a planning and a control perspective. From a control perspective, operating a small part of the grid as a stable island grid is difficult due to low system inertia, high R/X ratio and limited short-circuit capacity (Farrokhabadi et al. 2019). These properties make it more difficult to ensure frequency and voltage stability and the operation of a grid under these conditions needs to be further investigated (Redmann et al. 2021). Regarding the planning perspective, it needs to be decided which parts of the grid should be connected to form an island grid and which should be disconnected, since the available generation in a distribution grid is potentially lower than the number of loads. This can be described as an optimisation problem that is highly combinatorial, non-linear, non-differentiable, constrained, and multi-objective (Sanches et al. 2014).

A feasible islanding approach must account for the interdependencies between power and ICT system restoration (integrated perspective) and be able to create island grids even with severely impaired communication. A popular approach is to use a distributed multi-agent system (MAS) (e.g. Sampaio et al. 2020; Torres et al. 2018; Rokrok et al. 2017). A MAS is not only capable of solving complex optimisation problems with only locally available information, it can also adapt to changing communication infrastructure, providing a flexible and easy to adapt restoration process that also works in situations where a central coordination from the control room is not possible, thereby avoiding single points of failure.

The authors in Yang et al. (2021) focus specifically on restoration of critical loads, using a consensus-based approach with a MAS. The agents can monitor critical loads, control DERs as well as switches, and can efficiently discover an optimal restoration strategy for critical loads. This approach is robust to unreliable communication conditions, which is described as probability of failure of one communication channel, leading to potential package loss. However, the focus here is only on local, temporary communication failures for an abstract ICT system. The concrete effects of large-scale blackouts or the power system restoration on ICT performance are not considered.

A more specific ICT system is defined in Pietsch et al. (2020). A direct relationship between the communication infrastructure and power system infrastructure is defined, with power system nodes being connected through a communication overlay. A communication link can only be active if both power system nodes connected by it are energised. However, this limitation is not applied for the initial black start but for the merging of different island grids after first restoration steps. Moreover, the problem of finding the optimal restoration sequence is solved centrally and not distributed, thereby relying on centrally available information.

In Volkova et al. (2019) it is investigated, how to perform a distributed parallel power and ICT system restoration under consideration of an impaired ICT system. Figure 12.7 shows an exemplary system architecture with the three different layers of power system, ICT system, and MAS along with the restoration process.

The goal of the ICT system is to enable communication between agents in the presence of multi-level incidents (failure of intermediate ICT nodes, such as routers and repeaters, and upper-level communication services, such as wide-area routing, etc.). Therefore, the ICT system has been enhanced to support self-backhauling and create a communication overlay based on available ICT nodes (Volkova 2019).

Individual agents represent DERs, loads, or switches, which they can monitor and control. Together they perform the restoration process, which is based on the so-called combinatorial optimisation heuristic for distributed agents that has already been demonstrated to work in similar use cases from the energy domain and is robust against message delays and a changing communication overlay (Hinrichs

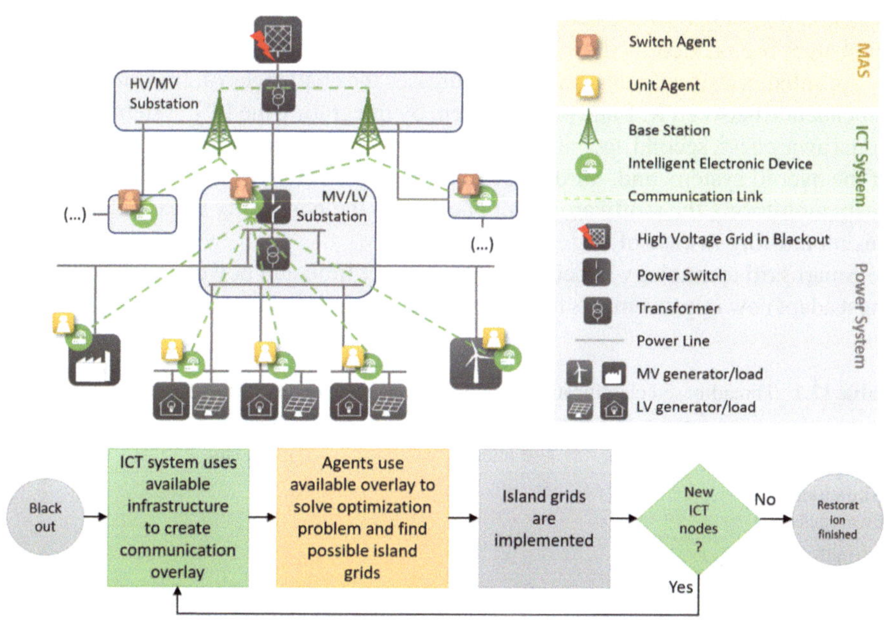

Fig. 12.7 An architecture for a blackstart-capable system and a resilient restoration process

et al. 2013; Holly and Nieße 2021). The resulting island grid configuration is then implemented into the power system, potentially activating more ICT nodes. The restoration is, therefore, performed in a cyclic manner, always using the currently available information to compute the optimal island grid configuration in a distributed fashion.

First results in Stark et al. (2021) confirm the importance of critical ICT nodes for a successful restoration, which supports the approach of explicitly considering ICT restoration explicitly in the objective function.

12.6 Discussion

In the previous section, an overview of approaches enabled by digitalisation and aiming at more resilient power systems was provided. The approaches, i.e. enhanced situational awareness, virtualisation, flexibility management, and automated recovery, are classified according to an attribute-based classification with anticipation, reaction (power system and ICT perspective), and recovery as classes. In this section, the characteristics of key challenges for digitalised power systems as well as of the resilience process are picked up and further classifications are discussed, i.e. a thread-based classification with respect to the key challenges and a time-based with respect to the resilience process.

Table 12.1 shows a thread-based classification by mapping the proposed measures to the identified main challenges they tackle, which were explained in Sect. 12.2. It is an *n:m* mapping because several approaches tackle the same challenge and several challenges are tackled by the same approach.

An enhanced situational awareness addresses the challenges of, first, mutual interdependence between ICT and power system by integrating the ICT system into situational awareness, second, technical system complexity by providing a better overview of the overall system, and, third, uncertainty about future developments by continuously monitoring the (software) components and the possibility to extend and adapt this monitoring in case of, e.g., new technologies.

Smart grid technology is being developed over time and power system operators must adapt new improvements to their systems. A virtualisation technique can bridge

Table 12.1 Thread-based classification: mapping of the approaches to the challenges they tackle

	Simultaneous behavior	Mutual interdependence	Technical complexity	Uncertainty
Situational awareness		X	X	X
Virtualization		X	X	X
Flexibilization	X		X	
Islanding with MAS	X	X		X

the gap of mutual interdependence by deploying virtualisation to the field devices. This technique also reduces technical complexity and uncertainty about future development for the entire life cycle by (re)allocating resources with control functions to the given locations quickly and economically.

Using the generation flexibilities from DERs or battery storages for balancing power in an automated way can help to reduce the technical complexity in system operation for the operators while at the same time ensuring the continuous functionality of an important ancillary service in the future power system. This also provides countermeasures for any unforeseen effects that lead to an imbalance and load and generation, caused by future system complexity. By actively using the flexibilities of small producers and consumers, the potential threat of simultaneous behaviour of small producers and consumers becomes a desirable measure to ensure power system stability and can also help to prevent unwanted simultaneous behaviour.

A MAS-based islanding concept can help to prevent unwanted simultaneous behaviour of small consumers and producers during the restoration process after a large-scale blackout. By closely monitoring the behaviour of loads and coordinating generation on distribution level, it can be ensured that they act in a grid-stabilising way. The mutual interdependence is considered in two ways. On the one hand, MASs are resilient against communication failures and still able to restore (parts of) the grid. On the other hand, they help to restore the ICT system in the area of the island grids, allowing the TSO to include more detailed information into their own restoration plan (Roggatz et al. 2020). Finally, it is a flexible and easy to adapt restoration process, where agents can be added or removed, objectives and constraints can be updated, thereby providing a good base for handling uncertainty.

Figure 12.8 shows the same bathtub curve as in Fig. 12.2 as base for a time-based classification. The classification is done by a mapping of the approaches to the different phases of resilience, explained in section Resilience of Power Systems. A solid circle indicates a main contribution of the respective approach while a dotted circle refers to an assistance. Approaches can not only assist a specific resilience phase but also each other, thereby creating interdependencies—e.g. generation flexibilities from DERs are needed during the recovery process. The approaches are enumerated in the order of the subsections in section Real-Time Measures for Resilient Digitalised Power Systems. It is, as for the thread-based classification, an *n:m* mapping because several approaches contribute to the same phase and an approach can contribute to several phases.

An enhanced situational awareness, first, can play an important role in the normal operation, because the approaches aim at detecting events that may reduce the functionality of a digitalised power system. Detecting such an event early enough can help to mitigate it without any major performance degradation. Second, a digitalised power system benefits also in the phase of stabilisation, since the stabilisation benefits from the knowledge which components and data are trustworthy. Third, it can assist in the other three phases because the improved situational awareness ensures that, for example, during the delicate phase of restoration the energy system only relies on the most trustworthy components and data.

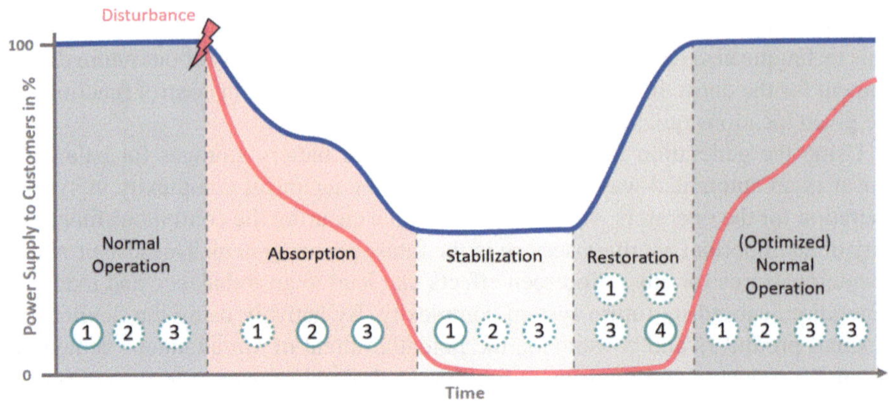

Fig. 12.8 Time-based classification: restoring system functionality (cf. Fig. 12.2) with the proposed approaches mapped to the different phases. 1 refers to anticipate: an integrative perspective, 2 to react: an ICT Perspective, 3 to react: a power system perspective, and 4 to recover: an integrative perspective. A solid circle indicates to a main contribution and a dotted circle an assistance

Preventing cascading failures requires ICT solutions to react promptly. A situation awareness system can help to detect the disturbances and virtualisation can be deployed quickly at a substation automation system with remedial control functions to reduce the severity. During absorption of disruption, virtualisation can assist power systems by reallocating computing resources with new remedial functions to the filed devices where disturbance occurs i.e. protective function. A modular engineering approach using virtualisation throughout the entire life cycle can pave the way toward a resilient power system by bridging the gap between technical complexity and uncertainty about future developments.

Generation flexibilities from DERs mainly help in the absorption of disruption after an incident and stabilise the system, e.g., by removing congestions, providing a manual frequency restoration reserve, or even load shedding. Balancing power, and in particular FCR is also constantly used to compensate for the small frequency deviations in the system caused by an imbalance between feed-in and consumption. It, therefore, enables the normal operation of the power system by contributing to the system's robustness, but also supports during the stabilisation and restoration of the system.

The automated recovery with MASs naturally plays its main role in the restoration phase by providing an initial local restoration of the power system and supporting the overall grid restoration.

Additionally, all approaches can potentially support in an (optimised) normal operation, for example by sharing data collected in the different phases with the system operator or by including methods of machine learning, which could enable an improved handling of future incidents.

12.7 Conclusion

The digitalisation of power systems requires a shift from classical robust design based on risk analysis towards resilient design. This necessity is motivated in this paper with four so-called basic causes that can result in incidents, for which classical risk analysis is not sufficient: a simultaneous behaviour of small producers and consumers, a mutual interdependence between ICT-related and energy technical resources, the technical complexity in system operation, and the uncertainty about future developments. With the concept of resilience, it is possible to deal with these challenges. A resilient system behaviour implies stabilising the system with a certain minimum of functionality, restoring the lost functionality, and bringing the system back to normal operation. Thereby, the latter can be optimised considering the lessons learned from the incident to improve the system's ability to handle incidents in the future.

For a resilient digitalised power system, anticipation, reaction, and recovery are important attributes, next to planning and system hardening mechanisms. Four solutions in these areas of contribution are presented in this paper. The solutions tackle different challenges and contribute to different phases of resilience: an enhanced situational awareness to anticipate, virtualisation and flexibilisation to react, and a distributed black start. The set of these four solutions tackles all identified main challenges and contribute to all phases of resilience directly or indirectly. Therefore, applying this set of solutions increases the resilience of a digitalised power system.

But, of course, the proposed solutions represent only a small portion of the possible measures. While there exist many individual resilience measures, a coherent resilience strategy is still missing. Solutions need to be combined into one portfolio of measures to assess overlapping requirements or interdependencies. This is especially relevant for the phase of optimised normal operation, where the whole interaction between all the solutions after an incident needs to be evaluated and adapted if necessary. Another open question is how to measure the resilience of the power system in a standardised way to be able to compare different sets of solutions with regards to their improvement of the system's resilience.

References

Acatech/Leopoldina/Akademienunion: The resilience of digitalised energy systems. Options for reducing blackout risks (Series on Science-based Policy Advice). Acatech/Leopoldina/Akademienunion (2021)

Akorede, M.F., Hizam, H., Pouresmaeil, E.: Distributed energy resources and benefits to the environment. Renew. Sustain. Energy Rev. **14**(2), 724–734 (2010)

Angenendt, G., Zurmühlen, S., Figgener, J., Kairies, K.-P., Sauer, D.U.: Providing frequency control reserve with photovoltaic battery energy storage systems and power-to-heat coupling. Energy **194**, 116923 (2020)

Asif, R., Ghanem, K., Irvine, J.: Containerization: for over-the-air programming of field deployed internet-of-energy based on cost effective LPWAN. In: 2020 First International Conference of Smart Systems and Emerging Technologies (SMARTTECH), pp. 65–70 (2020). https://doi.org/10.1109/SMART-TECH49988.2020.00030

Bag, G., Lednicki, L., Landernäs, K., Ericsson, N.: Experiments on approaches of virtualization for industrial internet of things applications. In: 2019 24th IEEE International Conference on Emerging Technologies and Factory Automation (ETFA), pp. 1226–1229 (2019). https://doi.org/10.1109/ETFA.2019.8868971
Basciftci, Y.O., Ozguner, F.: Trust aware particle filters for autonomous vehicles. In: 2012 IEEE International Conference on Vehicular Electronics and Safety, ICVES 2012, pp. 50–54 (2012). https://doi.org/10.1109/ICVES.2012.6294259
Brand, M., Ansari, S., Castro, F., Chakra, R., Hage Hassen, B., Krüger, C., Babazadeh, D., Lehnhoff, S.: A framework for the integration of ICT-relevant data in power system applications. In: Proceedings of the 2019 IEEE Milan PowerTech, pp. 1–6. IEEE, Milan (2019)
Brand, M., Babazadeh, D., Krüger, C., Siemers, B., Lehnhoff, S.: Trust assessment of power system states. Energy Inform. **3**(1), 1–11 (2020)
Brand, M., Babazadeh, D., Lehnhoff, S.: Trust in power system state variables based on trust in measurements. In: 2021 IEEE Madrid PowerTech. IEEE (2021a)
Brand, M., Castro, F., Hage Hassan, B., Krüger, C., Logemann, T., Siemers, B., Weller, D., Wolff, T., Lehnhoff, S.: Demo abstract: a platform to assess the trust in power system components, data, and services. In: Abstracts of the 10th DACH+ Conference on Energy Informatics (2021b)
Brand, M., Engel, D., Lehnhoff, S.: Assess: anomaly sensitive state estimation with streaming systems. Energy Inform. **6**(Suppl 1) (2023)
Brange, L., Englund, J., Lauenburg, P.: Prosumers in district heating networks – a swedish case study. Appl. Energy **164**, 492–500 (2016). https://doi.org/10.1016/j.apenergy.2015.12.020
Case, D.U.: Analysis of the cyber attack on the Ukrainian power grid. In: Electricity Information Sharing and Analysis Center (E-ISAC), vol. 388 (2016)
Ciapessoni, E., Cirio, D., Pitto, A., Sforna, M.: Quantification of the benefits for power system of resilience boosting measures. Appl. Sci. **10**(16), 5402 (2020)
Cortekar, J., Groth, M.: Adapting energy infrastructure to climate change–is there a need for government interventions and legal obligations within the german "energiewende"? Energy Procedia **73**, 12–17 (2015)
Council of European Energy Regulators (CEER): CEER benchmarking report 6.1 on the continuity of electricity and gas supply - data update 2015/2016 (2018). https://www.ceer.eu/documents/104400/-/-/963153e6-2f42-78eb-22a4-06f1552dd34c. Accessed 02 Aug 2021
Dayabhai, S., Prestwich, J.: A substation automation solution that uses virtualization to reduce cost while ensuring redundancy and security compliance. In: Power and Energy Automation Conference (2018)
Deutsche Energie-Agentur (dena) - German Energy Agency: dena Ancillary Services Study 2030 (2014). https://www.dena.de/fileadmin/dena/Dokumente/Themen_und_Projekte/Energiesysteme/dena-Studie_Systemdienstleistungen_2030/dena_Ancillary_Services_Study_2030_-_summary.pdf. Accessed 02 Aug 2021
Deutsche Energie-Agentur (dena), G.E.A.: Ancillary services study 2030 - security and reliability of a power supply with a high percentage of renewable energy (2014). https://www.dena.de/fileadmin/dena/Dokumente/ThemenundProjekte/Energiesysteme/denaStudieSystemdienstleistungen2030/denaAncillaryServicesStudy2030.pdf. Accessed 02 Aug 2021
Dominguez-Garcia, A.D., Hadjicostis, C.N.: Coordination and control of distributed energy resources for provision of ancillary services. In: 2010 First IEEE International Conference on Smart Grid Communications, pp. 537–542. IEEE (2010)
Du, Y., Tu, H., Lu, X., Wang, J., Lukic, S.: Black-start and service restoration in resilient distribution systems with dynamic microgrids. IEEE J. Emerg. Sel. Top. Power Electron. (2021)
ENTSO-E: Commition Regulation (EU) 2017/2196 - establishing a network code on electricity emergency and restoration (2017). http://data.europa.eu/eli/reg/2017/2196/2017-11-28

Economic Regulation Authority Western Australia: Annual performance report - energy distributors 2017/18 (2019). https://www.erawa.com.au/cproot/20349/2/FINAL%20FOR%20PUBLICATION%20-%20Annual%20Performance%20Report%20-%20%20Energy%20Distributors%20201718.PDF. Accessed 02 Aug 2021

European Network of Transmission System Operators for Electricity (ENTSO-E): Continental Europe synchronous area separation on 8 January 2021 - interim report (2021). https://eepublicdownloads.azureedge.net/clean-documents/Publications/Position%20papers%20and%20reports/entso-e_CESysSep_interim_report_210225.pdf. Accessed 02 Aug 2021

Farrokhabadi, M., Canizares, C.A., Simpson-Porco, J.W., Nasr, E., Fan, L., Mendoza-Araya, P.A., Tonkoski, R., Tamrakar, U., Hatziargyriou, N., Lagos, D., et al.: Microgrid stability definitions, analysis, and examples. IEEE Trans. Power Syst. **35**(1), 13–29 (2019)

Gautier, A., Jacqmin, J., Poudou, J.-C.: The prosumers and the grid. J. Regul. Econ. **53**(1), 100–126 (2018)

Habib, S., Khan, M.M., Abbas, F., Sang, L., Shahid, M.U., Tang, H.: A comprehensive study of implemented international standards, technical challenges, impacts and prospects for electric vehicles. IEEE Access **6**, 13866–13890 (2018)

Hadjicostis, C.N., Charalambous, T.: Asynchronous coordination of distributed energy resources for the provisioning of ancillary services. In: 2011 49th Annual Allerton Conference on Communication, Control, and Computing (Allerton), pp. 1500–1507. IEEE (2011)

Hillberg, E., Zegers, A., Herndler, B., Wong, S., Pompee, J., Bourmaud, J.-Y., Lehnhoff, S., Migliavacca, G., Uhlen, K., Oleinikova, I., et al.: Flexibility needs in the future power system (2019)

Hinrichs, C., Lehnhoff, S., Sonnenschein, M.: Cohda: A combinatorial optimization heuristic for distributed agents. In: International Conference on Agents and Artificial Intelligence, pp. 23–39. Springer (2013)

Holly, S., Nieße, A.: On the effects of communication topologies on the performance of distributed optimization heuristics in smart grids. INFORMATIK **2020**, 783–794 (2021)

Holly, S., Nieße, A., Tröschel, M., Hammer, L., Franzius, C., Dmitriyev, V., Dorfner, J., Veith, E.M., Harnischmacher, C., Greve, M., et al.: Flexibility management and provision of balancing services with battery-electric automated guided vehicles in the Hamburg container terminal Altenwerder. Energy Inform. **3**(1), 1–20 (2020)

Huang, P., Lovati, M., Zhang, X., Bales, C., Hallbeck, S., Becker, A., Bergqvist, H., Hedberg, J., Maturi, L.: Transforming a residential building cluster into electricity prosumers in Sweden: optimal design of a coupled PV-heat pump-thermal storage-electric vehicle system. Appl. Energy **255**, 113864 (2019)

Hunt, R., Flynn, B., Smith, T.: The substation of the future: MOVING toward a digital solution. IEEE Power Energ. Mag. **17**(4), 47–55 (2019). https://doi.org/10.1109/MPE.2019.2908122

I.S.G.A.N.: Spotlight on smart and strong power T&D infrastructure. T&D Case Book 2 (2016)

Janssen, K.L., Dankbaar, B.: Proactive involvement of consumers in innovation: selecting appropriate techniques. Int. J. Innov. Manag. **12**(03), 511–541 (2008)

Kandaperumal, G., Srivastava, A.K.: Resilience of the electric distribution systems: concepts, classification, assessment, challenges, and research needs. IET Smart Grid **3**(2), 133–143 (2020)

Kariyawasam, S., Wickremasuriya, A., Rajapakse, A.: Teaching IEC 61850 based substation automation through hands-on experiences. In: 2017 IEEE Electrical Power and Energy Conference (EPEC), pp. 1–6 (2017). https://doi.org/10.1109/EPEC.2017.8286217

Knezovic, K., Martinenas, S., Andersen, P.B., Zecchino, A., Marinelli, M.: Enhancing the role of electric vehicles in the power grid: field validation of multiple ancillary services. IEEE Trans. Transp. Electrif. **3**(1), 201–209 (2017). https://doi.org/10.1109/TTE.2016.2616864

Li, S., Gao, Z., Zhu, Y., Wang, T., Rui, M.: Research on designing methods of the secondary system in digital substation. In: 2016 IEEE PES Asia-Pacific Power and Energy Engineering Conference (APPEEC), pp. 289–293 (2016). https://doi.org/10.1109/APPEEC.2016.7779514

Lin, Y., Bie, Z., Qiu, A.: A review of key strategies in realizing power system resilience. Glob. Energy Interconnect. **1**(1), 70–78 (2018)
Lindstrom, A., Hoff, S.: Average U.S. electricity customer interruptions totalled nearly 8 hours in 2017 (2018). Accessed 02 Aug 2021
Liu, T., Sun, Y., Liu, Y., Gui, Y., Zhao, Y., Wang, D., Shen, C.: Abnormal traffic-indexed state estimation: a cyber-physical fusion approach for smart grid attack detection. Futur. Gener. Comput. Syst. **49**, 94–103 (2015). https://doi.org/10.1016/j.future.2014.10.002
Liu, Y., Ning, P., Reiter, M.K.: False data injection attacks against state estimation in electric power grids. ACM Trans. Inf. Syst. Secur. (TISSEC) **14**(1), 1–33 (2011)
Ma, J., Wang, T., Wu, J., Wang, Z.: Integration of protection and control systems for smart substation. In: 2010 5th International Conference on Critical Infrastructure (CRIS), pp. 1–5 (2010). https://doi.org/10.1109/CRIS.2010.5617481
Mahzarnia, M., Moghaddam, M.P., Baboli, P.T., Siano, P.: A review of the measures to enhance power systems resilience. IEEE Syst. J. **14**(3), 4059–4070 (2020)
Müller, F.L., Szabo, J., Sundström, O., Lygeros, J.: Aggregation and disaggregation of energetic flexibility from distributed energy resources. IEEE Trans. Smart Grid **10**(2), 1205–1214 (2017)
NERC: Technical analysis of the August 14, 2003, blackout: what happened, why, and what did we learn? (2004). https://www.nerc.com/docs/docs/blackout/NERCFinalBlackoutReport071304.pdf
Petit, F., Vargas, V., Kavicky, J.: Grid modernization: metrics analysis (gmlc1.1)-resilience. Technical report, Pacific Northwest National Lab. (PNNL), Richland, WA (United States) (2020)
Pietsch, M., Klein, A., Steinke, F.: Merging microgrids for optimal distribution grid restoration under explicit communication constraints. In: 2020 Resilience Week (RWS), pp. 48–54. IEEE (2020)
Pillitteri, V., Brewer, T.: Guidelines for smart grid cybersecurity. NIST Interagency/Internal Report (NISTIR), National Institute of Standards and Technology, Gaithersburg, MD (2014). https://doi.org/10.6028/NIST.IR.7628r1
Redmann, F., Ernst, A., Orlik, B.: Black start capability and islanded operation of power converters with virtual synchronous generator control. In: PCIM Europe Digital Days 2021; International Exhibition and Conference for Power Electronics, Intelligent Motion, Renewable Energy and Energy Management, pp. 1–8. VDE (2021)
Rinaldi, S.M., Peerenboom, J.P., Kelly, T.K.: Identifying, understanding, and analyzing critical infrastructure interdependencies. IEEE Control. Syst. Mag. **21**(6), 11–25 (2001)
Roggatz, C., Power, M., Singh, N.: Power system restoration: meeting the challenge to resiliency from distributed generation. IEEE Power Energ. Mag. **18**(4), 31–40 (2020)
Rokrok, E., Shafie-khah, M., Siano, P., Catalao, J.P.: A decentralized multi-agent-based approach for low voltage microgrid restoration. Energies **10**(10), 1491 (2017)
Sampaio, F.C., Leao, R.P., Sampaio, R.F., Melo, L.S., Barroso, G.C.: A multi-agent-based integrated self-healing and adaptive protection system for power distribution systems with distributed generation. Electr. Power Syst. Res. **188**, 106525 (2020)
Sanches, D.S., Junior, J.B.A.L., Delbem, A.C.B.: Multi-objective evolutionary algorithm for single and multiple fault service restoration in large-scale distribution systems. Electr. Power Syst. Res. **110**, 144–153 (2014)
Shen, L., Cassottana, B., Heinimann, H.R., Tang, L.C.: Large-scale systems resilience: a survey and unifying framework. Qual. Reliab. Eng. Int. **36**(4), 1386–1401 (2020)
Shiwen, Y., Hui, H., Chengzhi, W., Hao, G., Hao, F.: Review on risk assessment of power system. Procedia Comput. Sci. **109**, 1200–1205 (2017)
Stark, S., Volkova, A., Lehnhoff, S., de Meer, H.: Why your power system restoration does not work and what the ICT system can do about it. In: Proceedings of the Twelfth ACM International Conference on Future Energy Systems. e-Energy '21, pp. 269–273. Association for Computing Machinery, New York, NY, USA (2021). https://doi.org/10.1145/3447555.3465415
Steghöfer, J.-P., Kiefhaber, R., Leichtenstern, K., Bernard, Y., Klejnowski, L., Reif, W., Ungerer, T., Andre, E., Hähner, J., Müller-Schloer, C.: Trustworthy organic computing systems: challenges

and perspectives. In: International Conference on Autonomic and Trusted Computing, pp. 62–76. Springer (2010)

TSOs of Austria, Belgium, Denmark, France, Germany, the Netherlands and Switzerland: TSOs' proposal for the establishment of common and harmonised rules and processes for the exchange and procurement of Balancing Capacity for Frequency Containment Reserves (FCR) in accordance with Article 33 of Commission Regulation (EU) 2017/2195 establishing a guideline on electricity balancing (2018). https://www.entsoe.eu/Documents/Network%20codes%20documents/NC%20EB/FCRProposalArticle%2033%1EBGL20180426FV.PDF. Accessed 06 Sept 2021

Torres, B.S., Ferreira, L.R., Aoki, A.R.: Distributed intelligent system for self-healing in smart grids. IEEE Trans. Power Deliv. **33**(5), 2394–2403 (2018)

Vardhan, H., Ramlachan, R., Szela, W., Gdowik, E.: Deploying digital substations: experience with a digital substation pilot in North America. In: 2018 71st Annual Conference for Protective Relay Engineers (CPRE), pp. 1–9 (2018). https://doi.org/10.1109/CPRE.2018.8349795

Volkova, A.: Blackout recovery: resilient NFV-enabled ICT infrastructure for the smart grid. In: Abstracts from the 8th DACH+ Conference on Energy Informatics, pp. 73–76. Springer, Salzburg, Austria (2019). https://doi.org/10.1186/s42162-019-0098-7

Volkova, A., Stark, S., de Meer, H., Lehnhoff, S., Bremer, J.: Towards a blackout-resilient smart grid architecture. In: International ETG-Congress 2019; ETG Symposium, pp. 1–6 (2019)

Wang, A., Iyer, M., Dutta, R., Rouskas, G.N., Baldine, I.: Network virtualization: technologies, perspectives, and frontiers. J. Light. Technol. **31**(4), 523–537 (2013). https://doi.org/10.1109/JLT.2012.2213796

Wang, Y., Chen, C., Wang, J., Baldick, R.: Research on resilience of power systems under natural disasters—a review. IEEE Trans. Power Syst. **31**(2), 1604–1613 (2015)

Wirtz, N., Monti, A.: A flexible framework to investigate cascading in interdependent networks of power systems. In: 2020 6th IEEE International Energy Conference (ENERGYCon), pp. 38–41. IEEE (2020)

Yang, L., Xu, Y., Sun, H., Chow, M., Zhou, J.: A multiagent system based optimal load restoration strategy in distribution systems. Int. J. Electr. Power Energy Syst. **124**, 106314 (2021)

Zhang, X., Hug, G., Kolter, J.Z., Harjunkoski, I.: Demand response of ancillary service from industrial loads coordinated with energy storage. IEEE Trans. Power Syst. **33**(1), 951–961 (2017)

Chapter 13
Using Optimising Infrastructure Planning Models for the Resilience Assessment of Future Integrated Energy Systems

Hans Christian Gils and Madhura Yeligeti

Abstract Optimising energy system models are often used to derive roadmaps for redesigning energy systems from climate targets. When considering large-scale, i.e. national or continental systems, these models usually focus on the minimisation of economic costs. As a result of the ongoing transformation, energy systems are increasingly characterised by the extensive and decentralised use of renewable energies, far-reaching digitalisation and the integration of the previously mostly separate sub-systems of electricity, heat and fuel supply. The associated structural changes also have an impact on the resilience requirements of the corresponding energy systems. This chapter addresses the effects of the transformation on the resilience of future energy systems. In doing so, it evaluates the possibilities and requirements for a structured resilience analysis with optimising energy system models. Based on this, it derives a modelling procedure for a systematic stress test analysis. Furthermore, this chapter discusses the limitations in the usage of such models in resilience analysis arising from structural as well as data-related aspects.

Keywords Energy systems modelling · Optimisation · Sector coupling · Stress testing · Resilience

H. C. Gils (✉) · M. Yeligeti
Institute of Networked Energy Systems (DLR-VE), German Aerospace Center (DLR), Oldenburg & Stuttgart, Germany
e-mail: Hans-Christian.Gils@dlr.de

D. van Doren et al. (eds.), *Towards Energy System Resilience*, Ethics of Science and Technology Assessment 55, https://doi.org/10.1007/978-3-032-03187-7_13

13.1 Introduction

13.1.1 The Transformation of the Energy System

Combating climate change requires a fundamental transformation of our current energy system (IPCC 2022). This goes hand in hand with the new construction, conversion or decommissioning of infrastructures such as power plants, electricity and hydrogen grids, and energy storage facilities. The system-analytical investigation of future scenarios for the restructuring of the energy system has identified great potential in the closer linking of the various demand sectors as part of so-called sector integration or sector coupling (Brown et al. 2018). Specifically, this refers to the increased use of electricity from renewable sources and fuels generated from this electricity, such as hydrogen, methane or synthetic hydrocarbons, in industry, buildings and the transport system (Fig. 13.1). This is accompanied by challenges in the transformation of infrastructures, such as the installation of charging stations for battery electric vehicles, electric heating systems and technologies for the generation, storage and transport of hydrogen. The implementation of these measures impacts local and regional energy systems as well as national or even continental ones. For example, the widespread expansion of photovoltaics, heat pumps, and vehicle charging stations may require the addition of battery storage or the expansion of distribution grids (Kippelt et al. 2017). If, on the other hand, we look at larger spatial units and thus higher grid voltage levels, the focus is particularly on the expansion of electricity transmission grids, the conversion of the gas infrastructure for the use of hydrogen, and the expansion of large-scale hydrogen storage facilities (Gils et al. 2021; Wetzel et al. 2023).

Due to the expected dominance of electricity generation by wind energy and photovoltaics, future energy systems are characterised by a strong decentralisation of supply. This contrasts the still dominating setup of mostly large power plants running on fossil fuels or hydro power.

The time-variable availability of wind and solar energy also creates a need for flexibility in the electricity system. To date, this need for flexibility has also been covered by large-scale gas-fired power plants and pumped hydro storage. However, if sector coupling is accompanied by the controlled or bidirectional charging of battery

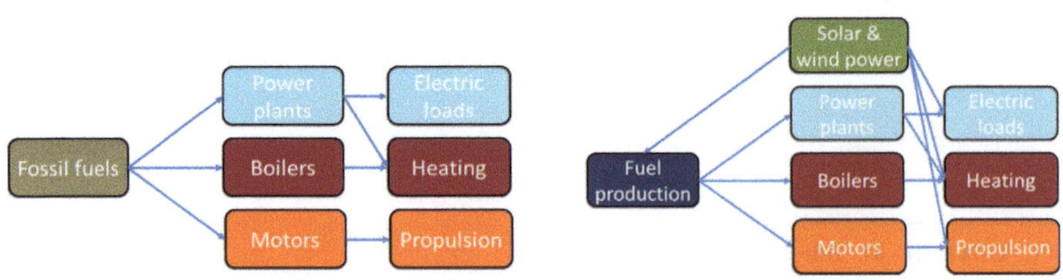

Fig. 13.1 Energy carrier conversion in traditional energy systems (left) and future systems with sector coupling (right)

vehicles or the use of thermal storage in combination with electrical heat generation, there may also be a shift to the decentralised level for the provision of flexibility. This is reinforced by the increasing installation of battery storage systems in residential buildings.

The decentralisation fundamentally changes the role of grids for the transportation of energy. Due to large amounts of distributed generation, power flows no longer occur only from the transmission grids to the distribution grids but also in the opposite direction. In addition, the shutdown of coal-fired power plants located on opencast mines or rivers and their replacement by wind power plants clustered in windy regions will change flow directions in the transmission grid. This also applies to existing gas pipelines in the transport network due to the possible switch to hydrogen and synthetic methane. In addition, the spatial concentration of good wind power potentials may increase the importance of electricity transmission over long distances. Due to the decommissioning of large power plants and their rotating mass, the reduced controllability of decentralised generation and the additional flexibilities in the system, ensuring grid stability must be reorganised.

For Germany as a country with high demand and population density, but limited expansion potential for renewable energy, energy imports are expected to continue in the future (Sensfuß et al. 2022). In this context, electricity and renewable fuels will replace coal, gas and oil, which goes hand in hand with the need for new energy infrastructures such as hydrogen pipelines and fuel import terminals as well as a strengthening of international power grid interconnectors. As it is foreseeable that not only the imported energy sources but also the corresponding import routes and countries of origin will change and diversify, new requirements for security of supply will arise.

Like many other areas, increasing digitalisation is also impacting the energy system. This comes with numerous opportunities for a more cost-efficient system operation, such as improved capabilities for monitoring plant and grid operations, targeted and predictive plant maintenance, and control of decentralised plants. The latter makes it possible, for example, to partially adapt the charging of electric vehicles or the operation of heat pumps to the current electricity supply. Thus, new sources of flexibility in the energy system can be tapped.

13.1.2 Impact of the Transformation on Energy System Resilience

The changes in power system design—renewable energies, decentralisation, digitalisation and sector coupling—are driving fundamental changes in the requirements for resilience, but also in the methods used to assess it.

At the heart of sector coupling is an increasing direct or indirect electrification of heat and fuel supply as well as mobility. This increases the potential impact of disruptions in the power system and thus the risk of cross-sector failures. The potential risk and complexity of damage propagation caused by the higher sector coupling may be partially compensated by redundancy of the available infrastructure. There are indeed certain redundancies with the parallelism of electricity and hydrogen transport infrastructure and the respective conversion plants. However, this is true even for today's system, in which infrastructures for the parallel use of numerous liquid, gaseous and solid fuels are in place. A decrease in redundancy can therefore only be avoided if a similar breadth of synthetic fuels is used in the energy supply in the future, or if parallel infrastructures are strengthened in some other way. In this context, diversification of power generation technologies is a key element to resilience. German energy system transformation pathways across numerous studies (Naegler et al. 2021, Lunz et al. 2016), indicate that solar photovoltaics and wind turbines will dominate with over 60% of all installed capacities in various future scenarios. For a carbon–neutral energy system, other renewable sources of electricity generation like hydropower, geothermal and solar thermal power generation are important, but have limited scope in Germany due to lower resource potentials. This results in a relatively high dependency on a small group of technologies with high variability highlighting a pronounced need for flexibility, which is accompanied by additional technical and regulatory challenges.

In principle, decentralisation of power generation favours the emergence of local self-supply, which offers the advantage of lower fault propagation in spatially restricted events. Nevertheless, due to the uneven distribution of demand and generation potentials as well as the temporal variability of renewable energies, it is rarely possible to achieve complete local self-sufficiency all year long. Further, it is not economically viable, since balancing effects between different renewable energy technologies and locations reduce costs for large scale systems (Neumann 2021). Nevertheless, decentralisation facilitates a wider distribution and higher disparity of energy conversion facilities that can contribute to the resilient design of energy systems. This applies, for example, to spatially concentrated extreme weather events such as flooding, but also to physical attacks on power system facilities.

Digitalisation also comes with potentials and risks for resilience. On the one hand, digitalisation enables improved control and monitoring of plants. Enhanced situational awareness can help identify early symptoms of cyber-attacks also contributing to system security. On the other hand, with millions of decentralised energy devices connected to the internet, the risk of attacks at unregulated or uncertified access points increases. Simultaneous infiltrations through numerous such channels can paralyse the entire energy system (Mayer and Brunekreeft 2020).

13.2 Modelling Future Energy System Infrastructures

13.2.1 Modelling Overview

To support decision makers in politics and industry in the planning of large-scale future energy systems, complex energy system models are made available by science (Fattahi et al. 2020). To consider the above-mentioned developments of the energy system transformation, these tools are continuously enhanced. However, due to the high complexity of the system as well as the variety of relevant temporal and spatial scales, these models can only ever represent a section of the system, and usually have to rely on extensive approximations of the physical processes. The trade-off is to use simplifications which, despite a reduction in complexity, provide a representation of the system that is as close to reality as possible (Cao et al. 2019). While earlier models focused on temporally and spatially aggregated observations, regionalised analyses in hourly resolution are now standard in the modelling of the German and European energy system. Regionalisation of the model often includes the representation of individual network nodes in the transmission network as well as the lines connecting them.

Energy system planning models are usually formulated as optimisation problems. Thereby, the total system costs are minimised from the point of view of an economic planner. There are differences in particular regarding the spatial, temporal and technological details and scope of such models (Gils et al. 2022). The central constraints of the optimisation concern the energy carrier and region-specific balances between demand and supply in each modelled time step. To satisfy these, different technologies for conversion, storage or transport of energy are available. Their technical characteristics are also formulated as constraints.

The energy system modelling framework REMix (Fig. 13.2) has been developed at DLR for over 15 years. It has been used to model and analyse numerous energy systems with diverse geographic, technological, and temporal scope and detail (Gils et al. 2021). REMix is characterised by an extensive representation of the different options for flexible sector coupling. This means that any conversion processes between different energy carriers (e.g. electricity, heat or hydrogen) can be modelled, as can the storage and transport of these energy carriers. It is also possible to explicitly allocate demand to different sectors of the energy system, such as industry or buildings, allowing flexibility within and between these sectors to be modelled in detail. This makes REMix very suitable for modelling multi-energy systems and sector coupling.

Up to now, REMix has been mostly used for planning national energy systems, with a special focus on the role of renewable energy sources. To reflect their time-variable availability, the system operation is optimised in hourly resolution. System design and operation are optimised together in one model run and with perfect foresight over the complete time horizon of usually one year. This means that all exogenous information, for example on renewable electricity generation and demand, is available to the model for all time steps right from the start. Due to the large variety

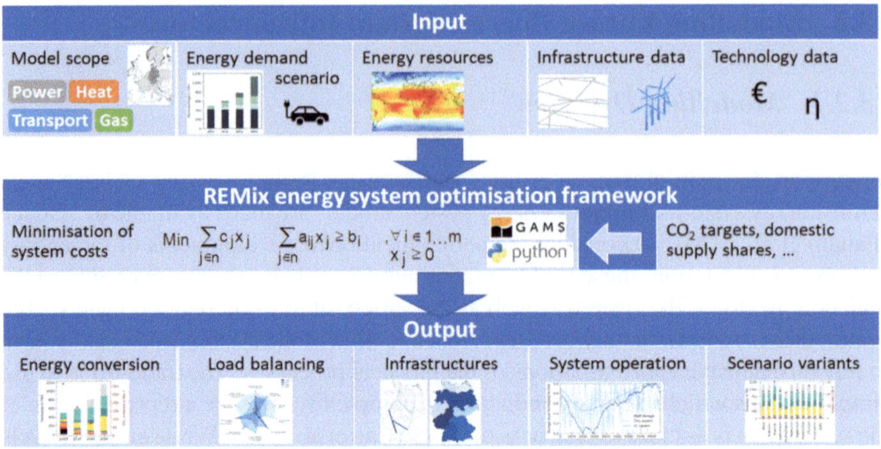

Fig. 13.2 Overview of the REMix modelling framework in its method and typical inputs and outputs (Gils et al. 2023)

of technologies and the many time steps, a consideration of the energy transport networks is often done in an aggregated way to limit computational complexity. In the following, REMix is used as an example for the class of linear optimisation models for power system planning.

13.2.2 Consideration of Security of Supply

As described above, optimising energy system models are usually designed to minimise supply costs. Other target criteria, such as sustainability or security of supply, can be considered as supplementary constraints. In the case of sustainability, this can include the specification of upper limits for carbon dioxide (CO_2) emissions or other environmental impact factors. In line with the political goals of climate protection, maximum amount of CO_2 emissions is often considered as another boundary condition. This usually leads to a deviation from the cost-optimal system, as consideration of the external costs of fossil fuels make them more expensive than renewable energy sources (Scholz et al. 2017). Complementary system requirements for security of supply translate, for example, into maintaining capacities of secured available power plant capacity, oversizing the power grid, or diversifying supply. These can be integrated into the optimisation model as additional constraints. This ensures that these additional system requirements are met. However, this is associated with higher supply costs compared to a system optimised purely for economic efficiency (Sasanpour et al. 2021). The aspect of resilience of energy systems has not yet been established as a standard procedure in the analysis of future infrastructure needs. As a result, aspects of security of supply in unforeseen extreme situations remain unconsidered. The consideration of security of supply within energy system models deals in particular with aspects of adequacy, i.e. the safeguarding of system operation. For example, the impact of stochastic power plant or power line outages on the probability of a load shortfall is investigated (Gils et al. 2018). This can be complemented by different scenarios of grid or power plant expansion as well as studies on the effect of different power generation profiles of renewable energies (Cao et al. 2023). Regarding the longer-term planning of power systems, the identification of robust infrastructure investment decisions is of particular relevance in order to hedge against possible trends and disruptions as much as possible (Gracceva and Zeniewski 2014). Concerning the introduction of additional constraints to meet security of supply requirements, it must be kept in mind that these constraints limit the solution space of the optimisation. The more extensive this is done, the more the optimisation approaches a simulation.

13.3 Requirements for the Resilience Analysis for Future Energy Systems

As technologies and structures in the energy system change, so do the requirements for resilience assessment tools. To enable resilience analysis with optimising energy system models, three prerequisites must be met. First, meaningful and consistent scenarios for the future energy system must be established and translated into model inputs. Second, the modelled system must consider the various trends in energy supply described in Sect. 13.1 so that their effect on resilience can be captured. And finally, model capabilities must allow for resilience analysis. This is specified further in the following for the scope of REMix-based models for national and continental energy system planning.

13.3.1 Scenario Modelling

A central pillar of the resilience analysis for future energy systems is the integration of consistent socio-technical scenarios. These include not only technical developments such as the expansion of electricity grids, but also place them in the context of social and political developments such as the acceptance of grid expansion or the progress of European integration.

These scenarios can be derived using the Cross-Impact Balance method and provide the context for numerous aspects of the system design by means of so-called descriptors.[1] By increasing the number and diversity of the scenarios considered, the informative value of both the structural and the incidence-based analysis can be strengthened. A clear reference of the descriptors to input variables of the model is

---

[1] See Chap. 4 in this book.

essential for the usability of the scenarios in the energy system model, such as the permitted expansion of the electricity grid or the added capacity of plants for the use of renewable energy. More specifically for the resilience analysis, the descriptors should also include relevant aspects such as the degree of decentralisation or technology plurality. The more precisely the descriptors and their characteristics quantify model input parameters or assumptions on the availability of technologies, the easier it is to incorporate the scenarios into the model.

13.3.2 Energy System Model Scope

Regarding the technological model scope, it is crucial that sector coupling is comprehensively considered, such that its impact on resilience can be analysed through the model application. Thus, a prerequisite for the model is the possibility of considering multiple energy carriers as well the various interfaces between the sectors, e.g. the use of electricity to generate heat and hydrogen, their possible conversion back into electricity, and the bidirectional charging of battery electric vehicles. Furthermore, the storage infrastructures must be considered for the various energy carriers. This allows the model-based examination of the propagation of failure through the sectors, from the electricity sector to the transport and heating sectors. This broad consideration of sector coupling results in the requirement to consider more than 50 technologies in models such as REMix.

A comprehensive analysis of the effects of decentralisation in the monolithic approach of a geographically large-scale examination partly collides with its reduced granularity. For example, decentralised deployed technologies such as photovoltaic systems or heat pumps can be modelled, but not their interaction with the power supply at the local and regional level. The optimisation approach can nevertheless be used in principle for analysing the effect of spatially limited stress cases, such as a local flood disaster, but it requires a higher granularity, which can be realised by limiting the geographic scope.

Due to the central role of energy transport networks for supply, but also their potential vulnerability to extreme weather events and physical attacks, it is important to ensure at least an approximate representation of network infrastructures in the model. This comes along with an appropriate spatial granularity of the model. For the example of Germany, this should at least reflect the grid bottlenecks between wind-rich regions and demand centres. However, a comprehensive resilience analysis for electricity grids requires the use of detailed dynamic and non-linear grid models, which in turn do not provide a complete picture of integrated energy systems.

In addition to the technological and spatial dimension, the temporal granularity is also important. To investigate the temporal course of the effect of extreme events, the temporal resolution should be as high as possible. Against the background of the trade-off between the model dimensions with regard to the size and solution time for the resulting mathematical optimisation problem, an hourly resolution is recommended.

The potentials and risks that digitalisation brings for resilience should also be considered to the extent possible. In doing so, the modelling of flexible decentralised consumers or prosumers is particularly relevant to consider a demand-side reaction to stress cases. This implies capturing the possibility of external control of decentralised systems. The extent to which this is technically possible and also accepted can be considered as an exogenous requirement in the model. The same applies to the improved possibilities of plant control in normal operation and during stress cases, as well as possible reductions in maintenance costs through predictive maintenance. Increased vulnerability to cyber-attacks and the possible multiplication of errors are also considered exogenously via the modelling of stress cases. The various software and hardware options for designing the Information and Communication Texhnology infrastructure and the associated differences in the impact of stress cases must therefore be analysed in advance.

13.3.3 Model Configuration

In addition to the aspects of model parameterisation explored, model capabilities must allow for resilience analysis. This includes the ability to model stress cases systematically and realistically and to represent various system properties required for the assessment of resilience (explained in Sect. 13.4). Since energy system optimisation models cannot explicitly simulate the development and consequences of extreme events and social response, the effects on the energy system have to be extrinsically estimated and fed to the model. However, the model runs can be tuned to represent stress response as closely as possible by limiting foresight of shock events, preventing unreasonable, sudden expansion of infrastructure and emptying of all storages. This can be enhanced by tailoring specific constraints within the optimisation to shift the system in a 'disaster-response' mode where annual emission or import limits are modified in the situation of a disaster.

13.4 Modelling Procedure for Resilience Analysis with Optimising Power System Models

Meeting the above introduced requirements, the model application for resilience analysis mainly consists of the stress case parametrisation and resilience measurement. How this can be realized will be explained again for the example of REMix.

13.4.1 Stress Case Characterisation

Resilience assessment can be distinguished between incidence-based and structural assessment. While the former deals with the effect of different, exemplary stress cases on a system, the latter focuses on the identification of system vulnerabilities. In terms of energy system modelling, incidence-based analysis involves modelling specific extreme events and testing one or more system configurations.

Stress events and their impacts on energy systems can be very different in their spatial and temporal dimension. This needs to be considered in both the model choice and model design and it determines the events suitable for investigation. For instance, the evaluation of localised extreme events such as heavy rainfall with subsequent flooding, can hardly be realised with the models used for large-scale system planning in the focus of this chapter. Instead, with a broad geographic scope such models are perfect for evaluating the impact of widespread meteorological effects like heatwaves or country-wide distributed cyber-attacks.

Different stress cases not only have differences in impact on the system, but also vary in their duration and predictability. Stress cases related to extreme meteorological conditions, such as floods and heat waves can usually be foreseen some days in advance in both their beginning and end. Other stress cases usually appear without warning time, as for example sudden weather extremes like tornados or human-induced events like hacker attacks. For these, it is also tricky to foresee an end of the impact, e.g. the return to functioning of affected infrastructures. An example for an intermediate case is the physical attack on a power line, which would come sudden, but with predictable repair time. For even other stress cases, an occurrence both with and without prior warning is possible. These include earthquakes and political unrest. Additional complexity can arise from combined stress cases, such as a hacker attack during a storm situation, which can have heterogeneous characteristics in terms of spatial extent, duration and predictability, which must be considered in the modelling.

13.4.2 Stress Case Modelling

The translation of the extreme events into concrete system damages is an important task in the modelling of stress cases. In the case of some events like earthquakes and hurricanes, vulnerability assessments in literature can provide probability of failure of electricity infrastructure and recovery time based on event intensity. However, for other stress cases, this can be approximated from historical events or predictive risk assessment studies. The effects also vary with the time course of the stress case, the definition of its location and propagation as well as the assumptions on available infrastructure. Quantification of infrastructure damage involves identifying the type and location of the affected infrastructure and estimating the extent and duration of failure.

Essentially, in the context of energy system optimisation models, the effect of any stress event can be grouped into three types of deviations from normal system operation:

(1) Infrastructure failure: depending on the type of stress case, certain technologies may not be available (spatially or temporally limited). This applies, for example, to power lines during severe storms or thermal power plants during severe drought.
(2) Change in demand (profiles): due to stress cases, the demand for certain energy sources may be different in total or over time. An example of this is the increased electricity demand for cooling and air conditioning triggered by a heat wave.
(3) Deviation in availability of renewable energy resource: extreme weather events are often accompanied by changes in solar irradiation, wind, or precipitation that deviate significantly from usual values, which impacts the facilities using these sources.

The consideration of potential failure propagation is also important in stress case parameterisation. Within its technological scope, the propagation of outages can be determined endogenously with an energy system model such as REMix. This applies, for example, to a shortfall in heat demand or restrictions in mobility caused by problems with the electricity supply. However, where significant technologies or operational effects are concerned that are not explicitly modelled, the propagation of failures must be extrinsically included while modelling the respective stress case. This may for example include the shutdown of equipment due to changes in the grid frequency or the failure of train operation caused by problems in the power supply. But, for technologies within the modelling scope, tools like REMix offer the possibility to investigate especially the effect of failures in one subsystem, e.g. power supply, on other subsystems, e.g. heat supply, for future systems with broad penetration of sector coupling. Equally, however, it can also be evaluated whether the technologies of sector coupling make it possible to avoid or at least reduce load shortfalls in the affected subsystem. This is made possible by the additional conversion facilities between different energy carriers.

A necessary adaption in the configuration of optimising power system models for resilience analysis is that the usual method of considering a perfect temporal foresight for the entire modelled period—usually one year—is unsuitable for resilience assessment. Depending on the type of stress case, there is no or only a very short warning time for the failure of infrastructures as well as for their recovery. From this follows that the model needs to be run without any temporal foresight so that the system is hit as unprepared as possible. This lack of foresight challenges storage operation, which is based at least in part on forecasts and expectations of the future supply situation. Besides, when events like heatwaves last for several weeks or months, there is no foresight at the start of the event, about how long the event lasts. For the considered model class, this results in the need for several model runs that build on each other, adding more knowledge and foresight with following model runs. For the case of a heat wave, this is illustrated in Fig. 13.3. Thus, the optimal system design with perfect foresight and endogenous system expansion is first determined. The occurrence of the stress case is then considered on this system for one or more points in time during the year. Subject to the characteristics of the stress case, this hits the system completely unprepared, or only with a foresight of a few hours or days. Thus, the second model run can start accordingly with the hour of the stress case or its earliest forewarning time, except here it is simply an optimisation of system operation under the constraints of limited infrastructure functionality, resource availability etc. The decision variables of capacity expansion for the modelled period and the operation until the event hits are fixed from the first model run. Depending on the further course of the stress case, the foresight then has to be adjusted even more times, which in turn requires new model runs starting accordingly after the occurrence of the stress case. In this way, the deviation from optimal system operation is captured and the energy system performance can ultimately be tested. This evidently requires extensive adjustments compared to the procedure of an integrated optimisation of all time steps in one model run and with perfect foresight. In addition, a specific design and sequence of model runs must be created for each type of stress case for a reasonable approximation of the stress case occurrence.

Depending on the background of the modelled stress case, e.g. a storm or a hacker attack, a variety of times and locations of an occurrence of the stress case are possible. For such cases, different combinations of time, location, durations and intensities of the events can be applied to the energy system for resilience assessment. In addition, synthetic stress cases that go beyond historical stress cases can offer testing the system for sudden shocks. To implement such stress cases stochastic methods and random sampling of different event parameters can be used. Thus, a wide ensemble of synthetic stress cases can be generated and integrated into the energy system model.

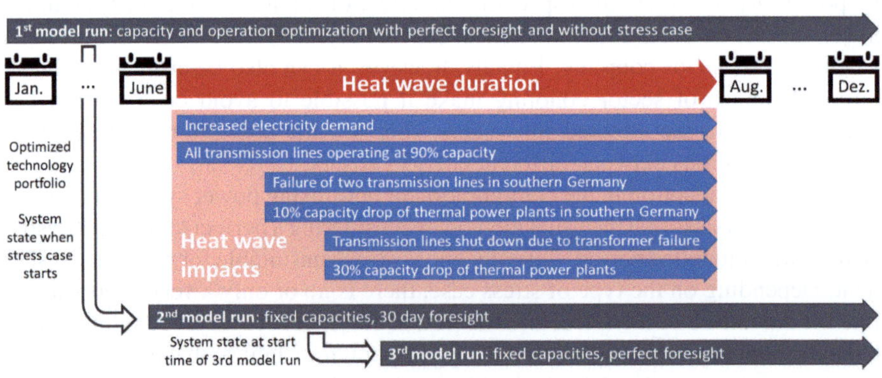

Fig. 13.3 Impact and modelling approach for a heat wave stress case. The central part illustrates the various impacts of the heat wave in their temporal sequence. Above and below, the different model runs are described. In the first model run, the stress case is not considered and the optimal design and operation of the system is assessed. The following model runs are then limited to optimization of operation and include the stress case effects. The second model run begins with the advent of the extreme event or shortly before it providing limited knowledge of the duration of the event. The prolonged effects of the later part of the event becomes available to the model only in the third model run

As a result, system characteristics that are particularly conducive to resilience can be identified, and conversely, particular system elements at risk (spatially, temporally and technologically) can be determined.

Identifying system characteristics relevant to resilience is linked to structural resilience assessment. For to structural resilience analysis, numerous scenarios need to be generated that create different energy system designs. This portfolio of designs combined with a wide spectrum of stress cases considering systematic variation of the spatial and temporal effects to obtain a comprehensive picture of the impact of different system configurations and design options for the energy transition.

But the assessment itself is enabled by metrics and indicators that are briefly explained in the following paragraph.

13.4.3 Resilience Measurement

The indicators used to measure resilience within the scope of energy system analysis are typically quantitative and can be either structural and performance-based. Structural or systemic indicators consist of various system characteristics, for example diversity of power generation technologies, and redundancy in the grid. In combination, they represent resilient system design and can be used to compare the resilience of different systems against different stressors.[2] Performance-based indicators or metrics focus on system operation and measure the fulfilment of relevant system services in response to the stressor. For instance, the (additional) demand not met due to a stress is particularly suitable as a measure of performance. For cross-sectoral modelling, this includes not only the energy carrier electricity, but also heat, hydrogen, and possibly others. Depending on the focus, these can be aggregated with or without weighting to serve as a measure of performance.

In the combination of the different elements mentioned, i.e. the preparation of a cross-sectoral system model, the implementation of scenarios and stress cases, and the evaluation of one or more performance measures, a complete incidence-based assessment of the resilience of energy systems can be performed. In this way, different system designs, represented by scenarios, can be evaluated for resilience against various stress cases. In addition, the temporally and spatially resolved operation behaviour of the considered technologies is also available as a result. This allows additional investigations of the individual roles of different technologies and flexibility options in the system, especially during the stress case. Thus, the model results can provide an extended basis for investment and operating decisions for energy system infrastructures. Furthermore, guidelines for regulation that favour a resilient design of the energy system can be derived.

[2] See Chaps 11 and 12 in this book.

13.5 Limitations of Resilience Analysis with Optimisation Models

The use of optimising multi-energy system models such as REMix for resilience analysis encounters some limitations. This concerns the electrical details of power system operation, which are highly relevant to both the effect of stress events and system recovery. Since the power grids are modelled only as a power transport option, details of system operation such as power system frequency cannot be captured. The required high aggregation and abstraction of technologies as well as the typical temporal resolution of one hour also limits the possibility to look into short-term ramping in the limitations therein. Related to this is the black start capability aspect. While this can be roughly approximated via boundary conditions such as the provision of certain technologies, it is ultimately also very closely linked to the grid topology and grid operation, both of which are only modelled in an abstract manner. In the case of a consideration of large areas of investigation, such as the German energy system, only spatially aggregated systems can be considered due to the high diversity of technologies and usually many time steps. Due to the high spatial aggregation, it is also not possible to evaluate the impact of small-scale stress cases or impacts on the local level. The analysis with an overall system model thus needs to be combined with a more detailed power grid model and/or models of decentralised energy systems.

From the abstracted representation of technologies also follows that effects of the penetration of ICT in the form of 'smart' buildings, industrial process automation etc. on energy consumption cannot be captured definitively. Decentralised energy trading, smart charging of electric vehicles, demand-side management and other energy system transformations facilitated by digitalisation can also only be captures in a limited sense as flexibilities within the energy system operation.

When it comes to the modelling of stress cases, the high abstraction causes that some impacts on different energy system components cannot be simulated but only estimated. The subsequent system response also does not necessarily represent societal behaviour after the shock.

The design principles for resilient systems[3] can only partially be incorporated as exogenous specifications or constraints in the design optimisation of energy systems. Generally, these principles can be classified into three types: those that either implicitly or can be explicitly considered with the optimisation model, those that can be interpreted in terms of energy system analysis but cannot be incorporated into the modelling, and those that cannot be interpreted within the considered scope of energy system analysis. Principles like modularity, networked design and flexibility are implicitly involved in current or future trends of energy system design. Diversity, multifunctionality, resource efficiency can also be explicitly incorporated in the optimal system designs. However, biomimetic principles such as homeostasis and punctuated equilibrium can be interpreted on scales and resolutions of energy system analysis beyond the modelling scope of optimisation models as REMix. This also

[3] See Chap. 3 in this book.

includes principles involving the role of actors in the systems (in our case people/society), which is not reflected in an optimisation model e.g. self-healing or self-organisation and learning. The third category of principles typically involves themes of evolutionary theory like autopoiesis, which find hardly any interpretation within the energy system analysis scope.

The focus on minimum cost system deployment in energy system optimisation models is in contradiction not only to the provision of safety margins in system design and operation, but also to the aspect of adaptivity. This can be levered by extending to a multi-objective optimisation or the usage of methods of multi-criteria decision analysis. However, the ability to learn from stress cases for the future is a central property of resilient systems. While the aspect of safety margins can be addressed via additional constraints, the aspect of the learning system can only be considered outside of optimisation.

To summarize, limitations essentially result from the necessary trade-off between spatial, temporal and technological scope and detail of large-scale energy system models as well as the optimisation logic. As a result, other modelling approaches and models that focus more on spatial, temporal or technological details need to be applied in addition to allow a comprehensive resilience analysis.

13.6 Conclusion

Optimising energy system models are valuable tools for providing orientation knowledge about the design of future energy systems as well as transformation paths to realise them. While costs and emissions are often the focus of scenario-based investigations with these models, resilience aspects are usually not considered. As shown in this chapter, optimisation models can make a significant contribution to the resilience analysis for future energy systems. This concerns in particular the system reaction to impacts of external stress cases such as the failure of infrastructures or changes in demand. The model allows to evaluate if deficits in supply can be avoided by an adjusted operation of the available technologies, and if this is not the case the amount of unsupplied demand can be quantified for each considered energy carrier. Furthermore, a spatial or sectoral propagation of failure can be identified within the scope of the modelled system. From this follows that optimising energy system planning models can be applied for both structural and incidence-based resilience analysis. In order to realise this, in addition to the selection of a suitable model parameterisation, above all the appropriate integration of possible stress cases is a central challenge, for which a solution approach was presented here. Furthermore, the comparative analysis of different system designs requires adapted resilience metrics that address the specifics of future energy systems.

References

Brown, T., Schlachtberger, D., Kies, A., Schramm, S., Greiner, M.: Synergies of sector coupling and transmission reinforcement in a cost-optimised, highly renewable European energy system. Energy **160**, 720–739 (2018). https://doi.org/10.1016/j.energy.2018.06.222

Cao, K.-K., von Krbek, K., Wetzel, M., Cebulla, F., Schreck, S.: Classification and evaluation of concepts for improving the performance of applied energy system optimization models. Energies **12**(24), 4656 (2019). https://doi.org/10.3390/en12244656

Cao, K.-K., Anderson, L., Böhme, A., Breuer, T., Buschmann, J., et al.: Evaluation of uncertainties in linear-optimizing energy system models - compendium. DLR research report (2023). https://doi.org/10.57676/w2rq-bj85

Fattahi, A., Sijm, J., Faaij, A.: A systemic approach to analyze integrated energy system modeling tools: a review of national models. Renew. Sustain. Energy Rev. **133**, 110195 (2020). https://doi.org/10.1016/j.rser.2020.110195

Gils, H.C., Bothor, S., Genoese, M., Cao, K.-K.: Future security of power supply in Germany – the role of stochastic power plant outages and intermittent generation. Int. J. Energy Res. **42**, 1894–1913 (2018). https://doi.org/10.1002/er.3957

Gils, H.C., Gardian, H., Schmugge, J.: Interaction of hydrogen infrastructures with other sector coupling options towards a zero-emission energy system in Germany. Renew. Energy **180**, 140–156 (2021). https://doi.org/10.1016/j.renene.2021.08.016

Gils, H.C., Gardian, H., Kittel, M., et al.: Modeling flexibility in energy systems - comparison of power sector models based on simplified test cases. Renew. Sustain. Energy Rev. **158**, 111995 (2022). https://doi.org/10.1016/j.rser.2021.111995

Gils, H.C., Wetzel, M., Gardian, H., Schmugge, J.: Realizing sector coupling in Europe and beyond: the future role of electricity and gas grids. In: 18th IAEE European Conference, July 2023, Milan (2023). https://elib.dlr.de/202959/

Gracceva, F., Zeniewski, P.: A systemic approach to assessing energy security in a low-carbon EU energy system. Appl. Energy **123**, 335–348 (2014). https://doi.org/10.1016/j.apenergy.2013.12.018

IPCC: Climate change 2022: mitigation of climate change. In: Shukla, P.R., Skea, J., Slade, R., Al Khourdajie, A., van Diemen, R., McCollum, D., Pathak, M., Some, S., Vyas, P., Fradera, R., Belkacemi, M., Hasija, A., Lisboa, G., Luz, S., Malley, J. (eds.) Contribution of Working Group III to the Sixth Assessment Report of the Intergovernmental Panel on Climate Change. Cambridge University Press, Cambridge, UK and New York, NY, USA (2022). https://doi.org/10.1017/9781009157926

Kippelt, S., Wagner, C., Rehtanz, C.: Consideration of new electricity applications in distribution grid expansion planning and the role of flexibility. In: International ETG Congress 2017, Bonn, Germany, pp. 1–6 (2017)

Lunz, B., Stöcker, P., Eckstein, S., et al.: Scenario-based comparative assessment of potential future electricity systems – a new methodological approach using Germany in 2050 as an example. Appl. Energy 555–580 (2016). https://doi.org/10.1016/j.apenergy.2016.03.087

Mayer, C., Brunekreeft, G.: Resilienz digitalisierter Energiesysteme. Blackout-Risiken verstehen, Stromversorgung sicher gestalten. Schriftenreihe Energiesysteme der Zukunft (München, 2020)

Naegler, T., Sutardhio, C., Weidlich, A., Pregger, T.: Exploring long-term strategies for the german energy transition - a review of multi-sector energy scenarios. Renew. Sustain. Energy Transit. 100010 (2021). https://doi.org/10.1016/j.rset.2021.100010

Neumann, F.: Costs of regional equity and autarky in a renewable European power system. Energy Strat. Rev. **35**, 100652 (2021). https://doi.org/10.1016/j.esr.2021.100652

Sasanpour, S., Cao, K.-K., Gils, H.C., Jochem, P.: Strategic policy targets and the contribution of hydrogen in a 100% renewable European power system. Energy Rep. **7**, 4595–4608 (2021). https://doi.org/10.1016/j.egyr.2021.07.005

Scholz, Y., Gils, H.C., Pietzcker, R.: Application of a high-detail energy system model to derive power sector characteristics at high wind and solar shares. Energy Econ. **64**, 568–582 (2017). https://doi.org/10.1016/j.eneco.2016.06.021

Sensfuß, F., Kleinschmitt, C., Köppen, S., Kiefer, C., Männer, W., et al.: Langfristszenarien für die Transformation des Energiesystems in Deutschland: Treibhausgasneutrale Szenarien T45, Webinar zum Energieangebot, 22.11.2022. https://www.langfristszenarien.de/enertile-explorer-wAssets/docs/LFS3_T45_Webinar_Angebot_Nov_2022_final_webinarversion.pdf

Wetzel, M., Gils, H.C., Bertsch, V.: Green energy carriers and energy sovereignty in a climate neutral European energy system. Renew. Energy **210**, 591–603 (2023). https://doi.org/10.1016/j.renene.2023.04.015

Chapter 14
Implementing Resilience Strategies in Future Digitally Interconnected Distribution Grids

Karen Derendorf, Peggy Bergmann, Ghinwa Harb, and Karsten von Maydell

Abstract One challenge in future digitally interconnected distribution power grids is that the digitalisation introduces the possibility of malfunctions, such as ICT network failures, which might lead to grid instabilities. Additionally, the rising complexity of the systems makes it harder to predict and respond to potential disruptions. Possible solutions for enhancing resilience in digitalized distribution grids are integrated fallback behaviours, local measurements and machine learning based operations implemented at single energy systems (assets) in the grid. In an example from the project IKTfree it was evaluated, how effective fallback behaviours in single assets can ensure stable grid operation in an event of ICT network failures. The analysed fallback behaviours range from keeping the latest power value to changing it to a pre-defined set point up to change the power level in accordance to the measured value at the connection point. The results show that these approaches can help to keep the voltage at house connection points and loadings of lines and transformer stations within operating limits. In conclusion, these results demonstrate the potential for fallback behaviours to enhance resilience in digitalized distribution power grids, even in situations where ICT control is lost.

Keywords Resilience · Stability · Distribution grid · Fallback · Disturbance in controllability · Disaster prevention · Disaster management

14.1 Introduction

In recent years, energy production in many countries has seen an increase in renewable energies within the energy mix. In Germany, the energy transition ("Energiewende") has led to the introduction of additional information technology

K. Derendorf (✉) · P. Bergmann · G. Harb · K. von Maydell
Institute of Networked Energy Systems (DLR-VE), German Aerospace Center (DLR), Oldenburg, Germany
e-mail: karen.derendorf@dlr.de

D. van Doren et al. (eds.), *Towards Energy System Resilience*, Ethics of Science and Technology Assessment 55, https://doi.org/10.1007/978-3-032-03187-7_14

networking of grid components and grid control. The widespread use of information and communication technology (ICT) in electricity supply brings both advantages and disadvantages. The growing decentralization and volatility of the generation side, as well as an increasing diversity of players, make coordination and robust control necessary. Additional future revenue opportunities, which require ICT-based retrieval of flexibility, continue to make the decentralized and grid-serving expansion of renewables attractive. At the same time, the system is becoming more vulnerable due to its interconnection with complex information. To an increasing extent, grid operational measures (measurement, metering, control) are dependent on a functioning ICT.

Distribution and transmission system operators can often rely on their own dedicated infrastructures, exclusive to the respective operator. These include, for example, the connection of substations or switchgear to the grid operator's control room. Large power plants are also integrated into the control and regulation systems in a similar way. These systems often have a reduced security risk due to their separation from the public communication network.

As an increasing number of small grid participants become interconnected through shared networks (like the internet), the controllability is enhanced resulting possibly in a higher amount of usable flexibility for the grid. At the same time, these connections exhibit a lower security compared to traditional communication networks used by grid operators, thereby introducing additional vulnerabilities that could potentially compromise critical infrastructure through disruption, damage or attacks.

These challenges will be described in the following section, followed by an overview on groups of participants in the distribution grid. Afterwards technologies are specified which are utilized in the distribution grid and which now and in future even more will be digitally interconnected. In a next part different possible disturbances in the grid are presented, which long from natural catastrophes to attacks which are only possible due to the enhanced digitalization. Measures implemented in single energy systems can help to withstand these disturbances. They are described afterwards, followed by an example from a research project which analyzed the effect of ICT outages for the distribution grid and how measures can enhance resilience against these.

14.2 Developments and Challenges in Future Energy Grids

The challenges in distribution grids arise from both changed and changing structures as well as new operational complexities. The power grid, which was originally planned and implemented unidirectional, with power generation at the highest voltage level and subsequent distribution of the power to voltage levels for the end consumer connected at medium or low voltage levels, is currently undergoing fundamental change.

The number of ***decentralized grid participants*** is increasing. They primarily include decentralized generators that produce energy at the lower voltage levels and feed it into the power grid "from below", for example photovoltaic systems connected as small rooftop systems at the low voltage level as well as larger solar parks at the medium voltage level. Likewise, wind turbines usually feed electrical energy into the power grid at the medium-voltage level. These two types of generators present two challenges. First, their generation output cannot be planned as with conventional generators (coal, nuclear, gas) because they are dependent on weather conditions. Second, their location and size are outside the scope of intervention of grid operators because investors and homeowners decide whether to install such plants and, as is the case of Germany, the grid operator is usually obligated to connect them to the grid. On the other hand, the grid operator is responsible for ensuring stable grid operation.

Another notable change in the electricity supply system is the increasing ***sector integration***. By transitioning the foundation of sectors like heat and mobility to electricity, new market participants emerge, whose behaviours and influence on demand and supply within the electricity system are not yet fully understood. For example, the electrification of the transportation sector increases the overall demand for electrical energy. At present, charging behaviour can be mapped well, especially at house connection points. However, the question remains as to how much and in what form usage behaviour will change in the future, particularly if there are new incentives that, for instance, encourage to not always fully charge vehicle batteries. It is also conceivable that the increased demand for electricity due to the increasing number of electric vehicles may exhibit less simultaneity than anticipated, with charging peaks potentially being distributed throughout the day and incentives like dynamic electricity tariffs that encourage charging when there is a lot of available power in the grid. Changes in the transportation system will result in other supply patterns that may vary greatly from household to household. It has not yet been conclusively determined to what extent ICT and corresponding regulation will help to ensure local coordination of local generation (e. g. through rooftop PV systems) and how they will incentivise flexible load operation at the lower grid levels where possible, and what consequences this will have for grid operation planning.

The integration of weather-dependent power generators leads to increased fluctuations in generation output compared to the previously very uniformly running, centralized power plants. To ensure the necessary ***balance*** between ***generation*** and ***consumption***, the number of controlling interventions in grid operation increases, especially when power grids are not fully equipped yet to deal with the changed supply side. On the one hand, this concerns the control of generation plants, e. g. by curtailment in the case of generation that is regionally not usable at the time (EnWG, §13a). On the other hand, the imbalance can be counteracted by controlling loads. Of special interest in the first step are industrial processes, because at this point a large amount of power can be changed by a single control command (Paulus and Borggrefe 2011).

The increased number of control actions can facilitate the ongoing safe and reliable operation of the power system. The challenge is to organize and coordinate the different control commands.

Locations for renewable energy are primarily determined according to meteorological and also geographical factors. The selected locations rarely coincide with the locations of energy demand. This creates the risk that grid resources such as lines or transformer stations can become overloaded. To prevent this, redispatch measures are implemented (***Redispatch 2.0***). In this process, generation plants located upstream of the endangered equipment are deactivated and generation plants located downstream of the endangered equipment are activated or their power output is increased. The increasing number of redispatch measures in Germany in the past years shows the necessity of this instrument. They keep the grid operation stable, but on the other hand they increase the cost of grid operation (ECO*Journal 2024), require increased monitoring and personnel deployment.

In addition to the traditional participants, new types of players are increasingly appearing in the course of the transformation of the power system. One group of players are ***aggregators***. They bundle a large number of decentralized generation assets, storage or consumption units and thus make the services of these plants available to various markets or other parties. In this way, even energy plants located in the distribution grid, which are typically too small to participate in energy markets, can take part in market activities. The minimal amount which has to be offered for control reserve is 1 MW (50Hertz Transmission GmbH, Amprion GmbH, TenneT TSO GmbH, TransnetBW 2024). To this purpose, for example, a large number of battery storage units are bundled ("pooling") by an aggregator and their combined output is offered as a virtual power plant or as flexibility in the balancing energy market as well as on spot markets (Lund et al. 2015). The aggregation and control of the individual components is often carried out via public technologies such as mobile networks or wired communication. Coordination is managed by a control system.

The challenge arises from the fact that energy quantities are traded on a German market and independently of the local supply situation. This can lead to energy imbalances between regions which is solved by redispatch of power generators, which rises the costs of power dissemination.

14.3 Digitised Energy Systems: Overview of Operational Areas

The ongoing changes in the electricity sector mentioned before go hand in hand with a steady integration of more information and communication technology. Because to make use of the potential capabilities and flexibilities of components, they have to be digitised. That means that they are no longer *on* and *off* with a specific power; instead they are equipped with integrated communication devices which can respond to external control signals and adjust or shift their power respectively.

14.3.1 Generation

The electricity generation of the past was integrated into higher voltage levels. Their control was done by the plant operators. Moreover, their production was easily plannable and—if there were deviations in the demand—the amount of generated electricity was easily changeable. Due to the integration of decentral wind and solar power plants, the number of generators increases tremendously, but their generation is dependent on weather conditions. The power plants are connected via remotely controllable and also non-controllable inverters to the grid, primarily at the middle and low voltage level. In case of grid overload the power of larger power plants is reduced.

14.3.2 Consumption

Now that electric power generation is not as plannable as in the past there is an increased need for electric power consumption to following the availability of (renewable) generation. To achieve this, it is essential to identify flexible loads and equip them with communication technology. These loads can be industrial and commercial facilities, but also heat pumps and storage systems in private households or electric vehicles which both might have the possibility to change their timing for receiving the required amount of energy to provide their service (heating and mobility) without compromising convenience. Especially for the components with a comparatively low power rating but a high prevalence, harmonized communication is needed which is able to connect to components from different manufacturers.

14.3.3 Grid Control

Grid operation and transportation of energy to the demand is becoming more complex as consumption and generation become more variable. Consequently, the operation of the grid must also become more flexible. On the one hand, this means that an increased number of control actions is needed in control centres, and operations must adjust to sudden weather changes. On the other hand, automatically switching devices will enable the grid to cope with these challenges effectively. One example of this are controllable substations like on-load-tap changer (in German *Regelbarer Ortsnetz-transformator* rONT), which might help to reduce the need for grid reinforcement actions like redimensioning of cables (Samweber et al. 2017).

14.3.4 Trading

Electricity trading in Germany takes place via the EEX exchange as well as through over-the-counter (OTC) transactions. At the exchange, electricity is traded on the spot market with intraday trading possible up to five minutes before the delivery quarter hour. For energy traders, accurate forecasting of electricity production and consumption is crucial. Ensuring an accurate forecast is even more important when prosumer households deviate from the standard load profiles, as they either manage their own energy systems to increase self-consumption or respond to a variety of incentive signals that introduce additional uncertainty and variability into load forecasts.

14.4 Examples of Digitally Connected Energy Systems Technologies

In the sectors outlined, several types of technologies are digitally connected to communication networks. They are able to participate in controlled energy generation and consumption. The number of contributing systems is strongly increasing. The following paragraphs will elaborate on some examples and specifications.

14.4.1 Photovoltaic Solar Systems

The power generation of photovoltaic systems is dependent on weather conditions and latitude but also on constructional parameters like azimuth and tilt angle of the installed panels. Incorporating these factors into the forecasting of PV system generation has improved in recent years. Nevertheless, the cumulative generation of PV systems typically exhibits a peak at noon. This peak might lead to overloading of lines or transformer stations, particularly when there are many PV systems in a neighbourhood and/or not enough consumers in the surrounding area. For those situations, external control of PV systems through ICT was made mandatory in some cases. The controlling player might be a direct seller (usually only for large, not household size systems) or the grid operator, with the latter only intervening in case of security risks. For example, in Germany, in the latter case, the system must be able to assume setpoints of 0, 30, 60 or 100% of the rated power(EEG, §9). The control signals are sent via ripple control (Schwab 2017). In addition, grid codes require the PV systems to be able to vary the active power as a function of the grid frequency when the grid frequency exceeds 50.2 Hz. This is not dependent on external control but relies on a local measurement. Regarding the reactive power for voltage maintenance, the grid codes in Germany allow three strategies:

- Adjustment of reactive power in response to voltage using a characteristic curve: Q(U).

- Adjustment of the power factor according to the actual power fed into the grid: $\cos \phi(P)$.
- A fixed power factor $\cos \phi$

14.4.2 Batteries as Parts of Home Energy Storage Systems

For rooftop PV systems, another approach to reduce peak feed-in is to store surplus energy on-site by installing a battery energy storage system and a battery management system that coordinates battery charging with available PV power. This can potentially also reduce the need for power transmission through the grid. However, a significant challenge arises from the differing objectives of homeowners and grid operators. Homeowners aim to minimize their own power consumption from the grid, as this enhances the profitability of their own small energy system. Currently, charging patterns often do not consider the overall grid situation, resulting in excess power to be fed into the grid, once the battery is fully charged. In contrast, the grid operators might benefit most from a reduction in the noon peak or during periods of high loading of lines and transformer stations.

Another way to utilise battery storage systems is by aggregating them into a pool, enabling them to participate in electricity markets. In Germany, battery systems contribute to grid stabilisation if the frequency is below 49.8 Hz or higher than 50.2 Hz or through the frequency containment reserve. Regarding reactive power for voltage maintenance, grid code regulations stipulate that when the battery system feeds power into the grid, it must adhere to the same options as those applicable to PV systems. If the battery system operates as a load, it must maintain a displacement factor between 0.9 and 1 (inductive).

14.4.3 Controllable Substations

The integration of decentralised energy sources feeding in at the low-voltage distribution grids necessitate further measures from grid operators, so that grid measurement values can be kept within acceptable limits according to grid codes.

In the low voltage grid in Germany, the specified nominal voltage range of $\pm 10\%$ according to DIN50160 for the end users only allows a maximum voltage rise of 3% for the integration of decentralised energy sources. The remaining bandwidth must be available for medium voltage grids and voltage drops.

In this context, grid operators are required to undertake expensive expansion activities in their grids. As an economical alternative to grid expansion, the controllable substation helps to adhere to the voltage range requirements by adjusting the voltage dynamically and automatically while being controllable remotely. By decoupling the voltages in the medium and low-voltage grids through controllable substations, more headroom is available to manage sudden voltage spikes that occur through feed-in in the low voltage grid (Schneider Elektrotechnik GmbH 2013).

14.4.4 Heat Pumps

Another new type of load connected to the electric power grid are heat pumps. In comparison to common household loads they have a high-power demand and—which is very important—their demand is not necessarily fixed to a special point of time at the day because they are usually paired with water storages. This flexibility offers a number of applications in the smart grid context in which they can be used, such as voltage regulation or power reserves through pooling of participants (Fischer and Madani 2017). Another application is based on an agreement with the grid operator. Based on the electricity and gas supply act (EnWG, §14a) it is permissible for the grid operator to specify shut-off periods for the operation of heat pumps during which the heat pump cannot be operated and the heat supply must be ensured by a buffer storage tank. For this purpose, the operator sends a shutdown signal in advance via ripple control technology. This measure is used, for example, to reduce the grid load at peak load times (VDE-AR-N 4100).

14.4.5 Smart Meters

A smart meter—a digital metering system—is an essential part of the digital infrastructure of the energy system. The addition of a communication unit, called smart meter gateway (SMGW), to a smart meter enables communication of the measured values to external stakeholders. Together they form an intelligent measurement system (in Germany so called *intelligentes Messsystem, iMSys*). The SMGW is a specially secured interface for communication between electricity consumers and generators (like PV) with the operators of the electricity grids. Beside the SMGW, there are also other communication interconnections in operation, for instance through cloud services, enabling aggregators to integrate home energy storage systems into the market without relying on metering data.

14.4.6 Electric Vehicles

Electric vehicles represent a new category of power consumers that draw electricity over extended periods at various locations, including homes, public charging stations and charging car parks. The power consumption typically ranges from kilowatts to several hundred kilowatts. Currently, the power flow is unidirectional. In initial research projects the integration of bidirectionally chargeable vehicles into the power system is being researched (Sakarwala et al. 2023) and demonstrated. This creates a new possibility for intermediate storage of electricity, where electric vehicles that are not currently used for mobility can act in the role of battery storage systems (see above) and respond to market signals to support grid operation.

14.5 Disturbance Events Affecting the Stability of Energy Systems and Their Potential Impacts

The digital connectivity of components offers numerous opportunities as described above. However, it also introduces new dependencies within the grid, potentially making grid operation more vulnerable to disturbances and less resilient in general. Extreme disturbance events might exacerbate and reveal the consequences of the dependency. How shock events can affect the power grid, especially when it has a high level of digitalisation, is described in the following.

14.5.1 Extreme Weather Events

Due to the climate change, the frequency and level of severity of weather events has been increasing. This directly leads to more frequent threats on power grid outages caused by damaged transportation lines or damaged transformer stations. To show the severe effect of weather events, some examples will be given here.

In 2008 a snowstorm extreme weather event hit South China and introduced over 129 line faults, which yielded to power outages for 14.66 million households. In 2012, Hurricane Sandy landed on the east coast of the United States and caused power outages for millions of people. In 2016 a tornado hit Jiangsu Province, China, which tripped over two 500-kV transmission lines, four 220-kV transmission lines, eight 110-kV transmission lines and caused power outages for 135.000 households.

14.5.2 Unpredictable Events like Large-Scale Staff Outage

It is imaginable that a strike or an area-wide disease might lead to shortage of personnel, potentially triggering cascading effects. Such an outage could halt production in manufacturing plants or disrupt processes in commerce and retail. In the next step of the cascade, this would lead to shrinking power demand which might influence the balance of consumption and generation in the grid when it was unforeseeable and the coordination of demand and consumption cannot sufficiently be adjusted on short notice.

14.5.3 *Outage of Grid Participants*

Especially for lower voltage grids, an outage of a grid participant might have a big influence. An example is a small manufacturer in an industrial park which usually has a predictable load curve. When having a lack of supply in articles it might behave very differently from its usual behaviour or even be turned off. Another reason for not taking any power from the grid might be software problems, as more and more commerce is dependent on software for their production lanes or working processes.

14.5.4 *Fail or Outage of Control Mechanism*

The possibility of monitoring the grid status and sending control signals by communicative interconnection of components also brings with it further possibilities for vulnerabilities. For a stable operation of the system based on control interventions, it is therefore always necessary to consider what will happen in the event of a failure or an outage of this controllability.

One potential issue is the outage of the communication system. If, just before the interruption, components were being controlled externally and supported the grid to be in stable operation, the question arises of how the components will behave if they do not receive new control signals. Potential behaviours could include to stay at the power level they were operating at, reducing the power level to zero or to use a fallback strategy whenever a missing ICT-signal is detected.

Another possible problem arising from the communication technologies might be incorrect signals. To recognize these, systems have to be installed which check the plausibility of measured values and estimate the grid status.[1]

An example for a problem in the power grid driven by the digitalisation of the system is the power outage in northeastern North America in the summer of 2003, when a software error on a transmission line caused an alarm signal to be transmitted out of time, leading to a cascade of faults and a widespread blackout (Dagle 2004).

14.5.5 *Hacker Attacks*

In addition to communication failures and erroneous measurements, intentionally manipulated signals can pose significant challenges. One example that has been discussed extensively in the literature is the attack on Ukraine's power grid in December 2015, in which, in a targeted attack, control was gained over distribution grid operators' grid control systems. Several substations were disconnected from the grid. The capital Kyiv and surrounding areas were affected by a power outage for three hours. Recovery was further complicated because of damage to IT

---

[1] This is described in more detail in Chap. 12 of this book.

infrastructure and an additional denial-of-service attack on call centres (Sullivan and Kamensky 2017; Styczynski and Beach-Westmoreland 2019; Lee et al. 2016).

14.5.6 Fail in Predictions

As mentioned, the variable power generation and consumption requires reliable forecasts. Despite the improvement in the last decades inaccuracies in predictions still exist and might lead to problems in the operation of the grid. Fast and spontaneous operation measures have to be executed to compensate for the errors which sometimes might depend on very local phenomena like clouds heading over a small town with a lot of PV systems.

14.5.7 Increasing Complexity Due to Increase of Controllable Components

In 2050 in Germany, there will be around 13–17 Million controllable components as active grid participants which will lead to an increase in the complexity of the future smart energy system.

This complexity is due to the diverse interaction of stakeholders like consumers, governments, grid operators and other institutions operating as independent decision makers with autonomous behaviours, goals and attitudes. The actors perform actions via distributed decision-making processes which impact the physically constrained grid via diverse electronic means (from control and command systems to smart meters). One example is the increased complexity of social behaviour and the interaction of individuals guided by narrow economic rationality (Joint Research Centre 2024).

14.6 Operational Measures to Deal with Stress Cases in Distribution Grids

To face the challenges arising with the transition in the energy system and the incorporation of more digitalisation, different measures are conceivable to keep the system stable and to additionally make it more robust and resilient against stress events.

These measures are meant to act only in critical situations where the grid operator is foreseeing some bottlenecks for stable grid operation. This is important to note, because it was found that acceptance for any measures can be enhanced if people generally have control over their own systems. If measures are only for emergency situations, the acceptance of those measures increases, as also their occurrence is supposed to be very low (Richard et al. 2018).

In the research project ReMoDigital, some of these measures have been studied in a simulation of the distribution grid to find out how they can for instance prevent grid sections from losing power supply. The system boundary was set at the transformer station to the transmission grid. This is mentioned here, because the definition of the analysed system always might influence the result.[2] For the analysis, different future scenarios where implemented and different stress cases where applied. For the results, a comparison was made between the effects with and without certain measures. The effects in different scenarios and stress cases are visualized together with analysis from other simulations also done in the project, like changes in transportation due to stress cases.[3]

14.6.1 Islanding of Feeders

One possibility to keep parts of the system running, even when another part is not functional anymore, is by forming power grid islands (Bie et al. 2017). In order to do so, specific technology has to be integrated which is capable to form a grid, which means that it imprints frequency and delivers a reference voltage to the islanded part of the grid. Such services can in principle be provided by inverters, e. g. in battery storage systems. A key issue is to ensure enough generation in the islanded area to support all loads. Moreover, the islanding operation must be organised in advance of the stress case. It needs to be specified which inverters should operate as grid-forming inverters and which inverters are grid-following. The islanded operation might also consider changing the power of grid-following inverters through a local control signal from a decentral control which manages the power flows in the island.

Technically, technologies to enable islanded grid operations are available. However, at the moment there is no regulation in Germany for the general implementation of those islanding possibilities. Only for critical infrastructures, like hospitals, an islanding operation mode is often implemented for this single connection point. Here, the service is typically delivered not by inverters but by diesel generators.

14.6.2 Hardening the Grid Operation by the n-1 Criteria

To mitigate the risk of a (regional) blackout in Germany the n-1 security criterion has been implemented, which ensures that the systems still will stay operational, even if **one** of the **n** lines fails. Each responsible grid operator adjusts their grid accordingly, so that the criterion is met.

---

[2] See Chap. 4 of this book.

[3] See Chap. 15 of this book.

At present, these measures are not implemented for lowest voltage distribution grids. Whether also grids in the medium voltage range should be hardened to be included into n-1 evaluations is still investigated and discussed.

14.6.3 Emergency Charging Stop for Electric Vehicles or Heat Pumps Through Grid Operator

In emergencies, the grid operator might benefit from the ability to stop components that can be assumed to not be critical but which are significantly large to influence the grid situation. Besides the downregulation of industrial plants, electric vehicles (EV) and heat pumps are loads that can be utilised for this purpose. In Germany this is governed by §14a in the EnWG (Energiewirtschaftsgesetz = law for energy economics) which states that the network operator may enter into a contract with the end-consumer allowing the network operator to reduce the power consumption of these systems for a limited period of time.

For instance, there are considerations to implement a "traffic light" concept in Germany in order to define what measures can be taken to deal with critical grid congestions (Zacharias 2017). In "yellow light phases", flexibility is used to reduce grid congestion and return the grid back to "green light" (normal operation). This could for example mean that a grid operator temporarily turns off a charging station for EV.

Another approach that is discussed, is to differentiate conditional and guaranteed power for the provision of electricity to flexible loads. The conditional power might be restricted to certain times or imbued with higher prices. This method might enable grid operators to restrict loads in stress cases like outages or control problems, while customers have transparency and are supplied with an amount of electricity that covers their critical needs.

14.6.4 Integrating Fallback Behaviours in Distributed Components

Another option to increase the resilience of digitalised distribution grids is to define and implement fallback behaviours in distributed components for situations in which control signals cannot be sent, for instance when ICT networks are not functional.

Depending on the stress case, it could be helpful for grid operators to know and define the behaviour of components when their connectivity is lost. Fallback behaviours could include the restriction of infeed and load to levels that can safely be assumed to keep the grid within operating limits.

A more dynamic approach would be to implement characteristics that adjust infeed and load according to the local measurement of voltage and/or frequency in case of emergency (see example below). Requirements can be specified according to the grid operator's needs and included in grid codes.

Furthermore, modern machine learning approaches can enable control systems to learn about their environment and detect grid situations also past their own connection point (Schlachter et al. 2024). This approach could allow components to independently adapt their behaviour in order not to put additional stress onto the grid.

First results as presented in the next paragraph show that these integrated measures can help the grid to stay in stable operation, even under adverse circumstances. At the same time, it remains uncertain whether individuals would accept their home components operate autonomously without their interaction. That underscores the necessity to investigate the question of acceptance. Furthermore, it is essential to discuss if and how these measures should be integrated, for example through legislation incentives.

14.7 Research Example from Project IKTfree

The project *IKTfree* (Derendorf et al. 2022) looked at future grid situations where ICT network hypothetically failed and therefore the intervention of the grid operator was no longer possible. One focus of the project was to evaluate how much each component could possibly contribute to keeping the grid situation within operating limits without any knowledge about what was happening in the grid. For that, different grids and different operating conditions were analysed in which the components (PV, EV, heat pumps, batteries) were outfitted with additional control circuits that were activated as a sort of safety mechanism when a loss of ICT control was detected (so called fallback).

The following example from the projects results shows how a P(U) characteristic (power in dependence on measured voltage) employed in PV systems is effective in keeping voltage and loading within operating limits. The effectiveness was compared in three different residential grid types and with different parameters for the fallback control.

Scenario: Summer midday where PV systems are installed at a high number of connection points in a residential area (80–100%). There is just a low load but a high infeed from the PV systems. If there are battery systems present, they are assumed to be fully charged. This represents a scenario that would lead to a critical situation in the grid in case of ICT loss, if the safe operation relied (hypothetically) completely on a working ICT connection.

Measurements without any control: The infeed from PV systems affects the local line loading and voltages. In the chosen critical scenarios under investigation, voltages reach a maximum of 1.09 p.u. (per unit) in smaller grid topologies and 1.16 p.u. in a larger real grid topology, where 40% of all voltage measurements would be out of operating limits. In one grid under investigation, 30% of all lines would be overloaded.

Measure: To counteract these situations, the implemented P(U) characteristic activates once a defined voltage threshold is reached and reduces the active power infeed from the PV systems depending on the currently measured voltage at each respective connection point. Different characteristics designs were compared to each other, where the reduction of power infeed activates earlier or later in regards to the voltage measurements.

Result: In the analysis, employing the P(U) characteristic was able to keep voltage and loadings of lines within bounds when ICT control was lost. The advantage of this strategy is that simultaneity effects are avoided because power is reduced according to the local demand. The characteristic is selective, since it activates only after a certain voltage is reached. However, this strategy can only consider the local grid situation and, depending on the situation, not all participants in a feeder contribute equally to the problem solution. To enhance overall resilience, it might be effective to implement a diverse range of measures that complement the effect of the localised power adjustments.

14.8 Conclusion

The simulation example analysed the challenges faced by future distribution grids in integrating more distributed renewable energies and communication technologies, focussing on the effect of interrupted ICT. A malfunctioning ICT could affect several existing and new participants in the distribution grid like PV systems, heat pumps or electric vehicles. The absence of ICT may cause grid instabilities or operational voltages out of range which in consequence might lead to outages of (parts of) the grids. The analysis showed that fallback behaviours implemented within the energy system technologies have the potential to improve resilience in digitalised distribution power grids.

These fallback behaviours could also be advantageous during other stress events affecting the grid. These events could be heavy rain or (ice) storms, which have become more frequent and severe in recent years due to climate change. Consequently, these fallback options will be further analysed to investigate their impact on improving resilience in distribution grids.

References

Bie, Z., Lin, Y., Li, G., Li, F.: Battling the extreme: a study on the power system resilience. Proc. IEEE **105**(7), 1253–1266 (2017). https://doi.org/10.1109/JPROC.2017.2679040

Bundesministerium der Justiz: Gesetz über die Elektrizitäts-und Gasversorgung (Energiewirtschaftsgesetz). EnWG, §14a. Online verfügbar unter (2024c). https://www.gesetze-im-internet.de/enwg_2005/__14a.html. zuletzt geprüft am 02.12.2024

Bundesministerium der Justiz: Gesetz für den Ausbau erneuerbarer Energien (Erneuerbare-Energien-Gesetz). EEG, §9. Online verfügbar unter (2024a). https://www.gesetze-im-internet.de/eeg_2014/__9.html. zuletzt geprüft am 02.12.2024

Bundesministerium der Justiz: Gesetz über die Elektrizitäts-und Gasversorgung (Energiewirtschaftsgesetz). EnWG, §13a. Online verfügbar unter (2024b). https://www.gesetze-im-internet.de/enwg_2005/__13a.html. zuletzt geprüft am 29.11.2024

Dagle, J.E.: Data management issues associated with the August 14, 2003 blackout investigation. In: IEEE Power Engineering Society General Meeting, vol. 2, pp. 1680–1684 (2004)

Derendorf, K., Bergmann, P., Yang, S., Engel, B., Rebak, E., Tiedt, F., et al.: Hochverfügbarer Verteilungsnetzbetrieb bei Störung der IKT-Infrastruktur im Smart Grid": Förderzeitraum: 01.10.2018 bis 30.06.2022. Abschlussbericht zum Verbundvorhaben IKTfree: Deutsches Zentrum für Luft-und Raumfahrt; Institut für Hochspannungstechnik und Elektrische Energieanlagen. Köln. Online verfügbar unter (2022). https://www.tib.eu/de/suchen/id/TIBKAT%3A1873048742

ECO*Journal (Hg.): Netzentgelte für Strom: Darum sind sie regional so verschieden. EnBW. Online verfügbar unter (2024). https://www.enbw.com/unternehmen/themen/netze/netzentgelte-fuer-strom.html#netzentgelte-steigen-2024-strom-wird-teurer, zuletzt geprüft am 29.11.2024.

Fischer, D., Madani, H.: On heat pumps in smart grids: a review. Renew. Sustain. Energy Rev. **70**, 342–357 (2017). https://doi.org/10.1016/j.rser.2016.11.182

Forum Netztechnik/Netzbetrieb (FNN): Technische Anschlussregeln Niederspannung (VDE-AR-N 4100). VDE-AR-N 4100. Online verfügbar unter (2019). https://www.vde.com/de/fnn/themen/tar/tar-niederspannung/tar-niederspannung-vde-ar-n-4100. zuletzt geprüft am 02.12.2024

50Hertz Transmission GmbH, Amprion GmbH, TenneT TSO GmbH, TransnetBW: Präqualifikationsverfahren für Regelreserveanbieter (FCR, aFRR, mFRR) in Deutschland. PQ-Bedingungen. Online verfügbar unter (2024). https://www.regelleistung.net/xspproxy/api/StaticFiles/Regelleistung/Infos_f%C3%BCr_Anbieter/Wie_werde_ich_Regelenergieanbieter_Pr%C3%A4qualifikation/Pr%C3%A4qualifikationsbedingungen_FCR_aFRR_mFRR/PQ-Bedingungen-Delta_2024_07_05.pdf. zuletzt geprüft am 29.11.2024

Joint Research Centre (Hg.): Complex system science for smart grids. European Commission. Online verfügbar unter (2024). https://ses.jrc.ec.europa.eu/complex-system-science-smart-grids. zuletzt geprüft am 02.12.2024

Lee, R.M., Assante, M.J., Conway, T.: Analysis of the cyber attack on the Ukrainian Power Grid. Defense Use Case. E-ISAC. Washington. Online verfügbar unter (2016). https://media.kasperskycontenthub.com/wp-content/uploads/sites/43/2016/05/20081514/E-ISAC_SANS_Ukraine_DUC_5.pdf. zuletzt geprüft am 02.12.2024

Lund, P.D., Lindgren, J., Mikkola, J., Salpakari, J.: Review of energy system flexibility measures to enable high levels of variable renewable electricity. Renew. Sustain. Energy Rev. **45**, 785–807 (2015). https://doi.org/10.1016/j.rser.2015.01.057

Paulus, M., Borggrefe, F.: The potential of demand-side management in energy-intensive industries for electricity markets in Germany. Appl. Energy **88**(2), 432–441 (2011). https://doi.org/10.1016/j.apenergy.2010.03.017

Richard, P., Limbacher, E-L., Engelhardt, T.: Akzeptanz und Vertrauen von Verbrauchern. Einflussgrößen, Herausforderungen und Handlungsempfehlungen für eine erfolgreiche Digitalisierung der Energiewirtschaft. Hg. v. Deutsche Energie-Agentur GmbH (dena). Berlin. Online verfügbar unter Complex System Science for Smart Grids, zuletzt geprüft am 02.12.2024 (2018)

Sakarwala, A., Beg, N.,Derendorf, K., Schuldt, F.: Multi agent control based energy optimisation of a prosumer household and a community with bidirectional electric vehicles. In: Proceedings of the International Renewable Energy Storage Conference (IRES 2022), pp. 434–450. Atlantis Press (2023)

Samweber, F., Köppl, S., Bogensperger, A., Böing, F., Bruckmeier, A., Estermann, T., et al.: Projekt MONA 2030: Abschlussbericht Einsatzreihenfolgen. Ganzheitliche Bewertung Netzoptimierender Maßnahmen gemäß technischer, ökonomischer, ökologischer, gesellschaftlicher und rechtlicher Kriterien. FfE, zuletzt geprüft am 02.12.2024 (2017)

Schlachter, H., Geißendörfer, S., von Maydell, K., Agert, C.: Load recognition in hardware-based low voltage distribution grids using convolutional neural networks. IEEE Trans. Smart Grid **15**(1), 1042–1051 (2024). https://doi.org/10.1109/TSG.2023.3280326

J. Schneider Elektrotechnik GmbH: Voltage Regulated Distribution Transformer. SmartActiveTransformer. Online verfügbar unter (2013). https://www.j-schneiderps.com/fileadmin/user_upload/j-schneider.de/J._Schneider/pdf/Prospekte/smartactive_e.pdf. zuletzt geprüft am 02.12.2024

Schwab, A.J.: Elektroenergiesysteme. Erzeugung, Übertragung und Verteilung elektrischer Energie. Springer Vieweg Berlin, Heidelberg (2017)

Styczynski, J., Beach-Westmoreland, N.: When the lights went out. A comprehensive review of the 2015 attacks on Ukrainian critical infrastructure. Hg. v. Booz Allen Hamilton Inc. Online verfügbar unter (2019). https://www.boozallen.com/content/dam/boozallen/documents/2016/09/ukraine-report-when-the-lights-went-out.pdf. zuletzt geprüft am 02.12.2024

Sullivan, J.E., Kamensky, D.: How cyber-attacks in Ukraine show the vulnerability of the U.S. power grid. Electr. J. **30**(3), 30–35 (2017). https://doi.org/10.1016/j.tej.2017.02.006

Zacharias, J.: Konkretisierung des Ampelkonzepts im Verteilungsnetz, 10.02.2017. Online verfügbar unter (2017). https://www.bdew.de/media/documents/Stn_20170210_Konkretisierung-Ampelkonzept-Verteilungsnetz.pdf. zuletzt geprüft am 02.12.2024.

Chapter 15
Assessing the Resilience of the Current and Future Transportation System

Tudor Mocanu, Sophie Nägele, and Moritz Bergfeld

Abstract The resilience of the transport system has been extensively researched in the past decades. Research has so far focused mainly on natural hazards as the main threat to the system. Current and future developments in the transport sector and beyond like automatization, digitalization, transition towards renewable energies and climate change will likely increase the complexity of the transport system and pose new challenges to ensuring the resilience of the system. Amidst these developments, the role of individual behaviour during disruptions will be crucial and needs to be better understood. In spite of the wealth of previous research on resilience, so far, no standard methodology for the assessment of the resilience at system-level could be established. In order to provide a roadmap towards such a methodology we discuss the intended functionality, structure and components of the transport system and propose several approaches on how to operationalize resilience metrics in a model-based framework.

Keywords Transportation · Transportation system · Transportation resilience · Digitalization · Sector coupling · Climate change · Travellers' behaviour · Assessment framework

15.1 Introduction

In light of the goal of many countries to reduce greenhouse gas emissions and reach carbon neutrality, the transportation system will have to undergo a comprehensive transformation. One of the key factors in achieving carbon neutrality in the transport sector is a shift towards direct electrification, i.e. the increased usage of electricity as energy source at the expense of fossil fuels (e.g. International Council on Clean Transportation 2020). The increased electrification will add new components to, but also new opportunities and challenges for the transport system, such as the charging infrastructure, vehicle-to-grid (V2G) systems and the sector coupling. This increases

T. Mocanu (✉) · S. Nägele · M. Bergfeld
Institute of Transport Research, German Aerospace Center (DLR), Berlin, Germany
e-mail: Tudor.Mocanu@dlr.de

D. van Doren et al. (eds.), *Towards Energy System Resilience*, Ethics of Science and Technology Assessment 55, https://doi.org/10.1007/978-3-032-03187-7_15

the connection to and dependency from the energy system. In the context of the energy transition, this connection is becoming even more important.

In parallel with the energy transition of the transportation system, the trend towards digitalization and an increasingly intelligent transport system will also be one of the main drivers shaping the future of this sector. Digitalization will not only manifest itself in the context of autonomous driving, but will also enable new mobility services that largely rely on digital technologies (e.g. shared mobility services such as ride- or carsharing) or improve existing services by digital traveller information systems (Bieser and Kriukelyte 2021).

The above-mentioned trends and developments will increase the complexity of the transport system and the interdependencies between its components and thus increase the importance of analysing the resilience of such a system. Although studying the resilience of the transport system is by no means a new topic and has been conducted extensively in the past couple of decades, the forthcoming transformation of the transportation system might pose some novel research questions that have not been adequately considered so far.

The aim of this chapter is to provide an overview of the current state of research on transportation resilience and transportation resilience assessment and monitoring methods. We will hereby focus on aspects related to electrification, energy transition and digitalization. We do not attempt to achieve an exhaustive review of all previous transportation resilience studies, but rather to offer a selection of papers and studies as starting point for readers interested in a more in-depth research. In light of the present knowledge regarding the resilience of the transport system and its transformation and future challenges we will also discuss the necessity for a more comprehensive assessment framework. Therefore, we will also address some of the key aspects and requirements of such a framework. Furthermore, we also attempt to identify research gaps and possible directions for future research in light of the above-mentioned trends.

This chapter is structured as followed: First we give an overview of the literature on transportation resilience research, starting with a generic definition and comprehensive review studies and continuing with a more detailed presentation of selected topics of special relevance. We then present a roadmap towards establishing a framework for transport system resilience assessment. Finally, we also identify the most relevant research topics in the context of the energy transition and digitalization and map out potential directions for future research in this field.

15.2 Previous Research on Transportation Resilience

15.2.1 Definition of Resilience in the Transport Sector: A Diverse Field

Research on the concept of resilience spans across a highly diverse set of disciplines and topics. Consequently, there have already been quite a few systematic reviews of how to define transportation resilience (see e.g. Zhou et al. 2019; Wan et al. 2018). The wealth of research so far has produced a great variety of different transportation resilience definitions and has made it difficult to find a generally accepted definition of resilience in the transport field (Wan et al. 2018). The latter authors wrote a comprehensive transportation resilience definition review and precisely describe the main differences between transportation resilience definitions. They show that different definitions have different scopes. While some focus on the whole transportation system, others focus only on specific aspects. Some describe resilience as the ability to return to a specific functioning, while others define it as a capability of the system itself. In their comprehensive review, (Wan et al. 2018) summarize transportation resilience as the ability to absorb disturbances, maintain the system functionality and recover to a certain level of service after being affected by disruptions (Wan et al. 2018).

We will subsequently use this rather generic definition of resilience as a starting point to explore several aspects in-depth in the following sections. Furthermore, in Sect. 3.2 we will give some insights into how these concepts can be operationalized and implemented.

15.2.2 Comprehensive Literature Review Studies

Previous research has investigated transportation resilience from different perspectives. For example, (Wan et al. 2018) parse the existing literature by reviewing both objectives and research methods. They show that resilience has gained an increased popularity in the transportation research field in recent years, but relatively late compared to other fields. The review shows that natural hazards are predominantly investigated as a source of disturbance and ports as research objects. The dominant research method in this field is mathematical modelling (57.3% of the selected works). This method is followed by simulation (more than one third) and conceptual works such as definitions (less than one fifth), surveys (about 10%) and case studies (about 5%). Furthermore, the authors show that 70.5% of the research is of quantitative nature and only 29.5% is qualitative (Wan et al. 2018).

Zhou et al. (2019) provide a comprehensive literature review focusing on metrics and mathematical models to measure resilience and strategies to enhance resilience. The authors show that resilience indicators can be divided into topological, attributes-based, and performance-based metrics. Measurement approaches can be divided into optimization, topological, simulation, probability theory, fuzzy logic, and data-driven models. The authors argue that the enhancement of resilience is interrelated to the previous methods, however, most of the studies only focus on one method and on only one phase of the resilience assessment (most of the time operationalized by the cycle of disaster: mitigation, preparedness, response, recovery). What is missing from the findings is research about the role of the human being in the recovery phase and the societal perspective of resilience in general, as most of the research focuses on technology. Furthermore, the authors describe another problem: previous methods can hardly be applied to complex (interdependent) transportation systems (Zhou et al. 2019).

Due to the wealth of literature other researchers focus on only one aspect of the transportation system. For example, Bešinović (2020) investigates the railway transport system resilience. She emphasizes that railway transport demand is rising, which will be followed by more congestion and more complex, interdependent systems. Due to this development as well as system failures and climate change, she expects an increasing number of disruptions. She emphasizes the need for a field specific definition of resilience and for a widely accepted method to quantify impacts of disruptions and the evolution of system performance (Bešinović 2020).

Other researchers focus on more holistic approaches. Tamvakis and Xenidis (2012) investigate key variables for a holistic design of transportation systems including resilience. They highlight the need for resilience engineering, which ensures a certain minimum operation level in case of sudden disturbances and a quick recovery potential. They also present first steps towards a new approach to assess transportation systems resilience (Tamvakis and Xenidis 2012).

15.2.3 Threats to the Transportation System

A large part of transportation resilience research focuses on various threats to the transportation system. Dominant threats are natural hazards. Among the different kinds of natural hazards investigated are hurricanes (see e.g. Yin et al. 2014; Wu et al. 2012), typhoons (Wang and Taylor 2015), or winter roads (Nævestad et al. 2021). Wu et al. (2012) investigate evacuation logistics, and more specifically the activities and resources needed to reach a safe location during hurricanes Katrina and Rita. They found that evacuees most likely choose the homes of friends or relatives as shelter, take multiple cars and rely on experience and traffic conditions to choose an evacuation route. Yin et al. (2014) use a slightly different approach, employing an agent-based travel demand model system for hurricane evacuation simulations. Wang and Taylor (2015) investigate the impact of typhoons on human mobility patterns. Even though human mobility can be influenced by typhoons, the authors show that the power law still applies to human mobility patterns, which indicates that fundamental travel patterns are resilient. Nævestad et al. (2021) focus on a less dramatic natural hazard: winter roads in Norway. They investigate how heavy goods vehicle drivers adapt to winter roads and compare foreign and Norwegian drivers. They show that foreign drivers feel less safe on Norwegian winter roads and are more likely to have accidents compared to Norwegian drivers, who also feel more confident. In summary, natural hazards threaten transportation systems and the investigation of these aspects improves the knowledge on how to make the transportation system more resilient.

Other unexpected events such as industrial disputes or strikes have also been investigated to analyse the resilience of the railway system (van Exel and Rietveld 2009). The authors investigate rail users' behaviour before and after a national strike in the Netherlands. They analyse associations between traveller and trip characteristics as well as anticipated and actual reactions. The results show that 44% of those who wanted to travel by train on the day of the strike cancelled their trip, 24% used the car as a driver instead, 14% took another mode as passenger and 18% took the train on another day. Such results can help to adapt to strikes and enhance the resilience of a specific, affected transport system. Finally, pandemic situations have also been investigated in the light of the resilience of transportation systems (Teixeira et al. 2021). The authors investigate whether the Corona pandemic influenced the use of bike sharing. They show that bike sharing is an asset for the resilience of urban transport systems especially during a health crisis as they are perceived to be capable of preserving social distancing.

15.2.4 Data Sources and Analysis Methods

Another important part of transportation resilience research concentrates on the data and methods that can be used to investigate the resilience of a transport system. Examples for data sources are mobile phone data to track down individuals (Peak et al. 2018), GPS data from probe vehicles like taxis (Donovan and Work 2017), location-based service data from Baidu Map analysed by GIS and spatial flow analysis (Chen et al. 2020), or social media data (Roy et al. 2019). These different data sources are used to investigate human mobility during disruptions in order to provide insights and knowledge for transportation system planners to prepare for potential hazards.

Different methods to quantify transportation resilience have been applied, such as pace deviations during normal versus unusual events (Donovan and Work 2017) or path length analyses (Morelli and Cunha 2021). Since many different definitions of resilience exist, just as many different quantification approaches have been developed. Consequently, Bešinović (2020) argues that there is a need for an agreed quantification of the impact of disruptions. This further illustrates the current wealth of transportation resilience research.

15.2.5 Resilience in the Context of Individual Behaviour

Disruptions require people to make quick decisions and can also trigger anxiety and fear. The evaluation of the respective situation decisively influences the individual ability to react adequately. Therefore, it is important to also investigate the effects of disruptive events on individuals.

One study investigates 15 destructive situations of five different kinds of natural disasters and their effects on mobility patterns before, during and afterwards (Wang and Taylor 2016). They show that mobility patterns are correlated during the different measurement points but not during very heavy events. This means that mobility patterns during normal state and disasters are somewhat related, but only up to a specific disruption threshold, after which the correlation declines. The authors emphasize the need for a deeper understanding of mobility during disasters to avoid negative effects for the population.

Marsden et al. (2020) emphasize the necessity to examine how the transport system and mobility patterns change when facing disruptive events. They investigate behavioural adaptations in reaction to disruptive events. The authors find seven coping strategies in response to three disruptive events, winter weather, flooding, and a bridge closure that were followed by road closures of several days to a few weeks: (1) Remoding: using a different mode of transport for at least the main leg of the trip; (2) Rerouting: taking a different route from that which was planned or would typically be taken; (3) Retiming: modifying the time at which a trip starts by either bringing it forward or pushing it back without altering where in the sequence of activities it occurs; (4) Rescheduling: changing when in the week a trip is made. This is distinct from retiming as the trip is moved in a sequence of activities; (5) Relocating: changing the destination of a journey such as shopping somewhere else; (6) Reallocating: passing over the responsibility for a journey to someone else (e.g. childcare pick up or caring trip); (7) Reducing: not conducting a trip at all but conducting the activity through ICT.

Another similar study by Parkes et al. (2016) shows that the most common behavioural responses as a reaction to overcrowding in public transport during the Olympic games in London 2012 were retiming and reducing journeys (33 and 32% of respondents respectively), compared to 19% rerouting and 14% remoding.

Zhu and Levinson (2010) find that the preference for driving a car was very strong after disruptions. Additionally, **travel experience** seems to be an important variable as new habits may emerge through experience. In another study about suburban train users, the authors identify coping mechanisms and aspects related to the decision-making process of suburban train passengers when service is interrupted or delayed (Adelé et al. 2019). They reveal that factors relating to the individual (e.g., **expertise**), the journey (e.g., **time constraints**) and also information (pre-trip or enroute communication of the disruption) affect the behaviour. They also show that travel behaviour is further influenced by **available transport services**, **car availability**, the situation in which the disruption occurs, perception of service recovery time, as well as the attitudes towards information services. For example, those with route expertise are more willing to wait, compared to those with no route expertise (van der Horst and Ettema 2005), or are more reluctant to use unknown routes (Chorus et al. 2006). If a car is available in the household, it is more likely to be used when a disaster happens (Adelé et al. 2019). Yet another study shows that **car access**, **travel time**, **travel cost**, **trip importance**, and **weather** influence whether someone will switch from public transport to car due to a public transport disruption (Nguyen-Phuoc et al. 2018).

Finally, there are also differences between the modes of transportation in terms of their robustness during disruptions. For example, **walking and cycling can be seen as more robust** compared to motorized individual transport and longer trips are more vulnerable compared to shorter trips. This raises the point that the focus on motorized individual transport in resilience planning could actually worsen the transportation system vulnerabilities (Morelli and Cunha 2021).

15.2.6 Future Trends in the Transportation System

15.2.6.1 Resilience in the Context of Automatization, Digitalization and New Transportation Services

Johnsen and Kilskar (2020) investigate the effects of **automatization** on the resilience of transportation systems. They find that automatization itself might increase vulnerabilities, for example due to human attention failures or poor human machine interaction design. The authors emphasize that one key resilience factor of automated systems is being able to handle unexpected situations. Furthermore, automated systems also have to be resilient in order to increase the trust in such new technologies. The authors point out that if automated functions are not perceived as safe and therefore resilient in face of countlessly different situations, users will not be willing to use such functions (Johnsen and Kilskar 2020).

Buinevich and Vladyko (2019) investigate the effects of **digitalization** on the resilience of transportation systems. They show that one important aspect of digitalization in the transportation sector are wireless communication technologies for intelligent transportation system applications, which represent a new potential target for cyber-attacks. Any interruption of these systems might endanger the safety of e.g. passengers and trains since such communication technologies are employed to warn drivers about obstacles, systems failures, and other interruptions. The authors emphasize the need for cyber-resilient transportation systems.

Mitusch (2021) analyses the resilience of **new transportation services**, e.g. car or bike sharing. The author concludes that, because of the more flexible usage options of these new modes they are more resilient compared to collectively used modes of transport. However, the author emphasizes that, as most of these new services are running on electricity, they are dependent on a well-functioning electricity supply. Furthermore, he points out that the optimal utilization of a pool of vehicles may imply a reduction in the total number of required vehicles compared to individual vehicle ownership. In case of a disturbance this may cause a lack of available vehicles since the number of shared vehicles is optimized for normal conditions.

15.2.6.2 Resilience in the Context of Climate Change

Markolf et al. (2019) discuss the transportation system resilience in face of climate change. Our transportation system is built on historical climate observations, but climate change is making such assumptions obsolete. The authors differentiate between direct and indirect threats. Currently, research is focussing more on direct threats and the goal is to make the system more robust. However, the authors conclude that the indirect threats are just as important to consider, and robustness does not protect the transportation system. The authors emphasize the importance of expanding the scope towards flexibility and agility above robustness to make the transportation system more resilient in face of climate change (Markolf et al. 2019).

One important aspect of a sustainable and therefore of a resilient transportation system is congestion (Afrin and Yodo 2020). However, until now research has focused on capturing "typical" circumstances, meaning the "normal" mobility (Marsden et al. 2020). The authors highlight that this leads to an overestimation of the rigidness and invariability of individuals' travel patterns. They emphasize how important it is to close this research gap, especially in face of climate change. Such research could lead to a better understanding of how flexible mobility patterns really are.

Adaptation strategies can also be observed in actions of transportation service and infrastructure companies. For example, the German national railway infrastructure operator, DB InfraGO (formerly DB Netz), has implemented a natural hazard strategy to adapt to the effects of climate change. They analyse dependencies and relations between different effects, economical aspects and derive the need for action (Meßenzehl 2019). However, due to climate change, weather will become less predictable and more variant, which will in turn lead to less accurate forecasts and forewarnings (Mitusch 2021).

15.2.6.3 Resilience in the Context of Energy Transition and Electrification of the Transportation System

The electrification of the transport sector and therefore the coupling to the energy sector brings various challenges and opportunities in relation to the resilience of both sectors. Brown and Soni (2019) survey members of the U.S. Department of Energy's Electricity Advisory Committee to assess the market potential and challenges from the integration of vehicles in the electricity grid. The authors examine the survey results in combination with a literature review, current policies and the ecosystem of stakeholders. They find that grid-to-vehicle integration is the most valuable mode. Furthermore, the management of transport and energy as complementary sectors could uncover the benefits for grid resilience. A coordinated policy approach involving multiple sectors and infrastructures can support complementarities across sectors and significantly contribute to network resilience. The main barriers to grid-integrated EVs are operational and administrative in nature (Brown and Soni 2019).

This knowledge raises the question of how electric vehicles could be integrated and utilized into the grid system during emergency events to enhance resilience. Hussain et al. (2020) investigate microgrids and energy management during emergency cases. Within their research they develop a communication-based energy management system for emergency cases. This approach provides interoperability and plug and play solutions. Because EVs are spatially distributed and portable storage systems, they are of high value in emergency situations. The energy stored in the vehicles can be used until help arrives. To enable different energy system components to work together in such cases, standards must be expanded and combined with smart grid or smart city concepts (Hussain et al. 2020).

If disasters can be predicted like extreme weather events, the benefits on resilience from electric vehicles can be enhanced. Roudbari et al. (2021) investigate the opportunities to enhance resilience by rescheduling resources and charging/discharging patterns before and during extreme weather events. The effectiveness could be demonstrated through several case studies.

On the other side it is necessary to consider the resilience of electric vehicles when it comes to emergency situations. Hussain and Musilek (2022) analyse existing literature to discuss the use of electric vehicles as resilience resource and the resilience of electric vehicles themselves. They propose different strategies and techniques to enhance the resilience of electric vehicles during disasters. Prior to an event, resilience can be increased by including storage capacity in the charging infrastructure or building renewably powered off grid charging stations. During disruptions the allocation of power to EVs and along evacuation routes has an impact on mobility and possible evacuation options. However, in addition to the V2G issue (see above), there are other issues that need to be addressed, such as coordination among stakeholders, standardization, installation of charging stations, and shaping public policy to include electric vehicles (Hussain and Musilek 2022).

15.3 Towards a Framework for the Assessment of the Transportation System Resilience

After having presented the main directions of research in the context of transport system resilience we now turn our attention to the question of how the resilience of this system can be assessed and quantified. Unsurprisingly, there is no readily available state-of-the-art or even state-of-practice methodology for this task, since assessing the resilience of the entire transport system has not been so far up on the agenda of policy makers and practitioners so far. However, as previously discussed in Sect. 15.1, the impending transformation of the transport system and the associated new challenges will render this assessment considerably more pertinent in the future.

The following sections will present several fundamental aspects that need to be considered in developing a roadmap towards a comprehensive and future-proof resilience assessment framework. We start by undertaking a more detailed examination of the transport system, followed by an analysis of conceptual transport resilience indicators and how they can be operationalized. Finally, we present a possible implementation of resilience assessment using model-based analyses.

15.3.1 Scope and Boundaries of the Transportation System

In order to assess the resilience of any system, it is imperative to have a clear understanding of the intended system functionality, system components and boundaries. The transportation system is surprisingly difficult to describe in unambiguous and comprehensive terms. For instance, one definition refers to the transport system rather generically as a set of multiple elements that interact with each other and produce both a certain demand for travel as well as the transportation supply (services) required to meet this demand (Cascetta 2009).

A system can first be described in terms of its intended functionality. In a nutshell, the purpose of the transport system is to ensure the mobility of persons and goods (and according to some wider definitions also of information) at an acceptable level of service and costs. Several implications for resilience analyses derive from this high-level definition. Firstly, it is important to consider both the supply and the demand and how well they match together. Secondly, the performance of the system should be assessed from the perspective of the end users, as there is no universal or objective definition of what are "acceptable" levels of service or costs associated with travel. Furthermore, demand for (passenger) transport has traditionally been considered as being derived from the necessity of carrying out activities at other destinations (Mokhtarian and Salomon 2001). An interesting question that arises from this consideration is whether or not the provision of alternative destinations, locations or substitutes for the originally planned activities in case of disruptions still falls within the boundaries of the transport system. A pertinent example for this discussion was the sudden increase in working from home in response to the onset of the COVID-19 pandemic. This drastically reduced the demand for commuting while still enabling the originally planned activities (working) to be carried out. While such a reaction is a clear indication of resilience, we consider that in this case it was not the transportation system itself but actually the "societal system" as a whole that has shown resilience, since enabling people to work is not the primary function of the transportation system. Nevertheless, this example demonstrates the extent to which the transportation system is integrated within other, broader systems.

Another way of describing the transportation system that is useful for resilience analyses is to consider its components and structure. (Tamvakis and Xenidis 2012) propose a two-layered definition of the transportation system where the inner layer consists of six physical components (infrastructure, vehicles, equipment, power systems, fuel, communication systems) and the outer layer consists of external stakeholders (e.g. government, general public, financial community etc.). This is a comprehensive definition that appears to include several relevant aspects in addition to the most obvious core elements (e.g. infrastructure and vehicles).

For the purpose of resilience analyses it might be more beneficial to expand the system definition proposed by Tamvakis and Xenidis (2012) by structuring the transportation system and its components according to different kinds of vulnerabilities and possible kinds of reactions with respect to that. For example, let's consider the physical component "Infrastructure". This component has several vulnerabilities, one example being that roads can be blocked by heavy snowfall. The factors and subsystems that influence the mitigation and adaptation reactions are equally important. In our example this could be how well-organized the snow plough operation is and how quickly the original functionality of the system component (e.g. a clean, snow-free road) can be restored. This part of the transportation system could be termed as "recovery support components". The concept can be applied to any of the key system components and its existence and performance is crucial in the assessment of the system resilience. As with the aspect of providing alternative destinations discussed above, the outer boundary of the transportation system cannot easily be defined and it is debatable up to which point these recovery support components should still be included in the definition. For the purposes of resilience system analyses, however, they should at least be considered.

Based on these considerations we propose an updated, end user-focused description and structure of the transportation system well suited for resilience analyses, which includes the following components (Fig. 15.1):

- Core:
  - *Individuals and end users*: the actual "beneficiaries" of the transportation system that need to get from A to B

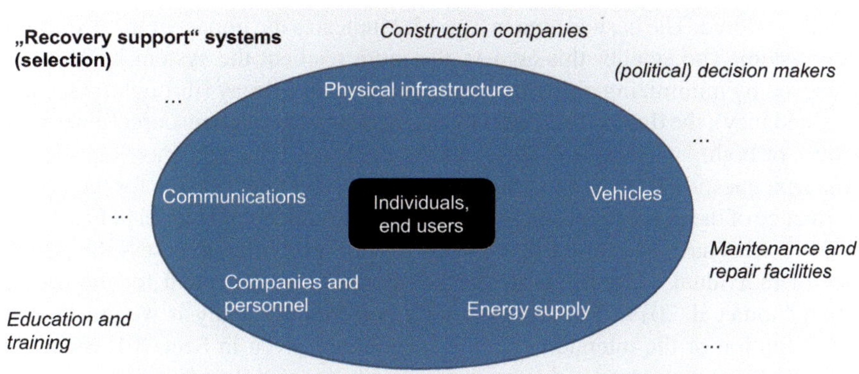

Fig. 15.1 Updated transportation system structuring for resilience analyses

- Key system components:
  - *Physical infrastructure*: the different transport networks (road, rail, waterways, stations, terminals, ports etc.)
  - *Vehicles*: the moving vessels (cars, trains, airplanes etc.)
  - *Companies and personnel*: the providers of transportation services (logistics companies, public transport companies, drivers etc.)
  - *Energy supply*: petrol stations, power plant operators, grid operators etc.
  - *Communications*: Mobile networks, 5G providers, internet platforms etc.
- Recovery support:
  - all systems (technical, procedures, companies, institutions) that act towards restoring the full functionality of any of the key system components in case of disruptions (e.g. construction companies repairing damage to the physical infrastructure or repair facilities repairing damage to vehicles etc.).

15.3.2 Operationalization of Resilience Concepts

With a better understanding of the purpose, scope and structure of the transport system we can now turn our attention to how its resilience can be quantified and assessed. Implementing the generic definition of resilience given in Sect. 2.1 into a framework for quantitative analyses is not straightforward. A useful, generic concept is the so-called "resilience triangle" that was originally proposed by Bruneau et al. (2003) for the evaluation of infrastructure resilience. This simplified conceptual model, as illustrated in Fig. 15.2, considers the time-varying performance of the (transportation) system before, during and after a disruptive event. In the immediate aftermath of the disruption (t0), the performance decays. The robustness of the system and its redundancies determine the extent of the decay. After the minimum of performance has been reached (t1), the system enters the recovery phase where the functionality is gradually restored. The hashed area in Fig. 15.2 indicates the impact of the disruption on the system. The smaller this area is, the more resilient the system is. This can be achieved by minimizing either the initial performance decay (through robustness and redundancy), the time to full performance recovery (through resourcefulness and rapidity), or both.

The next question is, then, what constitutes an appropriate measure for the overall performance of the transport system (the y-axis of the conceptual model in Fig. 15.2). Various metrics have been used both at sub-system levels (i.e. to assess the performance of individual components of the transportation system) and for the overall system (Zhou et al. 2019). For a system-wide performance analysis we must return to the definition of the intended system functionality given in Sect. 3.1. Since the purpose of the transportation system is to provide accessibility at minimal cost, it is precisely this metric which should form the basis for the resilience assessment.

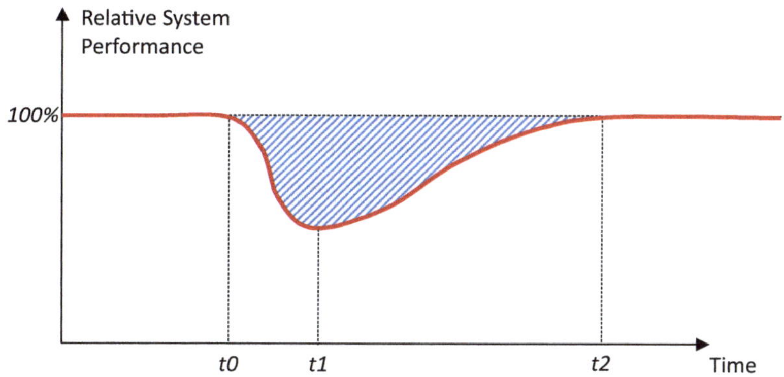

Fig. 15.2 Two phases of resilience measurement, according to Zhou et al. (2019)

This can be implemented into two relatively straight-forward metrics based on the underlying travel demand:

1. Percentage of serviced demand, i.e. ideal performance of the system is achieved when all trips and movement can be carried out as originally planned, in spite of any disruptions.
2. Impedance index, i.e. a comparison of the current total impedance in the network with a best-case (undisrupted) scenario (see e.g. Twumasi-Boakye and Sobanjo 2018). The most relevant impedance metric is travel time, but measures of generalized cost including monetary and/or other types of impedances are also possible.

For sub-components of the transportation system, other indicators are required. For example, access and availability of charging infrastructure for battery-powered electric vehicles is an indicator worth considering in the light of electrification of transport. Since the charging infrastructure depends on the power grid cross-sectoral effects must be considered.

It should be noted here that this operationalization of resilience analysis is an event-based assessment framework, i.e. it requires consideration of possible disruptions to the system. This is a useful approach in many cases, but there are also some drawbacks to consider here. Firstly, the assessment results will depend heavily on the exact disruption(s) considered so care should be taken when selecting the stressors. Ideally, for a comprehensive system analysis a wide-ranging set of disruptions should be investigated in order to avoid selection bias. Furthermore, when assessing future system configurations that greatly differ from the current one (e.g. a fully automatized and digitalized transport system as indicated in Sect. 2.6.1), it might not be so obvious at a first glance which disruptions are the most relevant. In some cases, resilience-enhancing measures for one disruption might also increase the vulnerability to other disruptions (e.g. digitalization might help in natural catastrophes but will also increase vulnerability to a hacker attack). Therefore, the event-based assessment could be supplemented by a structural one.

Structural indicators are generally employed to describe a system's robustness and redundancy. In transportation system analyses, such indicators have been developed mainly to describe the topology of (road) networks using graph theory concepts and include metrics such as the network diameter, average shortest path, average node degree, size of giant component etc. (Zhou et al. 2019). Another relevant structural indicator might be the total excess capacity, i.e. a network-wide aggregation of the difference between the maximum capacity of each link and the regular (undisturbed) traffic volumes. However, such indicators are difficult to apply in the context of multi-modal, system-wide analyses investigating disruptions more complex than singular link outages in simple networks.

One promising theoretical approach comes from microeconomics and has been employed in transportation in the context of choice modelling. The expected maximum utility, or logsum term, is a measure of consumer surplus in choice situations (see e.g. Jong et al. 2005). The greater the number of alternatives included in the choice set, the higher the logsum will be. From the perspective of the end user (e.g. a traveller), the transportation system can also be seen as a series of choices (e.g. destination, travel mode, route etc.). It should thus be possible to calculate logsums for each choice situation and aggregate them. By comparing the (aggregated) logsum to the actual utility of the (modelled) best or preferred alternative, it appears theoretically possible to evaluate the redundancy of the system. A more redundant system will have a higher difference between the logsum term and the best singular alternative. However, further research is required to operationalize this concept.

15.3.3 Model-Based Analyses for Resilience Assessment

Building upon the aspects discussed in Sects. 3.1 and 3.2, it becomes clear that establishing a framework for the comprehensive assessment of the resilience of current and future transport system is not a trivial task. This framework should include all relevant parts of the system (see Fig. 15.1) and enable the time-dependent, dynamic assessment of their performance. Based on this, indicators for both the event-based and the structural assessment discussed in Sect. 3.2 should be derived. In the following we will attempt to lay out a roadmap towards establishing such a framework.

Firstly, a proper model-based tooling should be developed. The starting point for this tool could be provided by already existing transport models and simulations. With such models it is already possible to compute the system performance metrics for any given timestep in Fig. 15.2. Conventional transport models usually consider both transport supply (e.g. the capacity and other characteristics of the road network, public transport scheduling etc.) and its impact on travel demand. If changes are made on the supply side, e.g. by reducing the capacities or removing certain links from the network altogether, then this could have an impact on demand at various steps in the transport model. Depending on the type of model and its sensitivity, such changes could be seen in terms of route choice and flow re-routing, mode choice, destination choice etc. By comparing the model results from disruption and undisrupted scenarios it is possible to calculate the impedance indices presented in Sect. 3.2. If the point in time with the lowest system performance (t_1) is selected and compared to the undisrupted system performance, the results could also be interpreted as an analysis of the system's robustness.

In more complex, so called "activity-based" travel demand models it is also possible to model changes in the activity patterns of individuals as a result of enhanced or degraded accessibility and thus also directly calculate the percentage of serviced demand. To assess future developments and cross-sectoral effects it may also be necessary to build interfaces between specialized models. Since the spatial and temporal resolution of different models might be very different a robust methodology needs to be established. Models with a higher temporal resolution might use a slice of temporal data from models with a lower temporal resolution. An interpolation methodology, as described later on, might help to build a sufficient interface. However, care should be taken when employing the models for such purposes because most of them have originally been developed and validated for long-term, strategic modelling.

The second step in the development of the model-based tooling should be to move from a static to a dynamic approach that is able to consider time-varying aspects before, during and after disruptions. Several approaches could make this possible. The first one is by employing dynamic simulations instead of conventional, static travel demand models. Dynamic simulations consider an explicit time axis and model the changes in the state of subjects or agents during the simulated period. Such simulations have been developed with the main focus on increasing the accuracy of the network assignment step, since there are obvious benefits in considering the dynamics of traffic flows and queues in a network. However, even in these dynamic simulations the travel demand (deriving from trip generation, destination, mode and time of day choice) is static and is not dynamically re-evaluated during the simulation. Understanding e.g. how travellers alter their activity plans spontaneously during a disruption is an open research question that warrants further investigation.

In transport models, the typical simulation period is one day (24 h), although there are also model implementations of week-long simulations (see e.g. Ordonez Medina et al. 2012). Depending on the disruptions to the transport system considered, it might be necessary to greatly expand this timeframe. This is especially relevant in the context of damages to the transport infrastructure, since repairing them could take months or even years. Performing continuous simulations for such extended periods of time is obviously not feasible. Instead, another approach could be to run multiple static models for several discrete points in time and then interpolate the results in between. This approximation would greatly simplify the modelling process and achieve an optimal trade-off between accuracy and computational efficiency by selecting the appropriate time intervals between the discrete simulations.

In both of the approaches mentioned above, it is important to also include the performance and behaviour of the relevant "recovery support" components of the transportation system (see Fig. 15.1). This goes beyond the scope of traditional transport models so new approaches will need to be developed for this purpose.

One obvious approach could be to make assumptions regarding the recovery support systems based on their observed performance during previous disturbances and stress situations. This simple approach could work well enough when there is sufficient historical data to base the assumptions on and when the interactions between different systems are negligible. For example, if a bridge has been destroyed by an extreme weather event, then the time required for reconstruction could be estimated from previous experience with such occurrences. However, if the construction companies (as an example of "recovery support" systems) are themselves affected by other disruptions, e.g. shortage of materials due to unavailable supply lines or budgetary and manpower constraints, then their performance might differ substantially from previous observations. Furthermore, the performance of recovery support actions for future technologies cannot be derived from historical data. This is illustrated by the example of electrification of transportation, where vehicles rely on electric energy, which is a different kind of infrastructure. Another example is autonomous driving, which could have a huge impact on travel patterns and vehicle movements. In these cases, it appears necessary to explicitly model the behaviour of the recovery support agents, e.g. by using (simplified) agent-based models or system dynamics models (Li et al. 2020).

Such a model-based tool for resilience analysis could primarily be used to assess the performance of the system by selecting relevant disruption scenarios and stress cases, simulating and analysing the results. By identifying those system features over which planners have an active influence on, the tooling can then also be used for design purposes. By finally running multiple simulations with different design parameters and options, it should be possible to identify some form of optimal system setup that maximizes resilience.

Figure 15.3 summarizes the proposed model-based resilience assessment framework. Starting from already existing transport models and simulations, an enhanced stress case evaluation tool will have to be established according the guidelines presented above. The utilization of this tool for the assessment of relevant stress cases will enable planners to monitor and better consider aspects of resilience when designing the future transportation system.

15.4 Conclusion

Research on the resilience of the transportation system has been conducted intensively during the last decades. There has been a specific focus on natural hazards and evacuations, but resilience has also been investigated from other perspectives. Latest developments in automatization, digitalization in the transport and energy transition and climate change have led to new research topics in the transport sector concerning resilience. Furthermore, these aspects are related to the individuals' behaviour, which is also highly relevant in the context of resilience.

While there is a wealth of research on current and future threats to the transportation system, up until now only a very limited focus has been placed on developing

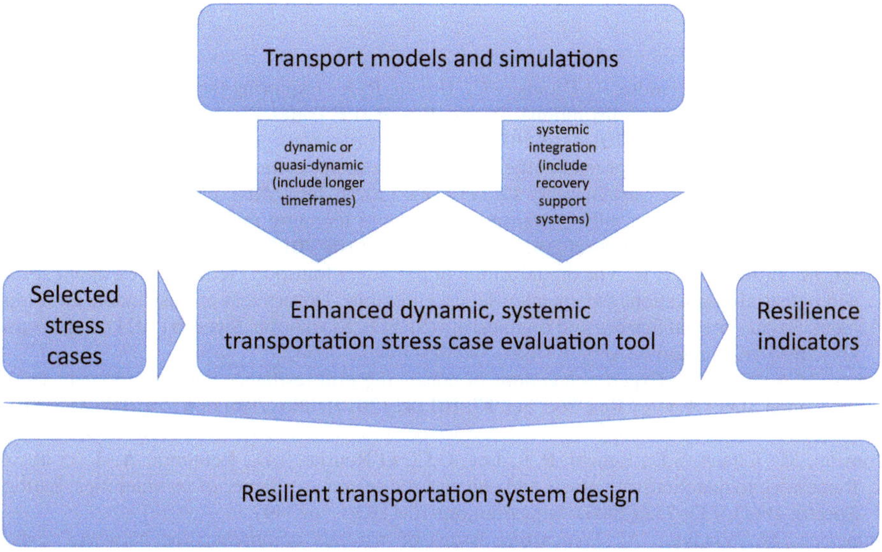

Fig. 15.3 Model-based resilience assessment framework

instruments and frameworks that allow planners and decision-makers to assess the resilience of the transportation system as a whole. Ideally, a comprehensive assessment framework should enable decision-makers to quantify both the costs and the benefits of a resilient transportation systems and take system design-related decisions accordingly.

This chapter has presented a vision of what such an integrated and comprehensive framework could look like in the future. For this vision to become reality and state-of-practice, research must now focus on developing practical and easily implementable tools that consider both short and long-term impacts on all resilience-related aspects of the transportation system, including the recovery support systems. The most obvious starting point in this endeavour are currently existing transport models and simulations, which can already be implemented to investigate the robustness and redundancy of several important parts of the transportation system, such as the transport networks. This should at least allow us to quickly identify the main vulnerabilities in our current system. Finding a quick solution for these vulnerabilities will, however, not be so straight-forward, especially in view of the often limited resources (financial, but also e.g. in terms of space in dense urban areas or shortage of skilled labour force available for construction work etc.) to expand our transportation systems. This will require an informed prioritization of policy interventions and optimal allocation of resources, for which a comprehensive resilience assessment framework, as proposed in this chapter, is an indispensable prerequisite.

References

Adelé, S., Tréfond-Alexandre, S., Dionisio, C., Hoyau, P.-A.: Exploring the behavior of suburban train users in the event of disruptions. Transp. Res. Part F Traffic Psychol. Behav. **65**, 344–362 (2019). https://doi.org/10.1016/j.trf.2019.08.009

Afrin, T., Yodo, N.: A survey of road traffic congestion measures towards a sustainable and resilient transportation system. Sustainability **12**(11), 4660 (2020). https://doi.org/10.3390/su12114660

Bešinović, N.: Resilience in railway transport systems: a literature review and research agenda. Transp. Rev. **40**(4), 457–478 (2020). https://doi.org/10.1080/01441647.2020.1728419

Bieser, J.C.T., Kriukelyte, E.: The digitalization of passenger transport. Technologies, applications and potential implications for greenhouse gas emissions. Department of Sustainable Development, Environmental Science and Engineering (SEED). Stockholm, Sweden (2021). https://urn.kb.se/resolve?urn=urn%3Anbn%3Ase%3Akth%3Adiva-305425

Brown, M.A., Soni, A.: Expert perceptions of enhancing grid resilience with electric vehicles in the United States. Energy Res. Soc. Sci. **57**, 101241 (2019). https://doi.org/10.1016/j.erss.2019.101241

Bruneau, M., Chang, S.E., Eguchi, R.T., Lee, G.C., O'Rourke, T.D., Reinhorn, A.M., et al.: A framework to quantitatively assess and enhance the seismic resilience of communities. Earthq. Spectra **19**(4), 733–752 (2003). https://doi.org/10.1193/1.1623497

Buinevich, M., Vladyko, A.: Forecasting issues of wireless communication networks' cyber resilience for an intelligent transportation system: an overview of cyber attacks. Information **10**(1), 27 (2019). https://doi.org/10.3390/info10010027

Cascetta, E.: Transportation Systems Analysis, vol. 29. Springer US, Boston, MA (2009)

Chen, Z., Gong, Z., Yang, S., Ma, Q., Kan, C.: Impact of extreme weather events on urban human flow: a perspective from location-based service data. Comput. Environ. Urban Syst. **83**, 101520 (2020). https://doi.org/10.1016/j.compenvurbsys.2020.101520

Chorus, C.G., Molin, E.J.E., van Wee, B.: Use and effects of Advanced Traveller Information Services (ATIS): a review of the literature. Transp. Rev. **26**(2), 127–149 (2006). https://doi.org/10.1080/01441640500333677

Donovan, B., Work, D.B.: Empirically quantifying city-scale transportation system resilience to extreme events. Transp. Res. Part C Emerg. Technol. **79**, 333–346 (2017). https://doi.org/10.1016/j.trc.2017.03.002

Hussain, A., Musilek, P.: Resilience enhancement strategies for and through electric vehicles. Sustain. Cities Soc. **80**, 103788 (2022). https://doi.org/10.1016/j.scs.2022.103788

Hussain, S.M.S., Aftab, M.A., Ali, I., Ustun, T.S.: IEC 61850 based energy management system using plug-in electric vehicles and distributed generators during emergencies. Int. J. Electr. Power Energy Syst. **119**, 105873 (2020). https://doi.org/10.1016/j.ijepes.2020.105873

International Council on Clean Transportation: Vision 2050: a strategy to decarbonize the global transport sector by mid-century (2020). https://theicct.org/wp-content/uploads/2021/06/ICCT_Vision2050_sept2020.pdf. Accessed 28 Oct 2024

Johnsen, S.O., Kilskar, S.S.: A review of resilience in autonomous transport to improve safety and security. In: Baraldi, P., Di Maio, F., Zio, E. (eds.) Proceedings of the 30th European Safety and Reliability Conference and 15th Probabilistic Safety Assessment and Management Conference. Proceedings of the 29th European Safety and Reliability Conference (ESREL), 01.11.2020 - 05.11.2020, pp. 5059–5066. Research Publishing Services, Singapore (2020)

de Jong, G., Pieters, M., Daly, A., Graafland-Essers, I., Kroes, E., Koopmans, C.: Using the logsum as an evaluation measure: literature and case study. RAND Corporation, Santa Monica, CA (2005). https://www.rand.org/pubs/working_papers/WR275.html

Li, G., Kou, C., Wang, Y., Yang, H.: System dynamics modelling for improving urban resilience in Beijing, China. Resour. Conserv. Recycl. **161**, 104954 (2020). https://doi.org/10.1016/j.resconrec.2020.104954

Markolf, S.A., Hoehne, C., Fraser, A., Chester, M.V., Underwood, B.S.: Transportation resilience to climate change and extreme weather events – beyond risk and robustness. Transp. Policy **74**, 174–186 (2019). https://doi.org/10.1016/j.tranpol.2018.11.003

Marsden, G., Anable, J., Chatterton, T., Docherty, I., Faulconbridge, J., Murray, L., et al.: Studying disruptive events: Innovations in behaviour, opportunities for lower carbon transport policy? Transp. Policy **94**, 89–101 (2020). https://doi.org/10.1016/j.tranpol.2020.04.008

Meßenzehl, K.: Das Naturgefahrenmanagement der DB Netz AG. Vorbereitung auf den Klimawandel. Deine Bahn **47**, 16–22 (2019)

Mitusch, K.: Resilienz von Verkehrssystemen unter besonderer Berücksichtigung des Klimawandels (2021). https://www.forschungsinformationssystem.de/servlet/is/543583/. Accessed 31 Oct 2024

Mokhtarian, P.L., Salomon, I.: How derived is the demand for travel? Some conceptual and measurement considerations. Transp. Res. Part A Policy Pract. **35**(8), 695–719 (2001). https://doi.org/10.1016/S0965-8564(00)00013-6

Morelli, A.B., Cunha, A.L.: Measuring urban road network vulnerability to extreme events: an application for urban floods. Transp. Res. Part D Transp. Environ. **93**, 102770 (2021). https://doi.org/10.1016/j.trd.2021.102770

Nævestad, T.-O., Sagberg, F., Levlin, G., Bjørnskau, T.: Competence, equipment and behavioural adaptation on Norwegian winter roads: a comparison of foreign and Norwegian HGV drivers. Transp. Res. Part F Traffic Psychol. Behav. **77**, 257–273 (2021). https://doi.org/10.1016/j.trf.2021.01.004

Nguyen-Phuoc, D.Q., Currie, G., De Gruyter, C., Young, W.: How do public transport users adjust their travel behaviour if public transport ceases? A qualitative study. Transp. Res. Part F Traffic Psychol. Behav. **54**, 1–14 (2018). https://doi.org/10.1016/j.trf.2018.01.009.

Ordonez Medina, S.A., Erath, A., Axhausen, K.W.: Simulating urban transport for a week time horizon (2012)

Parkes, S.D., Jopson, A., Marsden, G.: Understanding travel behaviour change during mega-events: lessons from the London 2012 Games. Transp. Res. Part A Policy Pract. **92**, 104–119 (2016). https://doi.org/10.1016/j.tra.2016.07.006

Peak, C.M., Wesolowski, A., zu Erbach-Schoenberg, E., Tatem, A.J., Wetter, E., Lu, X., et al.: Population mobility reductions associated with travel restrictions during the Ebola epidemic in Sierra Leone: use of mobile phone data. Int. J. Epidemiol. **47**(5), 1562–1570 (2018). https://doi.org/10.1093/ije/dyy095

Roudbari, A., Nateghi, A., Yousefi-khanghah, B., Asgharpour-Alamdari, H., Zare, H.: Resilience-oriented operation of smart grids by rescheduling of energy resources and electric vehicles management during extreme weather condition. Sustain. Energy Grids Netw. **28**, 100547 (2021). https://doi.org/10.1016/j.segan.2021.100547

Roy, K.C., Cebrian, M., Hasan, S.: Quantifying human mobility resilience to extreme events using geo-located social media data. EPJ Data Sci. **8**(1) (2019). https://doi.org/10.1140/epjds/s13688-019-0196-6

Tamvakis, P., Xenidis, Y.: Resilience in transportation systems. Procedia Soc. Behav. Sci. **48**, 3441–3450 (2012). https://doi.org/10.1016/j.sbspro.2012.06.1308

Teixeira, J.F., Silva, C., Moura e Sá, F.: The motivations for using bike sharing during the COVID-19 pandemic: insights from Lisbon. Transp. Res. Part F Traffic Psychol. Behav. **82**, 378–399 (2021). https://doi.org/10.1016/j.trf.2021.09.016

Twumasi-Boakye, R., Sobanjo, J.O.: Resilience of regional transportation networks subjected to hazard-induced bridge damages. In J. Transp. Eng. Part A Syst. **144**(10), Article 04018062 (2018). https://doi.org/10.1061/JTEPBS.0000186

van der Horst, R.S.N., Ettema, D.: Use of travel information and effects on mode choice for recreational trips. In: Transportation Research Board 84th Annual Meeting (2005)

van Exel, N.J.A., Rietveld, P.: When strike comes to town…anticipated and actual behavioural reactions to a one-day, pre-announced, complete rail strike in the Netherlands. Transp. Res. Part A Policy Pract. **43**(5), 526–535 (2009). https://doi.org/10.1016/j.tra.2009.01.003

Wan, C., Yang, Z., Zhang, Di., Yan, X., Fan, S.: Resilience in transportation systems: a systematic review and future directions. Transp. Rev. **38**(4), 479–498 (2018). https://doi.org/10.1080/01441647.2017.1383532

Wang, Q., Taylor, J.E.: Resilience of human mobility under the influence of typhoons. Procedia Eng. **118**, 942–949 (2015). https://doi.org/10.1016/j.proeng.2015.08.535

Wang, Q., Taylor, J.E.: Patterns and limitations of urban human mobility resilience under the influence of multiple types of natural disaster. PLoS ONE **11**(1), e0147299 (2016). https://doi.org/10.1371/journal.pone.0147299

Wu, H.-C., Lindell, M.K., Prater, C.S.: Logistics of hurricane evacuation in Hurricanes Katrina and Rita. Transp. Res. Part F Traffic Psychol. Behav. **15**(4), 445–461 (2012). https://doi.org/10.1016/j.trf.2012.03.005

Yin, W., Murray-Tuite, P., Ukkusuri, S.V., Gladwin, H.: An agent-based modeling system for travel demand simulation for hurricane evacuation. Transp. Res. Part C Emerg. Technol. **42**, 44–59 (2014). https://doi.org/10.1016/j.trc.2014.02.015

Zhou, Y., Wang, J., Yang, H.: Resilience of transportation systems: concepts and comprehensive review. IEEE Trans. Intell. Transp. Syst. **20**(12), 4262–4276 (2019). https://doi.org/10.1109/TITS.2018.2883766

Zhu, S., Levinson, D.M.: A review of research on planned and unplanned disruptions to transportation networks. In: Transportation Research Board 89th Annual Meeting (2010)

Chapter 16
Systems Consideration for Resilience Monitoring

Bert Droste-Franke

Abstract The task which is tackled with designing future energy systems is to build up a combination of technical facilities which can be used in future to provide the demand of usable energy predominantly from renewable energy sources. One major key to do this successfully is systems knowledge of intercorrelations between technical elements, environment, and society. Against this background in this chapter insights from systems analyses needed for policy support, basic theoretical systems views, derivation of fundamental systems characteristics for resilience, empirical findings from the Ahr-flood disaster, thoughts for a meaningful combination of systems knowledge, and practical requirements for the provision of applicable knowledge for action and for evaluating alternative options are gathered and put together. The results are an in-depth analysis of the field of energy systems analyses for resilience modelling and basic recommendations for building up a stress test tool as a nucleus for further endeavours in this direction.

Keywords Energy systems · Transport systems · ICT systems · Systems analysis · Resilience modelling · Systems theory · Stress test tool · Ahr-flood disaster

16.1 Introduction

The purpose of analyses in this project is to provide support for designing the digitalisation of future energy systems in a way that they are resilient or hardly vulnerable in catastrophic situations.

This endeavour comes with various challenges. In order to be applicable for monitoring, the analyses should best be carried out in parallel to the development of future energy scenarios which is continuously done anyway. Second best solution is that they rely on scenarios which have already been prepared, primarily, with respect to aim at the reduction of greenhouse gas emissions or a reduction in impacts on

B. Droste-Franke (✉)
IQIB - Institut für qualifizierende Innovationsforschung und -beratung GmbH, Bad Neuenahr-Ahrweiler, Germany
e-mail: bert.droste-franke@iqib.de

D. van Doren et al. (eds.), *Towards Energy System Resilience*, Ethics of Science and Technology Assessment 55, https://doi.org/10.1007/978-3-032-03187-7_16

the environment, resource availability and society. These are usually already available, regularly updated, and used to derive policy recommendations for the design of future energy systems. Thus, such scenarios need to be applicable, at least as fundamental basis, in the modelling exercises to analyse their performance in catastrophic situations to derive meaningful results.

Against this background in this chapter, first, theoretical requirements are discussed which need to be considered for the design of meaningful analyses. Second, the multiple characteristics of systems analyses are evaluated with respect to their design to derive expressive systems knowledge with respect to resilience of energy related systems. This part starts with theoretical understanding of systems and their modelling, stressing out the need to look at socio-technical systems with specific challenges, analysing important energy related systems, resulting with a proposal for a meaningful combination of systems and their models for resilience analyses. Third, practical requirements for the applicable knowledge and evaluation of alternatives supported by systems analyses are elaborated, before, in a fourth step, workflows and integrative tools are outlined. Finally, the findings are concluded in a set of recommendations for the realisation of supporting systems analyses for resilience monitoring.

16.2 Theoretical Requirements for Meaningful Analyses

In order to be valuable in the design process of future energy systems, recommendations drawn from analyses need to be applicable for policy decisions. This aim comes with several challenges. In order to bare the most important challenges in mind, the requirements for systems analyses for policy support elaborated by Droste-Franke et al. (2015) will shortly be discussed in the following.

It is previously pointed out that the analysed systems of interest are not only technological, but socio-technical systems which show large interconnections to their environment.[1] Thus, societal aspects including economics as well as impacts on the natural environment as source and sink need to be considered beside the technological system. Furthermore, physical infrastructure and technical material characteristics and laws of natural science may need to be considered in several contexts which are often represented only in a simplified way.

The quality of analyses and, thus, their results depend on the one hand on the consideration of all problem relevant knowledge (epistemic robustness) and on the other hand on the consideration of all important valuations (social robustness). In order to elaborate problem solutions, available knowledge needs to be problem-related. This is not automatically the case for scientific knowledge. While academia thrives with understanding certain circumstances often mono-disciplinary, isolated in laboratories, and without influence from the environment, problem solutions need

[1] See Chap. 2 of this book.

to deal with multiple, often uncertain or even unknown influences from the environment. Thus, knowledge from multiple disciplines fitting to the purpose is required, including even practical knowledge to derive reliable solutions. In some cases, even hardly qualifiable local empirical knowledge is key to develop applicable problem solutions. The success of solutions will, furthermore, be evaluated with respect to non-epistemic societal categories like efficiency, sustainability etc., which needs to be considered. But, to set the right focus, also the analyses themselves need to concentrate on problem solving which often requires the scientists to be aware that they may need to refrain to a certain extent from disciplinary epistemic aims of academic research. Furthermore, the actors in the process need to be aware that they may work under close scrutiny of politics or even general public. Thus, they need to be prepared to defend own or overall results differently than in the scientific process, being cautious about unintentionally compromising other scientists or even undermining credibility of results from science in total, which could be the effect, only because the public is not used to scientific discussion culture and the press uses controversies to raise attention to its work. Providing results in understandable adequate formats and transparency in argumentation chains are the key here. A sharp distinction between descriptive and normative elements could help, as well as a distinction of knowledge for action, system knowledge and knowledge about the societal valuation of relevant impacts. An additional focus needs to be laid on, partly implicit, normative elements to be adequate, particularly setting preferences on false-positives and false-negatives in a way which is consistent with the major societal preferences in the problem context. For instance, if we set the clear focus on protecting the environment in a certain situation, we need to set standards on the acceptance of a hypothesis which prefer falsely rejecting a potentially harmful action or substance (false negatives) and not falsely accepting them (false positives).

16.3 Ensuring Adequate Systems Knowledge by Systems Analyses

16.3.1 Systems, Entities and Modelling

For the resilience analyses in the area of energy systems, systems knowledge is an important component. This is the reason why we thoroughly analyse related systems or subsystems here, starting with a very general concept of systems.

We follow here the definition by Ropohl (2012). According to this concept, a system is defined as the form in which arbitrary circumstances of reality can be described and explained. "Entities" in this context can be described as coherent parts in reality (object language) and can be differentiated from the "system" as a meta language about entities.

In practice, all systems can then be defined as set of elements and set of relations (or functions) between the elements.[2]

Taking a closer look at this definition reveals that systems are, thus, a form of description which can be used to explain changes of elements or their characteristics due to changes in other elements mediated by interconnections which can be experienced and described as cause-effect chains.

One important issue in the definition of specific systems is to fix their extension or coverage and to distinguish them from their environment. Such a differentiation represents also the basis to define open and closed systems. Real entities stand out in comparison to controlled entities in laboratories with regard to the requirement to be described as open systems which are in certain exchange with their environment.[3]

Often not even the relations between elements can be formally described in a proper way (e.g. due to high heterogeneity and high number of elements) or these are non-linear, elements or their characteristics change in a way the description cannot capture. In other cases individual interconnections can be explained, but unknown or uncertain conditions of the elements or the framework lead to the situation that within these uncertainties different circumstances can realise. Furthermore, such a large number of interconnections can be the reason why the final consequences cannot be assessed with certainty. In this case we speak of "complex systems".[4]

Modelling of entities via the definition of system elements and relations is the usual way to approach systems in order to generate explanations of real circumstances. This can be done in a rough model approach assigning only rough interconnections of elements in a mathematical way, defining strict or statistical interrelations with fixed formulas. By doing this, we usually take one or multiple combined perspectives with respect to the description of relations between the systems elements. Each of the perspective usually reconstructs the circumstances by defining the relation in form of an operation describing the change of characteristics of one element as consequence of a change in characteristics of another element in the system. Such a description is of course limited, as it can only describe the defined states and changes of considered characteristics of elements in their interconnectedness. Other changes of characteristics of entities which are not describable by the defined characteristics of elements, cannot be captured by this perspective and are, thus, invisible for the modelling process. Defining such a formal description of a relation as an "operation" which affects the elements of the model, this limitation in the descriptions can then be interpreted as "operational closeness" of the system. Only those elements or their real counterparts are part of the operationally closed system which are affected by the defined operation (cf. Droste-Franke et al. 2015). Some examples may clarify this view on systems:

- For instance, the financial system can be described by financial states of defined elements and respected flows between them. A flow of one monetary unit from

[2] See Chap. 2 of this book.

[3] See Chap. 2 of this book.

[4] See Chap. 2 of this book.

one to another element will change the financial state of both elements. However, here the description is also limited to financial flows and states, concentrating on a single operation. The description of real entities which are represented by elements in the system like computers and banknotes or coins is not important for its modelling at a first glance. In a refined system such elements can be considered, e.g. with respect to the flow velocity of monetary units. In this way differences in the potential speed of exchange on various markets can be distinguished which may be important for modelling dynamics in the system.
- A geographical map designed for navigation represents also a system clearly restricted to one usage, navigation with a car, as pedestrian or with a bicycle. It is a certain description clearly dedicated to the purpose of a respective specific task of navigation. It is nearly impossible to use such a map for other geographical purposes, for instance, the analysis of land use or ecosystems.
- In case of energy or energy exchange as operation we result in a system description which comprises all known entities in the world. This is the case, although the system description itself is very limited and neglects many aspects of the real world. The statements which can be drawn from the analyses are purely restricted on statements about the energy content and exchange, similarly to monetary units in the financial system described in the first example.

With a closer look it becomes clear that these systems, consisting of their elements and relations, as well as many other systems can only be used to explain certain preselected dimensions of interrelations which are determined by the chosen description.

Additionally to this natural limitation of systems/systems descriptions and modelling to the extent of elements affected by the chosen operation, a further aspect shaping systems needs to be considered: the intention for choosing a specific description. Taking the intention into account, a system is an abstracted description of actual real entities and circumstances, reduced to the most relevant aspects with respect to the chosen intention or the specific pragmatical interest of addressees of the descriptions (cf. Ropohl 2012, 55; Droste-Franke et al. 2015). Reducing the elements to the ones which are most relevant may lead to a specific reduction of elements which need to be considered in the system to fulfil the intention.

With respect to the selected examples:

- In case of the geographical map in a navigation system it is sufficient if it consists only of roads which could become relevant to reach the selected target. These can be dynamically chosen or even be preselected by the user by choosing a certain map extension (e.g. only roads in Germany). Other geographical elements are not needed for the intention of navigation.
- In the context of the energy supply system an intentional limitation to relevant elements and their interconnections is key for the descriptiveness or practical applicability of the system for the explanation of real circumstance. Reducing it to the energy states and flows which can be used for the supply of energy finally usable by the consumer, it is strongly reduced to the resources and technical infrastructure of energy supply systems. Energy flows to other entities leave the system and are, thus, lost for the intention of usable energy supply. Often cited

"energy losses" which do not exist if you look at the complete "energy system", can, thus, be framed as "losses in the technical system to provide usable energy to the consumer".

A further challenge in modelling systems exists if the intentional reduction of the system does not result in a sufficient reduction of system elements, to be applicable for its analysis e.g. via computer models. Often a reduction in spatial or temporal dimensions is used to be able to handle the amount of data and computer power needed. However, at least roughly, interconnections to elements outside the chosen dimensions need to be considered. For instance, a reduction to the energy supply in Germany does not mean that we can simply cut out Germany from a map in the internationally interconnected energy supply system. Additionally, energy flows from and to other countries are relevant for meaningfully modelling the system. In this case, at least simpler approaches should be applied to model energy flows from and to Germany. In building up such cascading or nested model constructions, the system description is supplemented by a hierarchical view (cf. Ropohl 2012, 57).[5]

In reality, however, entities are not only influenced in one dimension, but also in others. Such influences lead to real events which cannot be explained with modelling in single or multiple dimensions. In this case, the simulations may miss the one dimension or perspective which has the decisive influence on the observed circumstances and, thus, is required to explain the observed effects.

16.3.2 Sociotechnical Systems

In order to model an energy supply system, we can start with one of the core elements, modelling the energy exchange, as described above. This part somehow needs to be considered in each energy supply model to a certain extent. However, driven by the focus of the specific simulations, further components needed differ largely. While energy technologies as means for the realisation of an energy supply system are considered with varying detail and coverage, differences occur, particularly, in other dimensions like economic or market aspects, social aspects, resource aspects, etc. (cf. Droste-Franke et al. 2015). A closer look shows that, in order to understand all processes, energy (supply) systems (including the demand) need to be modelled as a socio-technical system, considering also the relevant social components of the system, depending on the specific focus. This is particularly relevant with future energy supply systems largely based on the usage of often decentralised renewable sources and production partly taking place at the consumer which is much more intermingled with social processes than the usage of conventional ones. As reducing the need of usable energy per consumer such as light, cooling, heating, and other energy related products can be part of the potential solutions, even the restriction to

[5] See Chap. 4 of this book.

the pure supply system, taking only assumptions on consumption of final energy like amount of gas or electricity delivered to households, falls short.

An indication with respect to relevant issues in the context of socio-technical systems is provided by analyses of Büscher et al. (2019). In their discussion of challenges as socio-technical problems, they distinguish three dimensions for the challenges in converging infrastructures (cf. Büscher et al. 2020):

- Problem of control: the mixture of technical and social elements leads to problems in the control of systems influencing "predictability, security, safety, efficiency, etc." (ibid., 12) (factual dimension)
- Problem of change: generally shared expectations change during transformation processes with changes in the constellations. An area of conflict arises in the mutual orientation of actors, particularly with respect to still existing needs for reliable services to meet the (changed) demand. (social dimension)
- Problem of action: problems of operation and decision making may occur due to increasing uncertainty and non-transparency, being in parts also a result of social processes substituting clearly defined and calculable technical processes. Elements like "distributing risks and responsibilities, legal protection, and informal mechanisms such as trust and confidence" (ibid., 14) are important keys. (temporal dimension)

The observations of Büscher et al. (2019) show the multidimensional characteristics of converging infrastructures, requiring the modelling of many more aspects and perspectives, in order to understand processes in energy systems with much higher share of decentralised renewable energy sources used than in conventional energy systems.

A promising simulation approach to capture processes in socio-technical systems is agent-based modelling. It can be applied, e.g., for comparative analyses of dynamics in socio-technical systems. However, with changing framework conditions and systems as well as future uncertainty, modelling of complex future systems and, particularly, their validation with respect to similar (not yet existing) situations have limitations (cf. Droste-Franke 2020). Thus, fitting the model to best possibly describe future processes is nearly impossible as we do not know future framework conditions and need to work with assumptions on relevant model elements like future behaviour and social structures here.

Additionally, Levins (1966) and Kaiser et al. (2022) discuss in a more general manner that modelling is always a trade-off between generality, precision and realism. This holds for all modelling approaches. For instance, models adapted to a specific situation are very realistic and precise for the specific circumstances, like the considered operations, effects and behaviour of the model is consistent with observations. However, this model cannot be easily transferred to another, even similar, situation. Looking against this background closer at theoretical models like e.g. physical laws, reveals that we make a realistic and general statement here, while we do not have any data or parameters to use it for a specific situation. Therefore, the formula itself does not provide any predictive statement without adequate data. What we actually need here for policy advice in our context are general statements with predictive power.

This can be realised by modelling on basis of, for instance, average values, typical probability distributions, and representative models, for types of technologies (e.g. conversion, grids, ict) or types of actors (e.g. charging/driving behaviour) (cf. Kaiser et al. 2022, 7).

Taking these challenges into account, modelling the individual systems by considering known, especially technical, interrelations in form of equations could be a good starting point which can later on be complemented by agent-based analyses based on average values and probability distributions where needed.[6]

Besides others, relying on the elaborations on "resilience and values",[7] one could add the following challenges for systems knowledge. Interdependencies between relevant interacting systems in terms of crisis need to be considered to an extent that the results of the analyses are reliable. Particularly, human behaviour needs to be considered if it is decisive which is particularly the case in the transport system. Furthermore, modelling and indicators should be meaningful for crises situations and their predictive potential should be elaborated where possible. In crisis management, but also in the preparation phase for crisis, heterogeneous human agencies need to be considered. Existing value-landscapes should be accounted for to increase compliance, robustness and co-responsibility in systemic changes and, thus, social robustness of solutions.

16.3.3 Basic Resilience Characteristics of Energy Related Systems

Expected outcomes of the system with respect to energy supply depends on various closely related subsystems, some of the most relevant are systems for: usable energy generation, electricity distribution (grids), transport tasks, and information exchange (ICT).

The subsystems we consider have very different characteristics even from the technical point of view. These have strong influences on the reliability of services in the individual system. A short analysis of the differences is done in the following as basis for a meaningful combination of systems or systems models.

According to Ropohl (2012) we distinguish for systems their elements and their connection. In order to characterise the systems we derive six categories, showing the complexity and flexibility or strictness of requirements of the individual systems.

We distinguish six categories in the following:

(1) *Complexity of entities to be modelled*—depending on the task of the technological system, various entities can be of high relevance and are, thus, considered in respective models. This characteristic comprises the variety and complexity

[6] See Chap. 2 of this book.

[7] See Chap. 6 in this book.

in relevant entities, including the multi-disciplinarity required to capture the elements.

(2) *Flexibility in the type of element*—the higher the flexibility, the more alternative options can in principle be made available in the system in case of disturbance.
(3) *Mutual interference of operations*—in case that operations interfere with each other or are influenced by the environment, more options for stress and a higher risk of disturbance exist.
(4) *Flexibility in interconnection infrastructure needed*—the higher the flexibility in required interconnection infrastructure, the more alternative options or redundancy are available in case of disturbance of interconnecting elements.
(5) *Option for alternative pathway* via *other systems/local production*—in case that alternative pathways are possible via other systems, maybe even including alternative sources, these can be applied as backup in case of disturbance of services.
(6) *Complexity and technical requirements for connection infrastructure*—the lower the complexity of the options, the less disciplines needed to establish and operate them and the lower the technical requirements are for connection infrastructure, the easier alternatives can be established, e.g. by changing/rearranging (parts of) the elements and building new connection infrastructure meeting minimum requirements for interconnections.

In the following, three systems are assessed based on these categories: (a) the energy system, (b) the transport system, and (c) the ICT system.

16.3.3.1 The Energy Supply System

Regarding the energy supply system, the distinguished categories can be assessed as follows:

(a1) *Complexity of entities to be modelled*: the central operation is transport of energy. the system is subdivided in many physical systems/options in which either material with chemical/physical energy content or electromagnetic waves can be used. The form of energy applied or energy carriers which are transported from source to users, in principle, can be flexibly changed. While the entities themselves are not complex. Due to the large amount of options, complexity rather enters in the processes of energy conversion and distribution—see complexity of connection infrastructure below.
(a2) *Flexibility in the type of element*: in principle high flexibility in the element used, only energy content provided is important, no direct connection via the transport of specific elements is needed, ensuring efficient energy exchange between elements is sufficient.
(a3) *Mutual interference of operations*: typical operations do not interfere with each other, but capacity limitations and specific technical requirements of interconnections require regulation, regulating total capacity/balanced flows is sufficient, no need to ensure specific individual flows.

(a4) *Flexibility in interconnection infrastructure needed*: infrastructure is specifically chosen, fitting to a certain systems element applied (gas, electricity, liquid fuels). The potential of exchanging/switching infrastructure for a certain element is low because of high effort and costs. With high effort a (temporal) switch to predefined and potentially prepared alternatives may be possible.

(a5) *Options for alternative pathway* via *other systems/local production*: local production is possible with various technologies from stored energy or existing alternative sources which need to be adjusted to each other to a certain extent. Specific interlinkage to ensure applicability is required. Thus, indirect provision of energy is possible, from stored energy, e.g. initiated via communication.

(a6) *Complexity and technical requirements for connection infrastructure*: Many technical disciplines are involved to develop and install suitable energy conversion and distribution options. Conditions of infrastructure need to be permanently ensured/operated in some of the major components: AC-grids, adequate long-term alternatives of connection infrastructure are in practice hardly available.

16.3.3.2 The Transport System

Regarding the transport system, the distinguished categories can be assessed as follows:

(b1) *Complexity of entities to be modelled*: the central operation is physical transport of a specific matter. Moving specific individual/heterogenous material and persons from one location to another. Some complexity lies in the varying characteristics of individual goods and persons to be transported which can be rated from unspecific to very specific and the inflexibility in the transport task itself. The activity is only successful if the specific matter is transported to the right location, as previously planned.

(b2) *Flexibility in the type of element*: low flexibility in the individual material and persons transported, sometimes no flexibility at all.

(b3) *Mutual interference of operations*: operations/elements can interfere in various ways with each other with large damages, a specific flow needs to be ensured, elements themselves need to be regulated on their technical functionality and compatibility, human action and interaction is of high relevance.

(b4) *Flexibility in interconnection infrastructure needed*: depending on the transport task various modes can be used. High flexibility in most cases.

(b5) *Options for alternative pathway* via *other systems/local production*: no local production possible, but some functionality can be used to transfer the demanded functions of the transported matter by other systems (e.g. communication can be ensured via ICT or information can be transferred and, thus, a copy with the same functionality can be generated (e.g. via 3D-printing)).

(b6) *Complexity and technical requirements for connection infrastructure*: for basic functionality, depending on the elements used, not very high preconditions necessary, even transport via air is possible, but alternatives may be costly and usually provide lower capacity. Potential interference requires regulation of flows with some requirements in control of connection infrastructure.

16.3.3.3 The ICT System

Regarding the ICT system, the distinguished categories can be assessed as follows:

(c1) *Complexity of potential entities to be modelled*: the central operation is transfer of information, with information carriers as elements, transfer is realised potentially via electromagnetic signals/sound/data carrier—depending on sender's/consumer's needs. Information can be available in very different forms.
(c2) *Flexibility in the type of element*: specific information is transported, provision of a direct connection between sender and receiver is required.
(c3) *Mutual interference of operations*: typically no interference of individual information, but capacity limits: regulation of information flows needed to ensure adequate flow velocities.
(c4) *Flexibility in interconnection infrastructure needed*: depending on the kind of information various modes can be applied with the same final result, if issues like speed of transfer are not crucial for the required services.
(c5) *Options for alternative pathway* via *other systems/local production*: no option for local production, pathways in principle very flexible.
(c6) *Complexity and technical requirements for connection infrastructure*: low level connection infrastructure possible, but partly high technical requirements on (fitting) sender and receiver facilities.

16.3.3.4 A Comparison of Systems

The descriptions of the different systems constellations reveal various differences in systematic strengths and weaknesses of the subsystems with respect to their flexibility in elements and infrastructures/pathways which can be used to establish connections between elements. These influence the reliability of systems and provide a first hint into more and less vulnerable components of the individual systems. A comparison of summarised characteristics is shown in Table 16.1. However, a more concrete and detailed analysis would be needed to assess concrete vulnerabilities of the systems.

Looking at empirical findings in stress cases illustrates difficulties with the discussed systems in such situations. We take the flood event in the Ahr-valley in 2021 as an example here, because it was an extreme event in which even large parts of the infrastructure were strongly damaged or destroyed.[8]

With respect to the energy supply, in the flood event the two main grid-bound infrastructures of the energy system, the electricity and the gas supply were damaged to a large extend. With respect to electricity, temporarily 64,400 households could not be supplied. A large number of provisory substitutes were installed. This started

[8] Findings without providing explicit sources are based on direct experiences of the author.

Table 16.1 Comparison of characterising categories for the three systems Energy supply, transport and ICT

Category	Energy supply	Transport	ICT
Complexity in relevant entities	Low	Low to mid	Mid to high
Flexibility in the type of element	High (energy content needs to fit)	None to low	None to low
Mutual interference of operations	None, only capacity restrictions	Potential interferences require high effort of control	Mostly none, only capacity restrictions
Flexibility in interconnection infrastructure	Element specific, usually high costs of switching	Mid to high	High
Alternative pathways/ sources	Possible, backup via storage possible	Possible to realise only some functions, backup with adequate substitutes	Possible, no option for local production
Complexity and technical requirements on connection infrastructure	Very high, even permanent monitoring needed	Low to mid	Low, but high requirements on sender-receiver compatibility

with small to medium size generators for emergency supply up to provisory grid operation facilities, lines, transformators and other technologies.[9] The fuel for the generators was provided via special tankers approaching central places in the affected municipalities. The provisory restoration of electricity grids took about one month. With respect to gas 8261 grid connections could not be supplied. With a large effort, provisory lines could be built so that gas could be delivered again in mid-December to all connection points.[10] In the meantime, numerous tanks with liquified gas were used, where possible.

The experience with the flood-event shows that a large advantage of the energy system is that other sources can be used for energy supply. In case of the electricity supply, generators or a renewable energy-based solution as pilot[11] could be used to provide a base supply as substitution to electricity from the grid or from damaged decentralised power facilities. With respect to heating of water for cooking and other uses, gas cylinder-based solutions or electric heaters could be applied as substitution to the damaged local heating systems and missing grid-based gas supply. Furthermore, the installation of new control units allowed to use solar thermal energy for warm water supply. Using liquified gas in tanks instead of gas from the grid was possible by small changes in settings of the restored heating systems. However, with

[9] https://dokumente.landtag.rlp.de/landtag/drucksachen/3437-18.pdf.

[10] https://dokumente.landtag.rlp.de/landtag/drucksachen/3437-18.pdf.

[11] Faber mobile Power System: https://www.faber-infrastructure.com/index.php/solarcontainer-das-faber-mobile-power-system/ (Solar panels with a 80 kWh battery in a container as emergency supply).

respect to the grid infrastructure, provisory lines where the first step in restoration and even these took some time. In case of electricity supply the sheer number of electricians and electrical fuses needed for restoration in combination with the need to secure all house connections led to the situation that about one month was required to supply nearly each household with electricity again. The challenges with the gas supply were even higher, as partly lines where completely destroyed and substitutes had to be built up. The example shows that in both cases scarcity of experts, material and facilities (e.g. small generators) together with organisational challenges can be reasons for shortage and delays in temporal provisory solutions and grid restorations. Particularly in places in which the lines and other infrastructures where destroyed, the installations were improved to be more robust against future flooding events. Even with the provisionally options service was close to normal. Due to large electricity demand of building dryers required to dry walls, the capacity even needed to be increased. One of the main questions in this context for future renewable-based energy supply systems will be to find the most suitable way to provide also in such emergency situations decentralised renewable-based energy supply with sufficient capacities.

Transport infrastructure shows a different picture. While many roads could be used again as soon as they were cleaned of mud and rubbish, including many trees and thousands of cars, some villages could only be reached on food or by air, with the associated limits in capacity. Six weeks after the flood event, all villages could were accessible again via provisionally restored roads.[12] Bridges pose special challenges. While, first, all bridges needed to be controlled, 30 temporary bridges could be installed by German Federal Agency for technical relief (THW) until March 2023[13] with 103 bridges being severely damaged or destroyed.[14] Railway tracks were destroyed in parts of the route from Remagen to Walporzheim and were completely destroyed in the Ahr-valley behind Walporzheim. Some bridges will only be restored in 2026 while the railway is planned to be put back into operation by end of 2025.[15] Cycle paths will in parts be restored in parallel to the railway tracks. In between, cyclists must use the normal roads as alternatives, a specific challenge in a tourist region and for daily routes of residents. However, the supply of energy to the transport sector is a further issue. Currently, the provision of gasoline can be provided by tankers. In future, we will need electricity to adequately charge vehicles in such a situation. Currently this cannot be provided by normal emergency power supply. Solutions need to be found here.

In the example we see the effect that of course some transport infrastructure could be used again after cleaning. However, not all roads could be restored so far and due to missing connections some ways still cannot be used (particularly cycle paths and

[12] https://dokumente.landtag.rlp.de/landtag/drucksachen/3437-18.pdf.

[13] https://www.thw.de/SharedDocs/Meldungen/DE/Einsaetze/Inland/2023/03_Brueckenbau_Walporzheim/meldung.html.

[14] https://dkkv.org/wp-content/uploads/2023/02/Anfrage_Opfer-_und_Schadensdaten_der_Flut_2021.pdf.

[15] https://www.ahrstrecke.de/.

railway tracks) or still are in bad condition, more than three years after the event. The disadvantage of the transport system is the large and expensive infrastructure needed, particularly to cross rivers. Furthermore, in the restoration process building transport infrastructure is competing to many other infrastructure measures of the municipalities. For instance, Bad Neuenahr-Ahrweiler identified in total more than 1400 individual restoration measures after the flood.[16] Service in case of transport will prospectively only be back to normal by about 2026—five years after the flood-event happened. However, developments due to the COVID pandemic—for instance enabling more people to work at home than before—some trips could be substituted by virtual exchange.

In some cases, ICT systems infrastructure posed similar challenges to those associated with transport infrastructure: information to be sent is unique and needs to reach the receiver in its complete state for a successful service. Furthermore, the information itself is coded in a complex way which requires a direct fit of sender and receiver formats. However, meanwhile, data can be transported via air with the same capacity as via cables. Also, mobile devices needed for this service are broadly available. This lead to the situation that a few days after the event, mobile services were close to be completely restored. In the area of the fixed network, the restoration required longer, but two weeks after the event more than half of the connection points could be served again.[17] For security of reconnection no pre-check of experts was necessary. In some cases even an exchange of hardware in the buildings was not necessary.[18]

ICT systems have the advantage that they can be rapidly substituted and even be restored without specific physical security measures, due to the low voltage levels applied. Thus, it is very flexible with respect to the grid-based infrastructure. Also, the energy supply of the most important devices is low so that emergency power supply is usually sufficient.

All in all, looking at the six categories identified and discussed above, in case of the energy system, high flexibility in the pathways and sources of service provision are the key for fast emergency supply with less capacity, but useful for base services. However, technical requirements and security issues of the technology lead to higher effort in replacing or restoring the original infrastructure with the same services as before. Improvements could be more decentralised facilities protected against impacts of such events and an increased number of available renewable based mobile solutions for emergency electricity supply. Furthermore, also robust decentralised renewables-based, maybe even mobile, heating systems could be part of the solution.

With respect to the transport infrastructure, high effort is needed to restore the infrastructure with the same quality of service and the competitive situation to other building measures lead to negative long-term impacts on services. To improve services, more alternative pathways can be chosen in order to replace transport needs

---

[16] https://www.bad-neuenahr-ahrweiler.de/hochwasser/wiederaufbau/massnahmenplaene-wiederaufbau/

[17] https://dokumente.landtag.rlp.de/landtag/drucksachen/3437-18.pdf.

[18] Authors own experience.

by other services with similar functions. Virtual meetings as alternatives for personal meetings are one option here. Other options are more decentralised production of goods, e.g. in agriculture, but also using innovative technologies like 3D-printing. Also, innovative small private or public air transport could improve the situation. However, transport services will still need an emergency energy supply fitting to future energy systems which needs mainly to be based on renewable sources. In order to provide such options, these need to be set up systematically.

With respect to the ICT system, the situation is already quite well. Due to large flexibility in pathways, alternatives can be installed quickly. However, communication infrastructure is an important key technology for critical services in catastrophic situations. Thus, these need a sufficient emergency power supply as well as the use of alternative long-reaching communication pathways, e.g. via radio technology or satellite communication. A broader usage of such technologies could improve the situation.

For modelling, of course we need to consider that also other stress cases could take place. Furthermore, modelling is to be designed specifically for the phases of such an event to be supported: starting with the stress event and direct impacts onto the systems, continuing with a clean-up phase, a phase of temporal solutions, transitioning to a restoration or improvement phase of original services. But, having also various heterogeneous value landscapes in mind[19] also for the phase of infrastructure planning and preparation for such events adequate support of modelling could be helpful.

As the first attempt is to protect systems from damage where this is possible and models for daily operation of the system are already available, we concentrate here on modelling the direct impact of stress cases on systems and their services. For such a kind of modelling stress tests procedures have been developed for various systems. For instance, stress tests are carried out in the banking system[20] and with individual hospitals as pilot activities.[21] For Telecommunication systems effects of threat scenarios are analysed and similar analyses are carried out in the water supply[22] and transport systems.[23] Also, stress test are carried out with the current energy systems, mainly with the focus on national energy supply and the operation of transmission networks.[24]

---

[19] See Chap. 6 of this book.

[20] https://www.bankingsupervision.europa.eu/banking/tasks/stresstests/html/index.en.html.

[21] https://www.bbk.bund.de/SharedDocs/Downloads/DE/Mediathek/Publikationen/PiB/PiB-02-risikoman-krankh.pdf?__blob=publicationFile.

[22] https://www.bbk.bund.de/SharedDocs/Downloads/DE/Mediathek/Publikationen/PiB/PiB-15-sicherheit-trinkwasserversorgung-teil1.pdf?__blob=publicationFile&v=14; https://www.lfu.bayern.de/publikationen/get_pdf.htm?art_nr=lfu_was_00317; https://www.bdew.de/media/documents/220733_Rheinland-Pfalz_Bericht_Benchmarking_Wasser_Abwasser_Erhebungsjahr_2019.pdf.

[23] https://www.bmdv-expertennetzwerk.bund.de/DE/Publikationen/TFSPTBerichte/SPT109.pdf?__blob=publicationFile&v=3.

[24] https://www.bmwk.de/Redaktion/DE/Artikel/Energie/stresstest-strom-2022-ergebnisse-langfassung.html.

Core of the developed methodologies is that relevant stresses need to be modelled on various levels of the systems. Usually, not only one, but different stress scenarios are constructed and tested. The situation with and without the stress case is compared to each other to identify the effect of the stress case. These features are taken up in this study to discuss how stress tests for future scenarios with the main focus on energy, transport and distribution grid systems can be designed and carried out.

16.3.4 Approach for Meaningful Combination of Systems and System Models for Resilience Modelling

Taking the results discussed above as basis for the combination of system with the purpose of stress-tests in the context of resilience monitoring means that we need a systems (model) combination which shows important interrelations, also in circumstances with severe change in framework conditions up to destroyed infrastructure, cut from usual sources and partly required temporary replacement without being too complex. The focus is on the service of providing usable energy as, on the one hand, very basic requirement for many societal processes nowadays and, on the other hand, a system which is currently undergoing a major transformation. The challenge is to maintain the energy supply service despite the successive replacement of its key components. As starting point we concentrate on the analysis of the target situation of the transformation processes. Furthermore, we start with already existing models, in order to generate a fast track to considering resilience issues already successively in current modelling procedures. Thus, our main focus is the influence of severe impacts on the systems services.

The core of the analyses of resilient energy supply is of course the analysis of the technical energy supply system itself. The basic systems operation is energy exchange. Thus, basic elements are all energy carriers which are suitable for the provision of usable energy supply. More specifically, these are energy conversion technologies applied in the energy sector with their energy sources and the transportation of final energy to the consumers which fits to the service awaited on the consumer side. For this purpose, we take a national model covering all relevant processes with a high temporal and spatial resolution which is used for the optimisation of future energy supply scenarios, REMix.[25] The model can cover energy conversion in very detail and basic functionality of transmission grids with a strong focus on electricity provision. It is supplemented by technical modelling of distribution grids, as many new challenges arise with the integration of decentralised energy conversion and electricity feed-in in the grid area. In this way, main parts of the complicated elements are covered by the analyses.

Another challenge is the digitalisation of energy system. The increased number of often decentralised energy converters in the system requires concerted operation of facilities which can only be accomplished with digital devices. Thus, the

[25] See Chap. 13 of this book.

importance of ICT systems for secure energy supply increases strongly.[26] Therefore, basic behaviour of ICT systems and concepts for digital resilience measures are analysed and considered, including influences of intentional hacker attacks. In this way, the main weaknesses of ICT systems, potential modification of information and disturbance of transfer are covered.

The energy transformation has also major influence on the elements in the transport sector. With modelling the basic transport tasks on national and local level and the energy provision for the prospectively dominating technology of electric vehicles, main aspects of the new challenges in the transport system are addressed. Depending on the changes in transport tasks, requirements for energy are derived as well as options and usage of charging infrastructure are modelled. With this approach, changing framework conditions because of destroyed infrastructure and the weakness of the transport system, that all substitutes can be realised only with large loss in transport capacity, are tackled by deviations of transport behaviour with a specific connection to barriers occurring due to the new dependencies on available charging stations and electricity.

Model combination is particularly important in terms of demand which cannot be met. Important elements are, on the one hand, the energy conversion, transportation and distribution of electricity. On the other hand, the change of demand and temporal and sectoral use patterns in charging electric vehicles is important which occurs due to changed transport behaviour because of damaged transport infrastructure like roads, bridges or private and public charging stations for some stress cases. A detailed analysis of ICT systems is carried out particularly for the distribution grid, because in this area many parameters need to be considered to make usable electricity available to the consumers. Furthermore, new chances of decentralised production of electricity can be captured in this way.

With the adapted models important interlinkages need to be considered adequately without establishing a too complex technical model combination. Therefore, in a first step, we analyse the vulnerability and robustness including aspects of dynamic flexibility of the services which could be implemented in the system itself. Thereby, we look at the systems in the stress situation itself, neglecting physical repairing and restoration measures for the moment.

16.4 Practical Requirements in Ensuring Applicable Knowledge for Action and for Evaluation of Alternatives

In addition to the theoretical requirements discussed above, the practical realisation of a resilience monitoring with the given framework conditions lead to further requirements. These practical requirements cover different aspects of which some of the most important are discussed in the following.

---

[26] See Chap. 12 of this book.

From the discussion in the previous chapters,[27], it became clear that such a complex decision situation as the resilient design of technological systems, specifically decisions in the context of future energy systems, require stakeholders and actors involvement to solve issues with genuine uncertainty and high ambiguity. Following an analytic-deliberative approach shows how stakeholders and analyses can interact in various ways. Here we discuss which challenges arise and need to be solved to realise such processes practically.

As discussed above, in resilience monitoring structural and incidence-based analyses can be distinguished. Furthermore, systems can be prepared with the aim to resist to whatever may come.[28] Following each of these approaches pose different challenges on analyses and modelling exercises.

Considering structural issues of the systems requires that the modelled systems can be characterised according to resilience indicators like for instance diversity in energy sources, redundancy in technologies etc..[29] Thus, the specification of the system characteristics of modelled systems needs to be describable and influenceable by the modeller in an adequate way and, ideally, practically in a decision situation, in order to be able address and adapt the systems structure as needed.

The requirements are similar with respect to the determination of model parameters to prepare the system for whatever may come. However, for a complete analysis this includes also measures for the operation of systems like technical monitoring systems interpreting systems parameters in a way to find out if the system status is trustworthy at a certain point in time and carry out adequate reactions if it is not.

In order to test predefined systems against severe stress situations, concrete situations need to be realised for the systems. Suitable assumptions for creating or selecting stress cases need to be made according to the focus of the analyses and decision situation. Stress cases can for instance either be randomly constructed or constructed as frequently occurring circumstances or as extreme situations.

In order to realise such analyses with systems models some further prerequisites need to be considered. In the following, some practical aspects are discussed which are again derived based on the analyses of Droste-Franke et al. (2015).

The first aspect to consider is that simulation models need to be adapted to be able to fulfil the above mentioned requirement to model interesting stress cases. Beside others, the temporal, spatial and technological resolution needs to be set in an adequate way to be able to show impacts of stress cases onto the system. The results need to be meaningful to be able to evaluate decisive impacts with respect to important societal needs. Furthermore, the results occurring in parallel in various systems need to be displayed in an integrated way. Also, important interrelations need to be considered. In order to realise this, it may be necessary to simulate impacts of stress cases iteratively with a set of different models.

Beside the results being meaningful for the decision situation, the analyses need to be understandable and should ideally allow for taking various perspectives on the

---

[27] See particularly Chaps. 2, 5, and 8 in this book.

[28] See Chap. 3 of this book.

[29] See Chap. 11 of this book.

circumstances in focus. Thus, premises for base scenarios and technological set-ups need to be made transparent and implemented in a way that they can be varied, e.g. to enabling sensitivity analyses, and ideally adaptable to the view taken by discussants. Furthermore, the influence of uncertainties on the analyses and, thus, their significance for the meaning of results needs to be discussable which requires at least qualitative statements in this context. Ideally, effects of changes in assumptions should be directly made visible in the decision situation. In case that model runs cannot be performed on short notice in the decision situation, because of high effort or long duration of modelling procedures, several runs need to be prepared in advance. However, in order to be prepared for unforeseen new arguments, a revision of prepared calculations, e.g. recalculations with other parameters, not yet considered in the variations, needs to be made possible, e.g. by dividing the entire decision process into various workshops and performing additional simulations in the interim period between them. The main purpose is to enable comparisons of design alternatives of the systems which can be influenced in the decision-making situation with respect to their meaning for the considered systems and society.

Some further requirements for the involvement of practitioners and lay-people, but also scientific experts from various disciplines which in most cases are laypersons in the other relevant disciplines, are characteristics which ensure practical applicability of the analyses. Low technical barriers and easy usage of tools encourage people to actively engage themselves in the process. In a workshop situation a centrally moderated utilisation of tools represents another solution. Furthermore, it is valuable if the participants at least have the option to use the tools by themselves to get enough time to better understand the analyses, in terms of assumptions, get a feeling for sensitivities and get answers to their individual questions. Another issue concentrates on the usability of tools in various decision situations. The tools ideally need to be applicable in various discursive formats and technical set ups. For this purpose, the chosen platform needs to be flexibly applicable, e.g. for one-screen and multiple-screen set ups. Another aspect concerns the presentation of results. A choice of various formats like maps, texts/descriptions, pictures/icons, column and pie charts, curves, histograms, etc. should be allowed for to best customise to the preferred visualisation and analyses habits of individual participants.

16.5 Design of Workflows (and Integrative Tools)

As discussed above, in order to provide a picture as complete as possible, elements of structural and incidence-based analyses need to be combined. While structural analyses are closely related to the design of systems and, thus, the setup of simulation models, incidence-based analyses represent a test procedure against stress cases with adequately established models. The way they are designed depend on the assumed framework conditions determining the roadmap and specifications of system development. The following discussion summarises a series of discussions and experiences within three years of realising the stress test tool while referring to the expert group being the main authors in this book for more general aspects.

In order to realise such analyses first, it needs to be chosen which kind of scenarios are used. Beside a scenario of a target energy system in one future year, e.g. in 2050, also development pathways to the target scenario can be addressed. However, in order to set the right goals, the most important issue is to analyse the target scenario which is aimed for by energy policy.

Such a future energy scenario with suitable framework conditions, is the starting point of the analysis. This scenario may or may not yet include resilience measures to a certain point. In a second step this scenario needs to be analysed with respect to incidence-related and structural vulnerability and resilience assessment. After comparison of results, the design can be improved or scenarios can be redesigned respectively in order to meet the requirements in a better way. Finally, all options can be compared by all relevant evaluation criteria. If no intermediate re-design or re-calculation of scenarios is possible, prepared scenarios with various optional resilience measures can be shown and discussed. In this process, framework conditions which cannot be influenced and manipulable factors like resilience measures need to be distinguished. While framework conditions can realise without any option to influence them, chosen resilience measures should ideally show good performance and low costs in case of each possible set of framework conditions.

The selected stress cases in the first line should represent extreme situations in order to be prepared for them. Nonetheless, tests with frequently occurring situation may complement the analyses. Randomly chosen situations have the potential to simulate whatever may come. However, with limited modelling capacity the difficulty is to assume really extreme events with this methodology.

A general workflow which can be followed is to first set up sets of options for the development of technical systems including digital resilience measures. These are assumed on the basis of various contexts. The result are technology options with matching framework conditions as scenarios. Furthermore, consistent scenarios are selected which represent twins, differing only in the status of digitalisation. In this way the full potential of digital measures can be assessed with different framework conditions and stress cases taking place. Thus, the effectiveness of measures in various circumstances can be analysed and compared to each other for different options of measures. On this basis decisions can be made as to which measures should be developed and implemented.

In the analyses various levels in the causal effect chain need to be considered. The analyses start with a stress case in the context of a scenario set including framework conditions as well as technology bundles. The first level of impacts are damage to infrastructure and to functions of the systems. These propagate then to other systems and to the environment (2nd and 3rd level impact). Effects on all those various levels need to be displayed in an aggregated and detailed way.

16.6 Conclusion

From the discussion above, it becomes clear that challenges of systems analyses in the context of resilience modelling reveal from various sides.

In order to be compatible to already existing work carried out periodically in energy system analyses on national level, the respective results like scenarios or the respective models, ideally, need to be applicable for this additional analysis. Otherwise, too much and even double effort is necessary to carry out resilience analyses in this context. Furthermore, the tools need to support stakeholder involvement for decision making. Another aspect is that various disciplinary perspectives need to be combined to be able to tackle real problems.

In order to allow for stakeholder engagement in decision making and showing meaningful results like structural and incidence related resilience aspects, further practical challenges arise. Existing systems models or results of systems modelling, ideally, have to be adapted to cover also resilience analyses. Premises and uncertainties of modelling need to be made transparent in their effect in order to be able to assess the reliability of estimates. With respect to uncertainties at least qualitative statements need to be provided. During the discourse with stakeholders, aspects may become relevant which have not been in focus before and have not been analysed in detail yet. Therefore, unforeseen arguments need to be treatable in the decision process by revision of calculations. This may mean to carry out a series of several workshops. In the decision situation the discourse should focus on components which can be influenced by the decisions. Finally, the usage of the tools should be made as intuitive as possible or a moderated format is needed to make their use simple for everyone involved. This also includes customisable visualisation modes.

From a more theoretical point of view, the task of resilience monitoring means to apply scientific expertise to provide meaningful insights of a certain systems realisation with respect to non-scientific evaluation criteria of circumstances. In the analyses a distinction between normative and descriptive elements is very helpful. Additionally, it should be clear that for a decision beside systems knowledge, also knowledge about acting or measures is necessary, as well as a consideration of evaluation of alternatives. Thus, it is one and not the only knowledge component needed in the process and needs to fit to the other ones, applicable measures and meaningful description of impacts which can be evaluated by stakeholders. For a decision situation it is important to identify general measures with their potential effects. Analyses needed for this purpose should provide general statements with high predictable power. However, this means that rather average values and statistical distributions are used and the simulations do not reflect real processes.

Looking closer into the elements and interconnections in the systems which are relevant for analyses in the context of the energy supply, reveals that the generation of useful energy, its distribution, ICT systems and the use of energy in the transport sector show all different strengths and weaknesses. Concentrating on the systems analysis of these weaknesses and considering important interconnections results in a set of models which can be used to provide valuable statements with respect to vulnerability and resilience. Important aspects to cover for situations under stress are: temporal variability and flexibility of energy supply, grid capacities and control options, ICT options for control of grid operation, change of transport behaviour, and change of charging behaviour.

In a decision situation a stress test tool is used in the framework of a workflow. Typical phases include setting up framework conditions, designing the systems, setting up digital resilience measures and designing or choosing stress cases. To be meaningful and to facilitate a vivid visualisation of a stress situation and its effects, impacts should be displayed on different levels: damage to infrastructure and functions of systems, propagation to other systems, impacts on environment and society. With the stress cases particularly extreme events need to be covered as a prerequisite to be prepared for such severe circumstances.

With these reflections a first step can be made in the analysis of resilience of digitalised future energy systems. However, the analysis of strengths and weaknesses of the systems in focus taking experiences from the Ahr-river flood in July 2021 as example, shows that much more has to be done in preparation of such events to be able to temporally repair and later on restore damaged or completely destroyed infrastructure. Issues are having enough material, components and facilities as well as experts available as well as dealing with large efforts (including administrative ones) to restore large infrastructures like roads, railways, and bridges. The methodology of analysing systems with respect to six central categories developed here[30] may help to analyse also other systems with respect to their resilience characteristics and to reflect on modelling needs and potentials for similar endeavours as designing a stress test tool.

References

Büscher, C., Schippl, J., Sumpf, P. (eds.): Energy as a Sociotechnical Problem: An Interdisciplinary Perspective on Control, Change, and Action in Energy Transitions. Routledge, London (2019)

Büscher, C., Ornetzeder, M., Droste-Franke, B.: Amplified socio-technical problems in converging infrastructures. A novel topic for technology assessment? Technikfolgenabschätzung Theorie und Praxis **29**(2), 11–16 (2020)

Droste-Franke, B.: Challenges and potential improvements of systems analyses for sector coupling. A discussion along the dimensions of socio-technical problems. Technikfolgenabschätzung Theorie und Praxis **29**(2), 45–51 (2020)

Droste-Franke, B., Carrier, M., Kaiser, M., Schreurs, M., Weber, C., Ziesemer, T.: Improving Energy Decisions. Towards Better Scientific Policy Advice for a Safe and Secure Future Energy System. Ethics of Science and Technology Assessment, vol. 42. Springer, Berlin (2015)

Kaiser, M., Buklijas, T., Gluckman, P.: Models and numbers; representing the world or imposing order? Perspect. Sci. **30**(4), 525–548 (2022)

Levins, R.: The strategy of model building in population biology. Am. Sci. **54**(4), 421–431 (1966)

[30] Six systems categories for resilience characteristics: complexity of elements, flexibility in the type of element, mutual interference of operations, flexibility in interconnection infrastructure, Alternative pathways/sources, technical requirements of infrastructure), s. 16.3.3.

Ropohl, G.: Allgemeine Systemtheorie. Einführung in transdisziplinäres Denken. Edition Sigma, Berlin (2012)

Part V
Synthesis

Chapter 17
Shaping and Monitoring Resilient Energy Systems—A Synthesis

Davy van Doren, Bert Droste-Franke, Michael Brand, Karen Derendorf, Gabriele Fohr, Hans Christian Gils, Matthias Kaiser, Jörg Knieling, Sebastian Lehnhoff, Karsten von Maydell, Tudor Mocanu, Sigrid Prehofer, Ortwin Renn, Arnim von Gleich, Wolfgang Weimer-Jehle, Martin Wickel, Henning Wigger, and Thomas H. W. Ziesemer

Abstract In the light of energy transition, existing systems have shown to be often characterised by fundamental challenges that could potentially threaten its functioning, particularly when exposed to extreme forms of stress. To overcome this, we argue that an inter- and transdisciplinary approach—aimed to integrate relevant stakeholder perspectives and acknowledge different spatial–temporal scales—can provide

D. van Doren (✉) · B. Droste-Franke · G. Fohr
IQIB - Institut für qualifizierende Innovationsforschung und –beratung GmbH, Bad Neuenahr-Ahrweiler, Germany
e-mail: davy.vandoren@iqib.de

B. Droste-Franke
e-mail: bert.droste-franke@iqib.de

M. Brand · S. Lehnhoff
Energy Division, OFFIS—Institute for Information Technology, Oldenburg, Germany
e-mail: michael.brand@offis.de

S. Lehnhoff
e-mail: sebastian.lehnhoff@uni-oldenburg.de

K. Derendorf · H. C. Gils · K. von Maydell · H. Wigger
Institute of Networked Energy Systems (DLR-VE), German Aerospace Center (DLR), Oldenburg & Stuttgart, Germany
e-mail: karen.derendorf@dlr.de

H. C. Gils
e-mail: Hans.Gils@dlr.de

K. von Maydell
e-mail: Karsten.Maydell@dlr.de

H. Wigger
e-mail: henning.wigger@dlr.de

M. Kaiser
University of Bergen—Centre for the Study of the Sciences and the Humanities, Bergen, Norway
e-mail: Matthias.kaiser@uib.no

D. van Doren et al. (eds.), *Towards Energy System Resilience*, Ethics of Science and Technology Assessment 55, https://doi.org/10.1007/978-3-032-03187-7_17

indications in relation to the development of resilience. In this context, the included contributions have highlighted that resilience remains a topic surrounded by promises and confusion, but also illustrate that improved conceptualisation and operationalisation of associated processes and structures can be realised. Through clustering these contributions—according to understanding, envisioning, analysing, and managing resilience of energy systems—a structured overview is provided. Based on obtained insights, various recommendations are given how resilience in energy systems can be realised. Also, some indications are given how policy advice could be improved by means of resilience monitoring.

Keywords Energy transition · Resilience · Inter- and transdisciplinary approach · Knowledge transfer · Policy advice

17.1 Introduction

In the history of humankind, the impact of energy systems on daily lives might have never been greater. Although global energy intensity has been showing a decreasing trend since the late 20th century, energy consumption remains on a rising slope (IEA

J. Knieling · M. Wickel
HafenCity Universität Hamburg, Hamburg, Germany
e-mail: joerg.knieling@hcu-hamburg.de

M. Wickel
e-mail: martin.wickel@hcu-hamburg.de

T. Mocanu
Institute of Transport Research, German Aerospace Center (DLR), Berlin, Germany
e-mail: Tudor.Mocanu@dlr.de

S. Prehofer · W. Weimer-Jehle
Center for Interdisciplinary Risk and Innovation Studies, University of Stuttgart (ZIRIUS), Stuttgart, Germany
e-mail: sigrid.prehofer@zirius.uni-stuttgart.de

W. Weimer-Jehle
e-mail: wolfgang.weimer-jehle@cross-impact.de

O. Renn
Research Institute for Sustainability—Helmholtz Center Potsdam (RIFS), Potsdam, Germany
e-mail: ortwin.renn@rifs-potsdam.de

A. von Gleich
University Bremen, Bremen, Germany
e-mail: gleich@uni-bremen.de

T. H. W. Ziesemer
Maastricht Economic and Social Research Institute on Innovation and Technology, United Nations University (UNU-MERIT), Maastricht, The Netherlands
e-mail: Ziesemer@merit.unu.edu

2022). This growing and subconscious dependence has led to strongly evolving energy systems, in which managers and engineers need to harmoniously tune short- and long-term supply–demand dynamic. Past events, however, have shown that this inherent reliance on energy can lead to strong systemic upsets, particularly in the light of rapidly changing environmental conditions.

Active ambitions regarding sustainable development have ignited a transition of the energy system towards an increased exploitation of renewable but often fluctuating resources, increased options for decentralised energy production, as well as a stronger socio-technical entanglement of energy systems with a new range of actors and associated roles. However, this has made energy generation also more sensitive and less controllable. This lack of control is also evident in the light of increased extreme weather occurrences resulting from global climate change, which can amplify the level of stress on socio-technical systems even further (Cronin et al. 2018). On top of this, the strong introduction of digitalisation into socio-technical systems has created, additional to various acknowledged benefits related to data management and systemic control, various vulnerabilities that could be prone to unintended failure or destructive attack (de Mello and Ter-Minassian 2020).

For energy systems to operate effective and reliable, energy supply needs to be constantly ensured. In this regard, a main aim of the project ReMoDigital was to develop a societal-embedded concept to analyse the effects of digital energy transition, as well as to create a monitoring instrument that can facilitate the assessment of systemic stress-cases and resilient behaviour. To realise this, an inter- and transdisciplinary approach was followed to address and integrate relevant stakeholders and perspectives, as well as to acknowledge different spatial and temporal scales that can provide indications in relation to resilience. The presented chapters in this book are an illustration of this heterogeneous approach and have shown that existing challenges towards realizing and monitoring systemic resilience—although far from being trivial—can be overcome.

In this final chapter, we take a holistic view through integrating key take-aways and provide distilled recommendations on various levels. For this, we revisit some of the discussions that review resilience from different angles, including associated conceptualisations, existing political challenges, available system-analytical methods, and pragmatic suggestions towards operationalisation. The chapter concludes with some recommendations how resilience could be approached, developed, and monitored within energy systems, as well as how policy advice could be improved by means of resilience monitoring.

17.2 Understanding Resilience of Energy Systems

17.2.1 A Theoretical Perspective

To understand resilience as a concept within the context of energy systems, taking a deeper theoretical dive is required. In the centre of this, one quickly finds aspects of preservation and recovery (Holling 1996; Rose 2007; Carpenter et al. 2012; Longstaff et al. 2010).[1]

Preservation generally regards certain characteristics or system services to avoid unintended and undesired changes. In this regard, resilience includes the ability to prevent, resist or absorb adverse occurrences that may cause harm, destruction, or loss. It also relates to systemic preparedness and sustaining system services. Recovery, on the other hand, regards how systems cope with destructive effects. It regards the ability to bounce-back or adapt to the impacts of adverse events, as well as the potential speed and effectiveness of such reactions. It relates to crisis-management situations where systems need to respond flexibly and effectively to environmental perturbations (Hutter 2013; Oliver et al. 2015).

This theoretical consideration of resilience provides various lessons. First, it shows that capacity in relation to diversification and flexibilization is key for making systems adaptable and prepared to recurrent surprises, particularly with respect to critical infrastructures (Nemeth et al. 2011). Furthermore, it emphasizes the role of learning and capacity building. Instead of mere restoration, resilience is associated with improving existing structures and services through evolution and transformation. Also, it relates to how resource allocation can be effective and robustly managed (Ali et al. 2004).

Another valuable lesson regards the strong emphasis on systemic and sub-systemic features inseparably linked to resilience. Although there are arguably basic principles and functions of resilience relevant at different levels of agency—including individuals, organisations, communities, networks, and institutions[2]—the operationalization of these basic principles might be quite different. Action and interaction can strongly affect the scale and scope of resilience, a phenomenon associated to the posed concept of 'Panarchy' (Gunderson and Holling 2003). This regards differences in reactivity and feedback between the lower and higher levels within nested systems, as well as to behavioural direction provided by different stimuli between sub-systems.

The theoretical perspective also highlights the significant influence systemic environments can have on the nature and extent of resilience. In this regard, the ambitious concept of biomimetic resilience draws inspiration from engineering, chemistry, and biology to develop systems with functions that mimic biological processes.[3]

---

[1] See also Chap. 2 for a review of resilience in the context of complex systems.

[2] See also Chap. 1 for a structured conceptualisation of resilience and overview of selected application-contexts.

[3] See also Chap. 3 for a review on design principles in relation to biomimetic resilience.

Biomimetic principles—i.e. the self-organization and autopoiesis of elements and structures, processes related to inductive regeneration and homeostasis, as well as the regulation of efficiency, abundance, and variability (Naik and Singamaneni 2017)—can provide in combination with existing risk management procedures valuable lessons for resilient system design. It highlights how system structure and system services are intertwined and as such fundamental for the development and management of resilience.

17.2.2 A Normative Perspective

However, system management and governance exclusively based on theoretical considerations will likely fail in being valuable and legitimate. There is growing attention towards normative and socio-technical related aspects to review the available solution space (Kersbergen and Waarden 2009).[4] Although systems are often assessed in terms of traditional descriptive performance indicators, there are also subjective goals that can be linked to resilience. These include how to master crisis situations, how to 'build-back-better', or how to improve the confidence of involved risk management actors (Commission and Innovation 2022; Renn 2005).

An important normative principle regards precaution. It is believed that the value of precaution is strongly associated to the three concepts of complexity, uncertainty and controversiality. Here, complexity associated to the consequences of threat situations, uncertainty associated to the consequences of possible countermeasures, as well as controversiality associated to prescribed risk reduction measures, seem to positively relate to the importance of precautionary risk management instruments.[5]

In this light, it has been argued that precautionary risk management and resilience can be obtained based on adaptive, coping and participative aspects of resilience (Lorenz 2013). This requires precautionary criteria in the design of critical infrastructures or other installations, as well as the elicitation of concerns from different stakeholders and experts that can make designs more robust. Discussions regarding building social resilience have stressed the importance of participative and transformative capacities, which implicates the importance of deliberation in decision making processes (Saja et al. 2021), as well as resonates with contributions on the relevance of transdisciplinary- and post-normal science-based approaches.

17.3 Envisioning Resilience of Energy Systems

To go from understanding to realising resilience, there is need for political guidance. However, the complex nature of resilience makes finding golden bullets unlikely.

[4] See also Chap. 6 on resilience and values.

[5] See also Chap. 5 on risk, resilience, and precaution.

Arguably, a critically central and sensitive political issue regards how to balance preservation versus recovery measures. On the one hand, the allocation of economic means to prevention and mitigation measures is generally not associated with governance focused on economic development (Briguglio and Moncada 2019). On the other hand, it could be argued that context specific policies focused on vulnerability reduction can also drive precautionary behaviour (Mahzarnia et al. 2020; Kandaperumal and Srivastava 2020).

17.3.1 Planning and Prioritisation

A key aspect for developing resilient systems is planning. Arguably, a central challenge in planning procedures remains the inclusion of external perspectives. In general, studies have shown that various forms of external perspectives and expertise are not always well translated into policies. This also relates to participative processes in defining and realising resilience. Within the European Directive on the assessment and management of flood risks, public access to information and associated participatory processes for planning processes have been considered essential (European Parliament 2007). However, insights regarding the operationalization of democratic and participatory approaches within existing legislative contexts remain limited.[6]

The challenging integration of external perspectives in planning also relates to culture (Friedmann 2005). Planning cultures are known to be very diverse regarding the use of regulatory instruments.[7] Variable and place-specific traditions must be considered when thinking about more coherent and adequate approaches to infrastructure planning (Knieling and Othengrafen 2016). As such, there is need for capacity building among actors that need to implement such instruments. Particularly in the light of European and other transnational infrastructure projects, a coherent planning process may well contribute to strengthening joint development. Overall, these challenges associated to the legitimate integration of stakeholder perspectives has emphasized the importance of evidence-synthesis and -brokerage within policy formation processes.

Two other challenges relate to organisational neutrality and top-down governance approaches. As resilient systems are often the result of interplay between various sub-systems, centralized planning can be ineffective when multiple dimensions need to be considered. A potential solution lies within integrating forms of spontaneous order or self-organisation (Sørensen and Triantafillou 2016). Through providing self-organisation with normative guidance, combined with coherent combinations of existing political measures, novel approaches could be developed that might drive the realisation of resilience. Regarding organisational neutrality, potential conflicts of interest—both of responsible and involved organisations in decision making

[6] See also Chap. 9 on resilience in the system of infrastructure law.

[7] See also Chap. 7 on the governance of energy infrastructures and the influence of planning cultures.

processes—need to be critically reviewed. This also relates to the extent early participation results are binding, as well as to how such results could be organised in planning procedures.

17.3.2 Governance and Transformation

To operationalize political visions associated with resilient energy systems, various issues need to be addressed. Issues close to the conceptual core of resilience include governance approaches based on decentralisation. Decentralisation is generally considered as a general transfer of power and resources away from a central entity or agency (Torrisi et al. 2011). Within the context of energy systems, spatial decentralisation refers here to the process by which the supply side and the demand side are made 'closer', which potentially improves the capture of individual preferences. From a functional perspective, decentralisation can be understood as the enabled coordination between sub-systems that can robustly operate independently from external support. Decentralization also strongly links to the type of connections developed within systems. From an energetic perspective, systems often favour connections that develop into networks that can be characterised as lean and linear. De facto indications from healthcare and agriculture have highlighted that to improve the robustness and resilience of such systems, more diverse and shorter supply lines might benefit resilience.

However, despite this advocation of decentralisation within the context of system transition and transformation—often strongly based on assumptions of required technological innovation—there is also much uncertainty regarding associated or lacking regulatory and legal constructs. This also strongly relates to the dominance of vested and conservative governance cultures. This particularly concerns the bouncing-back dimension of resilience. Despite a general awareness that destroyed structures should be 'built-back-better' post-catastrophic, this often proves difficult to implement in practice (Wisner 2017). Bureaucracy generally operates in terms of exploiting available solutions that fit to existing 'governance' situations, thereby indirectly hindering the diffusion of new ideas. Strongly related to bureaucratic culture regards mindset paradigms of involved actors in the operationalization of resilience, particularly when tended towards risk aversion and stability.

A final challenge regards the availability of potential crucial capabilities in the context of stress. It has been shown that in times of crisis, different capabilities are required compared to those during 'business-as-usual'. However, due to the strong context dependency of most systemic stress cases, it remains challenging how required competencies could be generalised and prioritised (Wooten and James 2008).

17.4 Analysing Resilience of Energy Systems

17.4.1 Indicators, Measurement and Normativity Representation

To get a better grip on the successful operationalization of resilience, being able to reliably assess situations is essential.

Although it is generally acknowledged that the definition of suitable indicators is crucial, it is still challenging how to identify and define adequate indicators that enable the assessment of socio-technical systems.[8] In addition, it remains unclear how indicators should be characterised for assessing the level and control of resilience in energy systems. Based on cross-country comparative studies, it has been emphasized that the selection of indicators, variables, and weights can be subjective (Briguglio et al. 2006).

To develop sound resilience indicators, various analytical challenges related to measurement should be considered. The measurement of uncertainty and data estimation issues, the valuation of observations and derivation of information, the modelling of potential complex and ambiguous relationships, as well as the assessment and management of composed strategies, are all relevant processes associated to developing and understanding legitimate empirical-based resilience indices (Watson et al. 2015).

A final aspect that needs consideration relates to how the normative aspects of resilience can be reflected. The societal embedment of resilience has been strongly advocated in literature (Davidson-Hunt and Berkes 2008). In this regard, research has indicated that analysis and scenario building through models alone might not qualitatively represent existing value landscapes.[9] As models could be perceived as being too complicated, impractical, and unfit for public and non-expert participation, analytic-deliberative approaches have been proposed as a potential solution. However, insights from practice have been limited so far.

17.4.2 Assessments, Models, and Scenarios

To legitimately assess resilience, one needs to go beyond the measurement of specific individual indicators towards more complex interrelations. However, appropriate cost–benefit and multi-criteria-based approaches to assess complex and resilient systems are currently lacking.[10] Debated policy concerning climate-change has made

[8] See also Chap. 11 on indicators for resilience monitoring in energy systems.

[9] See also Chap. 8 regarding a discussion on the value of analytic-narrative-deliberative approaches for realising (social) resilience.

[10] See also Chap. 10 on economic aspects of resilience and the development of adequate indices.

clear that comparative assessments of costs—i.e. options of intervention versus not-intervention—can be vital in the legitimization of resilience-enhancing measures (Chenet et al. 2021). The cost–benefit dilemma associated to precaution versus inaction has often been discussed within the context of innovation (Hemphill 2020), as well as reignited in recent times within the context of COVID-19 (Gao and Yu 2020). Comparative assessments can potentially be combined with multi-criteria evaluation and optimization procedures that build on normative assumptions and prerequisites regarding systemic services.

This complex consideration of resilience strongly relates to system modelling. Optimisation models are believed to make a significant contribution to the resilience analysis for future energy systems when applied appropriately (Beyza and Yusta 2021). This includes both structural and incidence-based resilience analysis. Also, due to the relevance of nested systems in understanding and realizing resilience, addressing these multiple systemic levels is believed essential in modelling exercises. In this regard, it has been argued that an ambitious vision is required that considers subsystems and acknowledges disciplinary detail by means of multisystem modelling.

The analytical consideration of framework conditions has also been stated crucial in modelling attempts (Ahmadi et al. 2021). For systems in transformation, resilience can only be comprehensively assessed if the future context is analytically scrutinised. However, since socio-technical systems—like energy systems—are generally characterised by a high level of complexity, the reliable assessment of their unknown future states is challenging. To overcome this, potential future trajectories can be modelled and analysed based on pre-developed scenarios that reflect our perception of future circumstances and appropriately address epistemic uncertainty. Compared to classical types of analyses—which generally capture co-transformation of framework conditions through assumptions or pointwise sensitivity analyses—context scenario-based analysis of system transformation can offer analytical advances.[11]

17.5 Managing Resilience in Energy Systems

17.5.1 Measures

To manage resilient energy systems de facto—also in relation to existing and future forms of transportation[12]—it is necessary to go beyond the understanding of resilience, to operationalise the chosen direction for future systems, and to support the systemic assessment of performance. To facilitate system resilience, there is need for tools that can be used in making reliable and legitimate decisions, that can

[11] See also Chap. 4 for a review on the analysis of systems and context in relation to resilience.

[12] See also Chap. 15 on resilience in current and future transportations systems.

break down complexity associated to energy systems, and that can prioritise suitable and available design interventions—assumptions that have already been tested in different contexts of resilience including supply-chains (Zouari et al. 2021) and agriculture (Jung et al. 2021). Although basic principles have been developed to analyse the quality of systems, principle-based procedures in which feasible design interventions in socio-technical systems can be experimentally tested, as well as that are scientifically compatible and practicable for innovation and transformation processes, are still lacking.

Despite these de facto limitations, various concepts have been developed that indicate potential for realising and managing resilience. For example, to face challenges emerging in the new structure of power grids—mainly caused by various energy transition and digitalization trends—different measures are conceivable to make the system stable and robust in the light of stress events.[13] For a resilient digitalized power system, planning and system hardening mechanisms—including situational awareness, virtualisation, flexibilization, and islanding—could improve the level of anticipation, reaction, and recovery in systems.[14] However, insights from related forms of experimentation are still limited.

In addition, the issue of digitalisation seems also relevant regarding vulnerability reduction policy in relation to precautionary behaviour. Although digitalisation can make systems more transparent and accessible, it can also introduce new forms of flexibility that can increase system complexity and introduce associated challenges.

17.5.2 Central Elements of Stress-Test Tools

Additional to experimentation, the availability of suitable resilience monitoring tools to aid the development and management of resilient energy systems is key. Digital participatory decision support tools can support local-scale energy transitions, as well as provide guidance for local-scale social innovation (Hewitt et al. 2020).

On a general level, all aspects associated to understanding, governing, and analysing resilience are fundamental in conceptualisations of a resilience monitoring tool. To develop an appropriate monitoring approach for energy systems, it has been argued that distinct monitoring steps need to be defined, including scope determination, stress case and scenario modelling, and the measurement and assessment of resilience in such models.[15]

More specifically, since specificities influence complexity and the associated level and dynamic of resilience, monitoring tools require a strong and detailed characterisation of assessed system and sub-systems. Within the context of energy systems,

[13] See also Chap. 14 on resilience strategies in future digitalized distribution grids.

[14] See also Chap. 12 on challenges and potential solutions in the context of resilient digitalized power systems.

[15] See also Chap. 13 on resilience in relation to infrastructure planning models of future integrated energy systems.

this need for specificity is particularly evident regarding issues related to power versus heat. As such, monitoring tools need to integrate analytical components and select suitable model parameterisations that represent scoped systems. Also, there is a need to appropriately integrate possible stress cases and perform comparative analysis of different system designs. This also relates to the need of integrating adapted resilience metrics that can address different characterisations of potential future energy systems.[16]

17.6 Realising Resilience in Energy Systems

It is evident that the development of resilient energy systems is a challenging vision, but also one that can be realised. To achieve this, various key factors have been highlighted that need to be considered in the design, planning and management of system processes. Based on gathered insights—particularly on how resilience in energy systems have been defined, analysed, and operationalised in practice—various recommendations have been extracted. Although the following recommendations—clustered based on (a) developing resilience, and (b) monitoring resilience (see also Fig. 17.1)—should not be considered as definite and all compassing, the contained messages might motivate various actors in bringing resilient energy systems closer to realization.

17.6.1 Some Recommendations for Developing Resilience

For those energy systems that are characterised by limited resilience, it is first imperative to grasp how resilience can be developed. For this, the following *six recommendations for developing resilience* have been formulated:

(1a) *Understand resilience*: One of the key challenges in developing resilience is the creation of a common and contextual understanding regarding what resilience should represent. It must be highlighted that the degree to which resilience can be defined is strongly dependent on a common understanding of systemic operation and required services. It also requires that intra- and inter-systemic relations need to be made as transparent as possible, including the consideration of various spatial and temporal scales. Also, the assessment of existing design principles for developing resilience is likely to be influenced by how resilience is understood or conceptualised. One basis for such assessments includes the engagement of various disciplines in discussing context-specific meanings and roles of resilience.

(2a) *Embed resilience in society*: A strong fundament to realise resilience of the energy system needs to be fuelled with all aspects relevant to the general

[16] See also Chap. 16 on systems consideration for resilience monitoring.

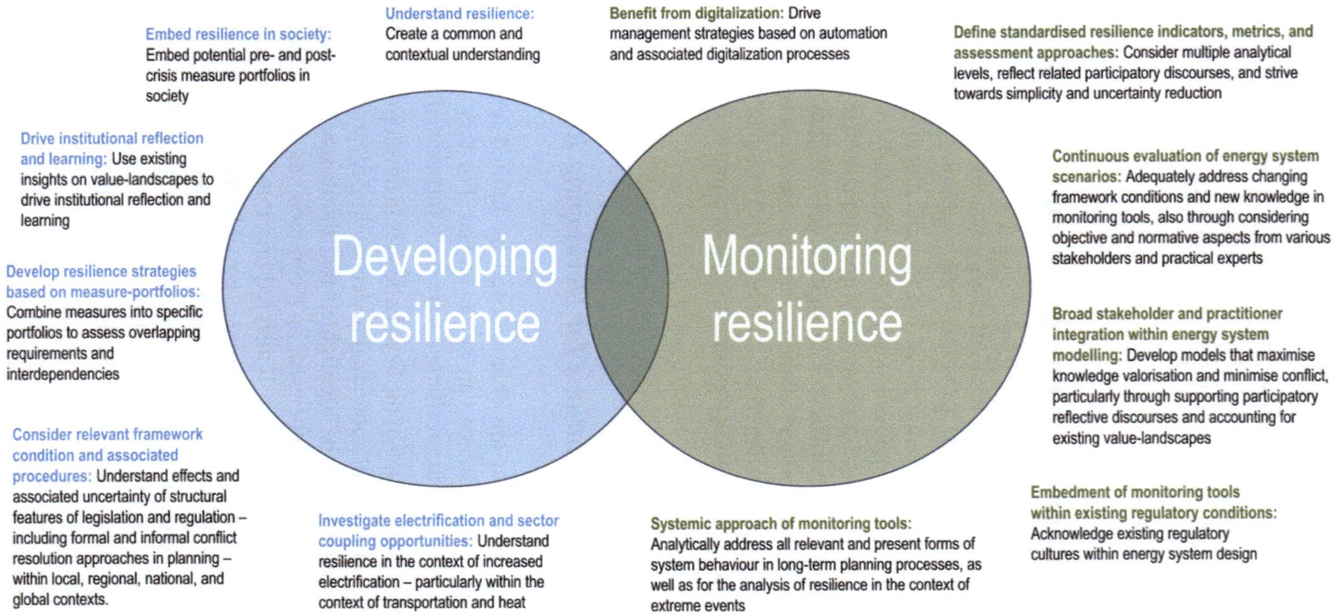

Fig. 17.1 Recommendations for developing and monitoring resilience

principles of recovery and prevention. However, to legitimately realise and strengthen system resilience, it is important to embed potential pre- and post-crisis measure portfolios in society. This requires an adequate grasp on the value-landscape of all relevant actors. To overcome challenging normative conceptualisations, one may gain insights into value-landscapes by utilizing pragmatic approaches and transdisciplinary co-creation, including combinations of bottom-up quantitative and qualitative approaches. Such broad involvement of actors—without limiting responsibilities of involved scientists and experts regarding enacted procedure and obtained output—should also be understood as a form for building social resilience in a continuous iterative fashion, through going beyond the pure analytic approach and introducing a deliberative element in scenario-building processes.

(3a) *Drive institutional reflection and learning*: To embed resilience in society, it is important to use existing insights on value-landscapes to drive institutional reflection and learning about how particular value-based behaviour patterns can be rewarded or encouraged. This also relates to stimulating societal experiments, particularly regarding the integration of local neighbourhoods and the identification of best-practice communities. Also, innovative measures can be communicated from selected networks that allow the comparison of daily routines in relation to alternative behavioural concepts. Incentives to inspire social learning—also for crisis management in communities and selected neighbourhoods—may serve to benchmark best practices and communicate existing moral standards.

(4a) *Develop resilience strategies based on measure-portfolios*: While many individual resilience measures exist, a coherent resilience strategy still seems missing. As such, measures need to be combined into specific portfolios to assess overlapping requirements and interdependencies. Such activity is especially valuable once combined solutions can be post-incident evaluated and potentially adapted. To realise this, a concretisation of guiding principles associated to resilient energy systems can facilitate the orientation within a potential solution space. To derive such principles, trade-offs should be made visible and societal ambiguity needs to be addressed through adequate public and stakeholder participation. This is also relevant to the argued importance of international coordination of research and implementation actions for developing resilient energy systems. This exemplary relates to international trade through investing in cross-border cables, international cooperation in the areas of grid control and standardization, as well as the strengthening of the national and international cyber-police in the context of digitalisation.

(5a) *Consider relevant framework condition and associated procedures*: To develop successful resilience strategies, it is imperative to understand the influence of framework conditions. There are various structural features of legislation and regulation that can be concentrated upon in the context of resilient energy systems. Although various instruments have been suggested for driving systemic resilience—including permit requirements, monitoring exercises, ex-post interventions, and planning approval procedures—more research is required to create a more generic and energy system-wide portfolio of options. This also relates to the integration of formal and informal conflict resolution approaches in planning procedures, including the specification of existing uncertainty of present and future conditions. Also, since planning approaches strongly dependent on the level at which framework conditions are assessed, there should be considerate attention towards the local, regional, national, and global context.

(6a) *Investigate electrification and sector coupling opportunities*: Another important future research avenue for energy system transition includes resilience in the context of increased electrification—particularly within the context of transportation and heat. This also strongly relates to the concept of sector coupling. Although sector coupling can be fundament in driving system resilience, more research is required to investigate how sector coupling can be most effectively operationalised. This also includes research on the evolution and complex dynamic of interacting system-wide production-supply processes, as well as the influence of such dynamic and associated uncertainty on the functioning of critical infrastructures.

17.6.2 Some Recommendations for Monitoring Resilience

From findings on how to develop resilient energy systems, more concrete recommendations can be drawn on how resilience of future energy systems—which are increasingly being set as political targets—can be monitored. Although resilience monitoring is arguably just one of various requirements within the context of sustainable and climate change robust energy systems, associated tools are likely to be important elements in the overall support of policy making. As such, the following *six recommendations for monitoring resilience* have been formulated:

(1b) *Benefit from digitalization*: The challenges posed by the energy transition cannot be managed without automation and associated digitalization processes. Although digitalization further increases system complexity and leads to new vulnerabilities, the potential of digitalization for a more resilient energy system must be harnessed. Digitalization enables direct monitoring, faster anticipation, and prompt response in the light of gradual systemic change and abrupt unforeseen events. Also, it allows for the improved utilization of flexibilities that exist within energy systems.

(2b) *Define standardised resilience indicators, metrics, and assessment approaches*: Research has highlighted that analysts, designers, and stakeholders should carefully choose indicator metrics. Due to strong influences of temporal, situational, and local differentiation during evaluation, metrics can yield significantly different results and recommendations. An important research avenue therefore regards how resilience assessment approaches of energy systems can allow the comparison between different solution-sets. Also, in the context of indicator selection, various analytical levels may need to be considered, as well as should reflect related participatory discourses. An important aspect here regards the argued need for simplicity and the reduction of uncertainty in indicator development and selection. Here, it is also important to critically assess the value of complex composite-indicators.

(3b) *Continuous evaluation of energy system scenarios*: To adequately address changing framework conditions and new knowledge in monitoring tools—e.g. regarding available technical options and insights from past socio-technical projects—a continuous and regular evaluation process of future energy system scenarios should be implemented. To derive valuable and legitimate input within such processes for scenario development, it is necessary to consider normative and empirical aspects from various stakeholders and practical experts. The development of transdisciplinary panels might be one of various potentially valuable approaches for obtaining representative input for scenario development.

(4b) *Broad stakeholder and practitioner integration within energy system modelling*: To provide a good basis for reflections for energy systems design, monitoring tools ideally require models which can reproduce the meaning of resilience for all associated sub-systems and impacts. To realise this, a transparent visualisation of these models that also addresses important interconnections is essential. Such monitoring tools should ideally be able to maximise knowledge valorisation and minimise conflict, particularly through supporting participatory reflective discourses and accounting for existing value-landscapes.

(5b) *Embedment of monitoring tools within existing regulatory conditions*: To make the uptake of monitoring tools by regulators as effective as possible, it is important that regulatory cultures are acknowledged within their design. This also regards to what extent modelled regulatory levers match with de facto regulatory approaches.

(6b) *Systemic approach of monitoring tools*: To be system-wide applicable, monitoring tools should be able to analytically address all relevant and present forms of system behaviour, existing interlinkages between sub-systems, and wider framework conditions influencing energy systems. This is needed for both long-term planning of systemic change, as well as for the analysis of resilience in the context of extreme events. As such, monitoring tools should be able to consider various analytical dimensions, including incidence-, structure-, and design-based approaches for resilient systems.

In relation to these recommendations, a stress-test instrument has been developed within the context of the project ReMoDigital, representing a potential starting point for resilience monitoring of envisioned and planned future energy systems. This developed tool—with a particular focus on various stress cases—concentrates on systemic correlations at the regional and national level, as well as focusses on the analysis of short-term catastrophic stress cases of potential future target systems. It addresses the effects of neglecting long-term developments and evolutionary trajectories, which could potentially introduce unobserved slow-burning but high-risk systematic anomalies. Future research could focus on complementing facets not yet addressed in the developed stress-test tool, including dependencies associated to international transmission grids or resources required for the development and operationalisation of future energy systems. The integration of existing and new insights within monitoring tools could significantly improve their operational value within future energy systems.

17.7 Conclusion

Recent history has highlighted that in situations where systems fail under stress, resilience associated concepts—like self-organisation, participation and dominant positive mindsets—can lead towards the light. Once the incapacity of existing organisations to adequately respond became evident during the Ahr-Valley crisis, extraordinary forms of self-organised support quickly sprung into life. To support all those being active with tractors, generators, and pumps during this abrupt collective challenge, oil traders freely delivered trucks filled with diesel slosh (von Hein 2021). A concerned citizen drove nearly 300 kms with his excavator to help others in need. A man in his mid-sixties, risking his life by unblocking a dam, likely saved thousands of a potential breach. Once wine cellars were flooded and machinery destroyed, growers supported affected colleagues with repairments, equipment, and donations from self-made products (ProWein 2021). When the elderly became high-up trapped in their apartments, the younger generation stood up to facilitate there escape from these prisons (Fischer 2022).

These examples of solidarity show that resilience—besides its conceptualisation in relation to objective, normative and pragmatic principles—can also just be considered as an inherent force to just keep going and a characteristic strongly embedded in individual and collective behaviour. If this drive to survive can be complemented with a supportive and harmonised system, the potential of resilience is likely to be strengthened. The past has shown that such complex and time-intensive designs of systems can be made (Simon 1977; Konrad Lorenz Institute 2005). When human rationality and inventive interaction patterns can harmonize with solidarity, when argued differences are overcome, and shared commonalities will be embraced, resilient systems can emerge.

References

Ahmadi, S., Saboohi, Y., Vakili, A.: Frameworks, quantitative indicators, characters, and modeling approaches to analysis of energy system resilience: a review. Renew. Sustain. Energy Rev. **144**, 110988 (2021). https://doi.org/10.1016/J.RSER.2021.110988

Ali, S., et al.: Measuring the robustness of a resource allocation. IEEE Trans. Parallel Distrib. Syst. **15**(7), 630–641 (2004)

Beyza, J., Yusta, J.M.: Integrated risk assessment for robustness evaluation and resilience optimisation of power systems after cascading failures. Energies **14**(7), 18 (2021). https://doi.org/10.3390/en14072028

Briguglio, L., et al.: Conceptualizing and measuring economic resilience (2006). https://doi.org/10.4324/9781315861104-4

Briguglio, L., Moncada, S.: The benefits and downsides of multidisciplinary education relating to climate change. In: Filho, W.L., Hemstock, S.L. (eds.) Climate Change and the Role of Education, pp. 169–187. Springer Cham (2019). https://doi.org/10.1007/978-3-030-32898-6_10

Carpenter, S.R., et al.: General resilience to cope with extreme events. Sustainability **4**(12), 3248–3259 (2012). https://doi.org/10.3390/SU4123248

Chenet, H., Ryan-Collins, J., van Lerven, F.: Finance, climate-change and radical uncertainty: towards a precautionary approach to financial policy. Ecol. Econ. **183**, 106957 (2021). https://doi.org/10.1016/j.ecolecon.2021.106957

Cronin, J., Anandarajah, G., Dessens, O.: Climate change impacts on the energy system: a review of trends and gaps. Clim. Change **151**(2), 79–93 (2018). https://doi.org/10.1007/s10584-018-2265-4

Davidson-Hunt, I.J., Berkes, F.: Nature and society through the lens of resilience: towards a human-in-ecosystem perspective. In: Berkes, F., Colding, J., Folke, C. (eds.) Navigating Social-Ecological Systems: Building Resilience for Complexity and Change, pp. 53–82. Cambridge University Press (2008)

European Commission Directorate-General for Research and Innovation: Strategic crisis management in the EU – Improving EU crisis prevention, preparedness, response and resilience. Publications Office of the European Union (2022). https://doi.org/10.2777/517560

European Parliament (2007) DIRECTIVE 2007/60/EC OF THE EUROPEAN PARLIAMENT AND OF THE COUNCIL of 23 October 2007 on the assessment and management of flood risks. Official Journal of the European Union, p. 8. https://eur-lex.europa.eu/eli/dir/2007/60/oj. Accessed 2 Oct 2025

Fischer, S.: Wenn die eigene Wohnung zum Gefängnis wird: Bad Neuenahrer Senioren können seit Flut ihr Haus kaum verlassen. Rhein-Zeitung (Kreis Ahrweiler) (2022). https://www.rhein-zeitung.de/lokales/kreis-ahrweiler/wenn-die-eigene-wohnung-zum-gefaengnis-wird-bad-neuenahrer-senioren-koennen-seit-flut-ihr-haus-kaum_arid-2384318.html. Accessed 2 Oct 2025

Friedmann, J.: Globalization and the emerging culture of planning. Prog. Plan. **64**, 183–234 (2005). https://doi.org/10.1016/j.progress.2005.05.001

Gao, X., Yu, J.: Public governance mechanism in the prevention and control of the COVID-19: information, decision-making and execution. J. Chin. Gov. **5**(2), 178–197 (2020). https://doi.org/10.1080/23812346.2020.1744922

Gunderson, L., Holling, C.: Panarchy: Understanding Transformations in Human and Natural Systems, Bibliovault OAI Repository. The University of Chicago Press, Chicago (2003). https://doi.org/10.1016/S0006-3207(03)00041-7

von Hein, M.: Solidarity after the floods, Deutsche Welle (2021). https://www.dw.com/en/showing-solidarity-after-the-floods-people-helping-people/a-58612983. Accessed 2 Oct 2025

Hemphill, T.A.: "The innovation governance dilemma: alternatives to the precautionary principle." Technol. Soc. **63**, 101381 (2020). https://doi.org/10.1016/J.TECHSOC.2020.101381

Hewitt, R. J., de Boer, C., Flacke, J.: Participatory development of digital support tools for local-scale energy transitions: Lessons from two European case studies. Glob. Transit. **2**, 138-149 (2020). https://doi.org/10.1016/j.glt.2020.07.003

Holling, C.S.: Engineering resilience versus ecological resilience. In: National Academy of Engineering (ed.) Engineering Within Ecological Constraints, pp. 31–43. The National Academies Press, Washington, DC (1996). https://doi.org/10.17226/4919

Hutter, G.: Organizing social resilience in the context of natural hazards: a research note. Nat. Hazards **67**(1), 47–60 (2013). https://doi.org/10.1007/S11069-010-9705-4/METRICS

IEA (2022) Energy efficiency 2022. https://doi.org/10.1109/FNWF55208.2022.00137

Jung, J., et al.: The potential of remote sensing and artificial intelligence as tools to improve the resilience of agriculture production systems. Curr. Opin. Biotechnol. **70**, 15–22 (2021). https://doi.org/10.1016/j.copbio.2020.09.003

Kandaperumal, G., Srivastava, A.K.: Resilience of the electric distribution systems: concepts, classification, assessment, challenges, and research needs. IET Smart Grid **3**(2), 133–143 (2020). https://doi.org/10.1049/IET-STG.2019.0176

van Kersbergen, K., van Waarden, F.: "Governance" as a bridge between disciplines: cross-disciplinary inspiration regarding shifts in governance and problems of governability, accountability and legitimacy. In: Clarke, T., Chanlat, J.-F. (eds.) European Corporate Governance, pp. 64–80. Routledge, London (2009). https://doi.org/10.4324/9780203875896-11

Knieling, J., Othengrafen, F.: Planning cultures in Europe: decoding cultural phenomena in urban and regional planning. In: Planning Cultures in Europe: Decoding Cultural Phenomena in Urban and Regional Planning. Taylor and Francis (2016). https://doi.org/10.4324/9781315246727

Konrad Lorenz Institute, Callebaut, W., Rasskin-Gutman, D. (eds.): Modularity: Understanding the Development and Evolution of Natural Complex Systems. The MIT Press(2005). https://doi.org/10.7551/MITPRESS/4734.001.0001

Longstaff, P.H., et al.: Building resilient communities: a preliminary framework for assessment. Homel. Secur. Aff. **4**(3), 1-2s3 (2010)

Lorenz, D.F.: The diversity of resilience: contributions from a social science perspective. Nat. Hazards **67**(1), 7–24 (2013). https://doi.org/10.1007/S11069-010-9654-Y/METRICS

Mahzarnia, M., et al.: A review of the measures to enhance power systems resilience. IEEE Syst. J. **14**(3), 4059–4070 (2020). https://doi.org/10.1109/JSYST.2020.2965993

de Mello, L., Ter-Minassian, T.: Digitalisation challenges and opportunities for subnational governments. In: OECD Working Papers on Fiscal Federalism (2020). https://doi.org/10.1787/822c0d6e-en

Naik, R.R., Singamaneni, S.: Introduction: bioinspired and biomimetic materials. Chem. Rev. **117**(20), 12581–12583 (2017). https://doi.org/10.1021/acs.chemrev.7b00552

Nemeth, C., et al.: Resilience is not control: healthcare, crisis management, and ICT. Cogn. Technol. Work **13**(3), 189–202 (2011). https://doi.org/10.1007/S10111-011-0174-7/METRICS

Oliver, T.H., et al.: Biodiversity and resilience of ecosystem functions. Trends Ecol Evol 30(11):673–684. Elsevier Current Trends (2015). https://doi.org/10.1016/J.TREE.2015.08.009

ProWein: Flood disaster strikes Ahr valley: a wave of solidarity from home and abroad (2021). https://www.prowein.com/en/Media_News/Press/Press_Releases/Flood_disaster_strikes_Ahr_valley_a_wave_of_solidarity_from_home_and_abroad. Accessed 2 Oct 2025

Renn, O. (2005) Risk governance – towards an integrative approach, white paper no.1. Geneva. https://doi.org/10.4135/9781412994064.n250

Rose, A.: Economic resilience to natural and man-made disasters: multidisciplinary origins and contextual dimensions. Environ. Hazards **7**(4), 383–398 (2007). https://doi.org/10.1016/J.ENVHAZ.2007.10.001

Saja, A.M.A., et al.: A critical review of social resilience properties and pathways in disaster management. Int. J. Disaster Risk Sci. **12**(6), 790–804 (2021). https://doi.org/10.1007/S13753-021-00378-Y/FIGURES/5

Simon, H.A.: The organization of complex systems. In: Models of Discovery and Other Topics in the Methods of Science. Boston Stu, pp. 245–261. Springer, Dordrecht (1977). https://doi.org/10.1007/978-94-010-9521-1_14

Sørensen, E., Triantafillou, P.: The politics of self-governance: an introduction. In: Sørensen, E., Triantafillou, P. (eds.) The Politics of Self-Governance, pp. 1–24. Routledge, New York (2016)

Torrisi, G., et al.: Defining and measuring decentralisation: a critical review (2011)

Watson, J.-P., et al.: Conceptual framework for developing resilience metrics for the electricity, oil, and gas sectors in the United States, SANDIA Report. Albuquerque (2015). https://www.energy.gov/oe/articles/conceptual-framework-developing-resilience-metrics-electricity-oil-and-gas-sectors

Wisner, B.: "Build back better"? The challenge of Goma and beyond. Int. J. Disaster Risk Reduct. **26**, 101–105 (2017). https://doi.org/10.1016/J.IJDRR.2017.09.027

Wooten, L.P., James, E.H.: Linking crisis management and leadership competencies: the role of human resource development. Adv. Dev. Hum. Resour. **10**(3), 352–379 (2008). https://doi.org/10.1177/1523422308316450

Zouari, D., Ruel, S., Viale, L.: Does digitalising the supply chain contribute to its resilience? Int. J. Phys. Distrib. Logist. Manag. **51**(2), 149–180 (2021). https://doi.org/10.1108/IJPDLM-01-2020-0038

Correction to: Energy Resilience as a Participative Process

Gabriele Fohr

Correction to:
Chapter 8 in: D. van Doren et al. (eds.), *Towards Energy System Resilience*, Ethics of Science and Technology Assessment 55, https://doi.org/10.1007/978-3-032-03187-7_8

In the original version of this chapter, citation text and copyright notices were omitted. These have now been added throughout the chapter, and the pagination has been revised accordingly.

The updated version of this chapter can be found at
https://doi.org/10.1007/978-3-032-03187-7_8

D. van Doren et al. (eds.), *Towards Energy System Resilience*, Ethics of Science and Technology Assessment 55, https://doi.org/10.1007/978-3-032-03187-7_18